Microwave Optics

The Optics of Microwave Antenna Design

S. CORNBLEET
Department of Physics
University of Surrey

1976

ACADEMIC PRESS

LONDON   NEW YORK   SAN FRANCISCO

A Subsidiary of Harcourt Brace Jovanovich, Publishers

ACADEMIC PRESS INC. (LONDON) LTD.
24/28 Oval Road
London NW1

United States Edition published by
ACADEMIC PRESS INC.
111 Fifth Avenue
New York, New York 10003

Library of Congress Catalog Card Number: 76-016960
ISBN: 0-189650-1

PRINTED IN GREAT BRITAIN BY
PAGE BROS (NORWICH) LTD
MILE CROSS LANE
NORWICH

Microwave Optics

Cassegrain Scanning Antenna. Polarization twist, phase corrected reflector with fixed polyrod feed.

Foreword

by
Professor John Brown
Imperial College, London

The design of microwave antennas has relied very extensively on optical principles and all designers have a working knowledge of elementary geometrical optical principles. Many designers have found that their work has duplicated that carried out much earlier by those whose interests have been primarily in extending geometrical optics theory. They must have wished—as I have done—for a comprehensive coverage, written in language with which they are familiar, of the wealth of ideas which can be tapped. Dr Cornbleet has now provided such a coverage. He is well-placed to do so—his mathematical background coupled with a long period in which he worked in industry as the leader of an active group designing antennas for airborne radar provides him with the two separate viewpoints which the task he has set himself requires. Perhaps of even greater significance is his enthusiasm—an enthusiasm which will be conveyed to the reader as he finds himself led on from detailed and illuminating treatments of the problems with which he will be familiar to parallels in other fields such as quantum mechanics.

Microwave optics is a subject in which practical design and theory go hand-in-hand. A casual glance at this text might suggest that it is a work only for the theoretician—it is to be hoped that this will not be the case for the designer can profit from knowing what theory has to offer, even if he is not interested in developing the theory himself. Dr Cornbleet has given complete proofs where these are helpful to an understanding of the material but also relates his material to practical situations. Even though some of the functions he uses, such as Zernike polynomials, may not be familiar, his treatment should not be beyond anyone who has studied mathematics in an engineering degree course. In a number of cases, particularly in the last chapter, he has very sensibly concentrated on describing the results which have been achieved giving the specialist, who wishes to explore the subject further, directions towards the very extensive literature which exists.

I have already mentioned Dr Cornbleet's enthusiasm. This comes through most clearly in Chapter 6 where he draws on materials from a wide range of topics to suggest the ways in which further developments in microwave optics might arise. My reaction on reading this chapter was to wish for a sabbatical year during which I might follow up some of these suggestions. Just over thirty years ago, network theory was revolutionised by the appreciation of the significance of functions of a complex variable in relation to network functions. The discussion in Chapter 6 tempts me to speculate whether we may be in sight of a corresponding breakthrough in our understanding of antennas. Perhaps one of the readers of this book may be stimulated into recognising the missing link in the chain.

May, 1976.

Preface

This book has its origins in a short intensive course on Microwave Optics that was presented at the University of Surrey in September 1974. The object of that course was to familiarise microwave engineers, and in particular those engaged in antenna design or application, with the basic elements of the optics of antennas, the variety of designs of such antennas and a ready means of assessing the resultant radiation patterns. This involves a heavy dependence on the first order theories, geometrical optics for the design process and the scalar theory of diffraction for the far-field patterns. Other subjects in which I have had some interest were included such as radome wall design, non-uniform lenses and polarizing gratings. The course was presented on an engineering handbook basis giving basic design data and formulae with the minimum theoretical background. Visiting lecturers rounded off the subject by discussing modern topics such as the geometrical theory of diffraction, fibre optics transmission and polarization effects in some of the less symmetrical reflector designs that can arise.

It appeared that sufficient interest was generated among the engineers and scientists who attended the course, to justify the inclusion of the theoretical background and to make the subject a complete study. The requirement of having to justify a great deal of assumed knowledge in this way has led to a far deeper reading into the available material and has created a more coherent view of the subject which now takes on the aspect of a science in its own right, well founded in the modern theories of theoretical physics. The result has been a reappraisal of the nature of the practical design process and has led to the discovery of wide ranging connections with other branches of physics and mathematics all in some way relevant to the one single subject which I have entitled this volume.

However, lest such a title appears to be too comprehensive, the rider was added that the parts of the subject dealt with are concerned with antenna designs and applications specifically. This has to be so since I have omitted any detailed study of other vital microwave optical subjects, such as fibre optics, optical and beam wave-guides, open resonator cavities, anisotropic or plasma media, optical wave-guide components and holography. Most of these subjects have been dealt with in recent volumes of their own.

The book now possesses three fundamental objectives. To the newcomer to the field of microwave antennas, it should provide fundamental design studies and a source of the relevant literature of the entire subject. The experienced designer, on the other hand, may find much of the familiar material cast in a new light which it is hoped will broaden his appreciation of the science of microwave optics. And the more theoretically minded student may find that the practicalities of antenna design give some substance to otherwise abstruse concepts in theoretical physics.

I have tried in the book to indicate wherever possible those fields where optics has had a long and valuable experience that could be usefully adapted to antenna design. On occasion the reverse process occurs and we find, for example, that we deal here mainly with the "in-the-large" problem of surfaces and wave-fronts as compared with the mainly ray methods used in optics, where the former method is only now being investigated.

It has long been a puzzle as to why the most beautiful mathematics that exists in the geometry of geometrical optics should not be available to those designers of optical systems that have most need of it, and I have attempted to correct that deficiency as well. A newcomer to the subject may very well be forgiven in thinking that the only designs that exist are the paraboloid and its Cassegrain variations and the Luneburg lens.

Chapter 1 is basically upon that theme. The basic laws of reflection and refraction are presented in their most fundamental form from which it is shown that virtually all optical designs can be obtained. This is illustrated by obtaining the design of most of the standard optical devices and showing that the method is available for application to some less familiar systems. Other consequences result, the methods of obtaining caustic curves, focal curves and curves of zero phase, for reflecting or refracting surfaces. All these entities play a part in the design or assessment of an optical system, curves of zero phase in the theory of inversions, focal curves in the assessment of distorted phase fronts and so on. The variation to the optical methods and designs that is created by the use of microwave lens media, leads to the concepts of wide angle lenses and reflectors and so to beam scanning. Wide field angle and few surfaces are the hallmark of the microwave system compared with the paraxial approximation with many surfaces of the optical system, and the accent has to be to make each surface play as valuable a role as possible in the system performance. This could be to correct an aberration, create aplanaticism, or tailor the field distribution for better radiation properties.

Chapter 2 extends the theory from discrete surfaces to the medium with continuous variation in refractive index. The fundamental law that arises is expanded completely for the usual basic coordinate systems and the derivation of ray paths in these media is shown to lead to the design of optical lenses. Here again the many possibilities are as yet not fully exploited, in particular the application of other more unusual coordinate systems. A new approach is given too to the standard problems in the spherically symmetrical media. The plane stratified medium is given separate attention due to its application in microwave practice as a transparent electromagnetic window. Here too new designs are presented as well as a new method aimed at dealing with a non-homogeneous layer. This gives a real physical comparison between the usual Born and W.K.B. approximations to the differential equation of the exact theory.

In the third chapter we maintain the first order approximation that constitutes the geometrical optics theory with a first order approximation to the diffracted fields created by an optical aperture. In all cases of relevance it is found that this scalar theory of diffraction gives results, and certainly qualitative effects, more simply and directly than the exact field theory. This even includes effects that would appear to require interference for explanation, such as the field distribution along the axis of a system. With the scalar theory a particularly pleasing closed form solution is obtained through the use of circle polynomials, which permits the inclusion of the basic aberrations that can occur in microwave systems as well as some more drastic field perturbations that can be used for pattern shaping. The symmetry of the method allows us to pose the question of the reversibility of the process so as to synthesize with a given aperture a required radiation pattern. This is the subject of the fourth chapter and the limitation of the procedure caused by having the continuous surfaces of the optical designs of the first chapters is soon established. It indicates however the advantages that are to be obtained by a discontinuous distribution and theoretical results for zoned circular apertures, or apertures containing multi-mode propagation of the correct form bear this out.

A departure is made in Chapter 5 where problems of polarization and its analysis and application are the subjects. Here, more than anywhere optical methods have outstripped those of microwaves and so a full optical treatment is presented. This is a matrix method and its applications are shown to include the design of polarization filters and rotators of several kinds at least one of which having had experimental confirmation at microwave frequencies. The theory in this case closely parallels that of particle spin and it is shown how even the very earliest of these studies was based on methods now being applied to quantum theory. In the context of this chapter this is the description of polarization by a quaternion form comparing with a particle spin described by spin matrix forms. Polarization has an intimate connection with transmission line theory and so a path is established between this subject and the higher theory of the rotation groups.

This and similar group aspects that have made an appearance throughout the book are followed up in the final chapter. We have found throughout

that analogues exist such as those of geometrical optics with particle trajectories, non-uniform media with non-uniform potentials, the polarization
with particle spin and the transparent window with the tunnelling effect.
This is completed by the derivation of a lens group acting upon a centred
system of rays closely analogous to the group of continuous transformations
in quantum theory. Specific examples of this are given and a new method for
the development of these transformations by means of a theory of functions of
a quaternion variable. Much of the theory is dependent upon a four dimensional study of optics which together with the theory of groups and differential geometry introduces new fields of study not usually associated with the
engineering design of antennas. In this case, and in a few other instances in the
preceding chapters, the subject is left in a state of conjectural continuity, or
as some would say suspended belief, which leaves open many new lines of
investigation and possibilities of new design methods and even practical
applications. If it serves to arouse interest in the wide scope that the subject
can cover then its purpose will have been achieved.

In each chapter beside the basic reference material, I have included a
selective bibliography containing what I have always appreciated as the
seminal papers on the subject and a variety of new references which the
experienced antenna designer will observe to be of a different kind from that
usually given in antenna literature. This is because I have tried wherever
possible to go outside the usual antenna and electronic engineering sources
into the field of the literature of optics, applied mathematics and physics.
To do this I have had to omit very many references that would otherwise be
included in a book of this title.

Several full and complete bibliographical lists already exist in the literature,
however, and it is understood that these would be familiar to a reader on this
subject.

The University of Surrey   S. Cornbleet
Guildford
May 1976

Acknowledgements

I am particularly indebted to Professor J. Brown of Imperial College, London, for contributing the foreword, to him and to Mr. S. H. Moss, Senior Mathematician at the Laboratories of the Marconi Company, Stanmore, Middlesex, for very many years of fruitful and valuable discussion, advice and encouragement and to Dr. R. C. Hansen whose suggestion began this venture. A debt is also owed to those very many people who contributed the dots and crosses that go on the figures marked as computed or experimental results This includes the computing staff of the Department of Physics at the University of Surrey, Guildford and their leader Mr. K. Knight and the experimental staff of the Antenna Group at the Laboratories of the Marconi Co. Ltd., Stanmore under Mr. W. G. Whyman and in particular to Mr. N. Dover and to many other friends in the field of microwave antennas. Gratitude must also be expressed to my colleagues, staff and students, in the Department of Physics for their encouragement of my studies in the theory of quaternions and mostly to the Microwave Group and their leader, Dr. K. W. H. Foulds, and to Dr. M. C. Jones, and to the generosity both in time and patience of the library staffs in both areas. Special thanks go to Mrs. Judith Smith for typing the manuscript with unfailing cheerfulness and at great speed thereby not allowing any decline in enthusiasm or of output at critical times, and to my wife for tolerating a period of seriously reduced social activity and other disruptions with great fortitude.

S.C.

Contents

1

Lenses and Reflectors

Optics, and in particular, geometrical optics, is among the oldest of the sciences, and the introductory chapters of any of the numerous texts on the subject would provide a suitable basis for subsequent discussion. Thus from the tract by J. L. Synge[1] we have

"Geometrical optics is an ideal theory and a useful one. The discovery that the propagation of light is an electromagnetic phenomenon made the subject of optics coextensive with electromagnetism. We may, however, study certain parts of the subject of optics without reference to electro-magnetism, always understanding that there is a limit to the physical accuracy of the results so obtained. It is customary to use the name 'physical optics' for the more complex and physically accurate theory, and 'geometrical optics' for the simpler ideal theory with which we shall be concerned. It is possible to justify geometrical optics as the limiting case of physical optics, the wave-length of the light in question tending to zero, but we shall be content with the development of geometrical optics on the basis of its own hypotheses."

What is not apparent in this quotation is that while the subjects of optics and electromagnetism are "coextensive", their application lies in essentially different areas. If in particular, we are concerned with the design of optical devices then geometrical optics provides the only method. This is, in fact, that certain part which can be studied without reference to electromagnetism. If, subsequently, we need to investigate the precise properties of the design so obtained, then physical optics has to be applied.

Consequently the development of geometrical optics "based on its own hypothesis" leads the way to the fuller understanding of current optical systems and to possibilities of new designs.

In this chapter we investigate the fundamental design of reflectors and lenses using the basic theory which is outlined in Appendix I. That is, the hypothesis of geometrical optics. This gives a first order design based upon the laws of reflection and refraction presented in what is thought to be a new setting. In the main, for the purposes of microwave antenna design, we will be concerned with an optical system of only a few surfaces, that collimates the radiation from a primary, nominally, point-source into a system of parallel rays, that is to a second point focus at infinity.

However the technique to be demonstrated for this purpose has a greater generality that makes it of interest as a subject in its own right. This not only leads to new methods of design but to the possibility of new optical systems whose practical application has yet to be fully assessed. It also presents a better appreciation of the role of geometry in geometrical optics, a field which is surprisingly little studied. A fuller understanding of the intimate manner in which the optics of rays is connected with the geometry of surfaces proves fruitful to both subjects.

Since we do not in practice use light of a wave-length approaching zero the collimating properties designed for cannot be perfect. The degree of this discrepancy becomes one of the major factors determining the use to which additional degrees of freedom in the system can be put. These additional degrees of freedom, as is well known, are obtained by the inclusion of additional reflecting or refracting surfaces. Other factors such as the reduction of the common optical aberrations, can also be achieved by the introduction of such additional surfaces. The primary concern of a geometrical optical design is to present a comparatively simple method, whereby the required additional surfaces necessary to contend with given aberrations or design criteria, can be derived.

Fortunately, in the field of microwave optics only few optical aberrations need to be so considered, and systems with more than three optical surfaces are rare. The large size and constructional complexity of some systems prevents the application of optimum designs. There is emerging a utilization of the higher frequencies of the spectrum allowing the possibility of smaller, more optical-like systems, for which a design procedure, along the lines presented here will be required. Once the methods have been specified and illustrated for systems of a few surfaces the extension to more complex designs should be easily accomplished and the way in which each additional surface can be used for a specific control purpose will be a matter for the individual system design.

1.1 SINGLE SURFACE FOCUSING REFLECTORS

The basic definition of the laws of reflection and refraction can be derived from the principle of least action (and vice versa) and are most simply given by the relation[2] (Appendix I)

$$\eta_r \, dr = \eta_\rho \, d\rho \tag{1.1}$$

in which from Fig. 1.1(a) η_r and η_ρ are the refractive indices of the media on the incident side with coordinates (r, θ) and on the transmission side with coordinates (ρ, ϕ) of a refracting surface.

For a perfect reflector we put the ratio $\eta_\rho/\eta_r = -1$ and Eqn (1.1) becomes the even more simple relation

$$dr = \pm d\rho. \tag{1.2}$$

The ambiguity in sign has been introduced to define the nature of the reflecting surface where the negative sign applies to a surface creating a real focus and the positive sign for a virtual focus.

From the elementary geometry of Fig. 1.1(b) a relation equivalent to that of Eqn (1.2) can be seen to be

$$r \, d\theta = \pm \rho \, d\phi, \tag{1.3}$$

where the sign in this case relates to the positive or negative values of the angle increment.†

We now consider the case where the surface concerned acts upon the entirety of a flat pencil of rays issuing from a single focus. This will then apply either to a two dimensional problem in which Fig. 1.2 is the cross-section of an infinitely long cylindrical reflector, or to a plane cross-section of a circularly symmetric system with axis containing the focal point. In these conditions, the most appropriate to the design of microwave antennas, Eqn (1.3) is integrable and can be used to derive reflecting surfaces with required focusing properties.

It is possible that because of these limiting conditions Eqn (1.3) has not

---

† The author is indebted to Mr S. H. Moss, Senior Mathematician, Marconi Space & Defence Systems Ltd for pointing out this vital relationship.

(a)

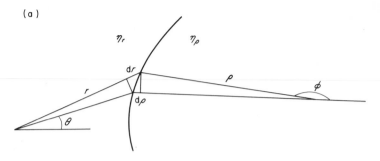

FIG. 1.1(a). Refraction.

(b)

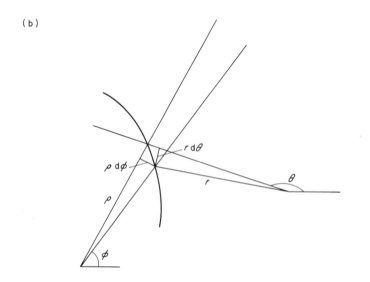

FIG. 1.1(b). Reflection.

received much attention hitherto, whereas nearly all fundamental optics has made much use of Eqn (1.2) since the time it, and its refractive counterpart Eqn (1.1), were presented by Hamilton.

Equation (1.3) joins together the optics and the geometry of surfaces in a most intimate manner in which the geometrical properties of curves can be derived from known optical systems as well as the reverse.

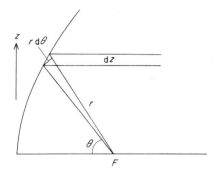

FIG. 1.2. Parabola.

We can illustrate the elegance of these seemingly elementary relations by deriving directly from them the optical properties of commonly known collimating and focusing devices.

The paraboloid

We require a surface (r, θ) which collimates all the rays from the origin into parallel rays. Then from Eqn (1.3) and Fig. 1.2

$$r\, d\theta = dz \text{ where } z = r \sin \theta.$$

Hence

$$d\theta = r \cos \theta \, d\theta - \sin \theta \, dr$$

or

$$r = \frac{\text{const}}{1 + \cos \theta}$$

the polar equation to the paraboloid.

The conic mirrors

A fundamental difference in approach between Eqns (1.2) and (1.3) can be illustrated by "designing" the conic mirrors. These are mirrors whose cross-section is a conic section and whose focusing properties are well established.[3] In this reference Brueggemann obtains the conic mirrors from the condition upon a reflecting surface to be perfectly anastigmatic, that is it focuses at the same point, rays in plane pencils which are at right angles, the tangential pencil and the sagittal pencil. It is shown by this treatment that the conic mirrors are the only surfaces with this property.

The two possibilities are illustrated in Fig. 1.3 where for 3(a) we have a real focus and in 3(b) a virtual focus. The appropriate forms of Eqn (1.2) are thus

$$dr = -d\rho \text{ for Fig. 3(a)}$$

and

$$dr = d\rho \text{ for Fig. 3(b).}$$

Integrating these directly gives

$$r \pm \rho = \text{const} = 2a.$$

This is the fundamental geometrical property of the ellipse (positive sign) and hyperbola.

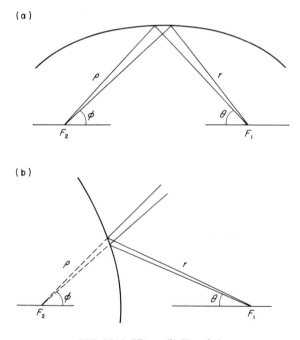

FIG. 1.3(a). Ellipse; (b) Hyperbola.

However, if we wish to derive the exact polar equations to the surfaces we must proceed from Eqn (1.3). Dealing with the situation of real focus we have from Fig. 3(a)

$$r \sin \theta = \rho \sin \phi$$

and

$$r \cos \theta + \rho \cos \phi = F_1 F_2 = 2a\varepsilon \qquad (1.4)$$

in addition to Eqn (1.3) $r \, d\theta = -\rho \, d\phi$. Applying relation (1.2) to these we find

$$\frac{d\theta}{d\phi} = -\frac{\sin \theta}{\sin \phi}$$

which integrates directly to give

$$\frac{\sin \theta}{1 + \cos \theta} \cdot \frac{\sin \phi}{1 + \cos \phi} = \text{const} = A.$$

Elimination of either θ or ϕ from this result by use of the triangle relation in Eqn (1.4) leads directly to the polar equation of the ellipse

$$r = \frac{4Aa\varepsilon}{(1 - A^2) - (1 - A)^2 \cos \theta},$$

from which we find that the constant of integration

$$A = \frac{1 - \varepsilon}{1 + \varepsilon} > 0,$$

where ε is the eccentricity of the ellipse.

Since the ellipse is described both by

$$r = \frac{l}{1 - \varepsilon \cos \theta} \qquad \text{and} \qquad \rho = \frac{l}{1 - \varepsilon \cos \phi}$$

Eqn (1.3) gives us the further relation

$$\frac{d\theta}{d\phi} = -\frac{\sin \theta}{\sin \phi} = -\frac{(1 - \varepsilon \cos \theta)}{(1 - \varepsilon \cos \phi)}.$$

For the hyperbola the related solution is

$$\frac{d\theta}{d\phi} = \frac{\sin \theta}{\sin \phi},$$

giving

$$\frac{\sin \theta}{1 + \cos \theta} \cdot \frac{1 + \cos \phi}{\sin \phi} = \frac{1}{A} = \frac{1 + \varepsilon}{1 - \varepsilon}.$$

We conclude furthermore that these (and the parabola of course) are the only single reflecting surfaces with the property of converting rays from a fixed point into rays converging onto a second fixed point. This agrees with the result of Breuggmann for anastigmatic reflectors.

2.2 TWO SURFACE FOCUSING REFLECTORS

In the same way that Hamilton (Ref. 1, p. 12) shows that Eqn 1.2 is summable over any number of individual reflecting (and later refracting) surfaces Moss points out[4] the piece wise summability of Eqn (1.3). So if, as in Fig. 1.4

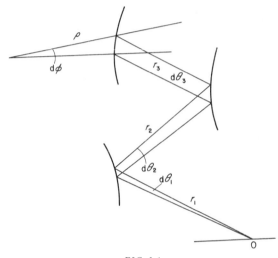

FIG. 1.4.

r_1 is the distance from the fundamental source at the origin to the first reflector, r_2 the distance along the ray from the first to the second reflector and so on and (ρ, ϕ) are polar coordinates about the final focus of the rays, we then have the two relations equivalent to Eqns (1.2) and (1.3) for n reflecting surfaces

$$dr_1 \pm dr_2 \pm \ldots \pm dr_n = d\rho$$

and   (1.5)

$$r_1 \, d\theta_1 \pm r_2 \, d\theta_2 \pm \ldots \pm r_n \, d\theta_n = \rho \, d\phi,$$

where the signs are determined by the same rules as for Eqns (1.2) and (1.3).

Cassegrain and Gregorian reflectors

The simplest illustration of this principle is the combination of a paraboloid and a secondary reflector to give the same result as for the paraboloid alone. If the two surfaces in Figure 1.5 have a common origin O, then from Eqn (1.5)

$$r_1 \, d\theta_1 \pm r_2 \, d\theta_2 = dz,$$

where the parabola is defined by $dz = r \, d\theta$.

If the polar equation of the secondary reflector is (ρ, ϕ) about the origin O, then $d\theta = d\theta_2$, and

$$r_1 \, d\theta_1 = (r \pm r_2) \, d\theta_2 = \rho \, d\phi.$$

But this is the defining equation of the conic mirror of the previous section and is thus a hyperbola (Cassegrain System) or ellipse (Gregorian system) depending upon the sign.

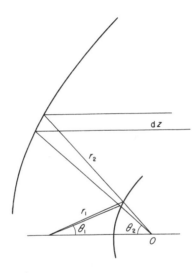

FIG. 1.5.

With the introduction of a second surface an additional degree of freedom has been created. This can be seen from the existence of an infinity of conics satisfying the above conditions. The unique system can then be obtained by establishing a second condition, the first condition being that of collimation alone. This can take several forms relating to aberrations or power distribution within the system. It was in computing a specific power distribution law requiring fundamental reshaping of the reflector surfaces that led Moss to the essentiality of including the condition that gave rise to the basic relation Eqn (1.3).

The cardioid of Zeiss

It can be seen from the nature of the relations in Eqn (1.5) that if all but one of the surfaces are specified then that surface alone can be determined from the given focal properties required. This means that of any two surface system where one surface has a preselected contour, a second surface can always be derived that will focus the rays again to a given point or to infinity. If we apply this to a spherical reflector there results two refocusing surfaces depending upon whether the reflection from the sphere is external (virtual focusing) or internal. The former of these as we now show, for a focus at infinity gives the cardioid reflector and sphere combination known as the cardioid of Zeiss.[5] Although the construction in the form of three dimensional reflectors is impracticable due to the double internal reflection, it is a further illustration of the method and has been constructed in a one dimensional form by methods which will be given later.

We have again the relation referring now to Fig. 1.6(a)

$$r_1 \, d\theta_1 - r_2 \, d\theta_2 = dz, r_1 = FM, r_2 = PM,$$

where, for a circle radius, $a \, dz = a \cos \theta \, d\theta$. From the properties of the reflection at P, that is a further application of Snell's law, we see that

$$\theta_2 = 2\theta,$$

and hence

$$r_1 \, d\theta_1 = (a \cos \theta + 2r_2) \, d\theta. \tag{1.6a}$$

The second statement of the principle of least action. Eqn (1.2) gives

$$dx = dr_1 + dr_2 = a \sin \theta \, d\theta,$$

which integrates as usual to give the constancy of the optical path

$$r_1 + r_2 - a \cos \theta = \text{const} = 2d_1 + d_2. \tag{1.6b}$$

A third relation is obtained from the geometry of the figure

$$r_1 = d_2 \cos \theta_1 + a \cos (\theta - \theta_1) + r_2 \cos (2\theta - \theta_1). \tag{1.6c}$$

The solution of these equations is dependent on the arbitrary choice of d_2, the required position of the eventual focus, and d_1. It is found by inspection that the simplest of these is obtained by putting $\theta = \theta_1$ and $d_1 = d_2 = a/2$. The resulting solution has $r_2 = \text{constant} = a/2$, and

$$r_1 = a(1 + \cos \theta_1).$$

This is the equation to the cardioid.

When the above values are incorporated into Fig. 1.6(a), we see as in Fig. 1.6(b) that the quadrilateral $FOPM$ is a trapezium with the three sides FO, OP, and MP of fixed length. A parallelogram linkage can thus be constructed which will describe the cardioid by the movement of the point M while the point P is confined to the circle. The linkage however is not "pure" since it requires a sliding member at N.

Further, since the intersection of the initial and final rays are at the point Q and $FOPQ$ is a parallelogram, Q will describe a circle centred at F. Hence this system automatically obeys the sine condition (see Appendix I) and thus the two surfaces and the two degrees of freedom are fully subscribed.

Other choices of focal position lead to a complex relation between θ and θ_1. These in general require computation for which the three equations given form a natural iterative process.

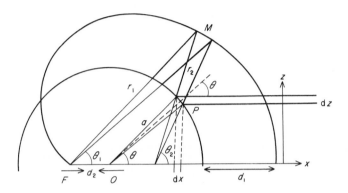

FIG. 1.6(a). *Cardioid corrector.*

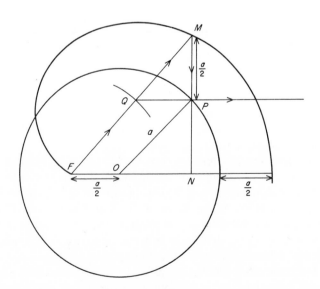

FIG. 1.6(b). *Zeiss-cardioid.*

Corrector for spherical mirror

When the first reflection occurs internally to the sphere, the second surface will give the required shape of reflector to correct the caustic of spherical aberration and to provide an accessible focal point in the manner of a Cassegrain system.

The defining relations are closely analogous to those for the cardioid geometry. We obtain from Fig. 1.7a $\theta_2 = 2\theta$ and hence

$$r_1 \, d\theta_1 = (a \cos \theta - 2r_2) \, d\theta,$$

where $r_1 = FM$ and $r_2 = PM$ and

$$dr_1 + dr_2 = dx; \qquad x = a \cos \theta.$$

The second of these integrates giving

$$r_1 + r_2 + a \cos \theta = 2d_1 + d_2 + a$$

the constant of integration being the limiting length of the ray as θ tends to zero.

The geometry of the problem provides the two further relations

$$r_1 \cos \theta_1 = r_2 \cos \theta_2 + a - a \cos \theta - d_2,$$

$$r_1 \sin \theta_1 = r_2 \sin \theta_2 - a \sin \theta.$$

Some simplification is obtained by assuming $d_2 = 0$ then squaring and eliminating r_2 gives

$$r_1 = \frac{a}{2} \left[\frac{2 - 2\beta + \beta^2 - 4\beta \cos \theta + (4\beta + 3) \cos^2 \theta - 4 \cos^3 \theta}{\beta - 1 - 2 \cos \theta + 2 \cos^2 \theta} \right],$$

where $\beta = 1 + 2d_1/a$.

In principle this can be converted into the proper polar equation (r_1, θ_1) by the use of the optical differential formula above relating $d\theta$ and $d\theta_1$, but this appears to be unjustifiably complex.

Considerations of symmetry show that the reflector has to be of a concave shape towards the incoming rays and thus has to be situated entirely within the region outside the cusp of the caustic curve (Fig. 1.7(b)). If this were not so two symmetrically placed points on the subreflector would receive rays at different angles of incidence and yet have to focus them at the symmetrical origin. Furthermore, all the rays between the cusp and the subreflector would be missed altogether thus limiting the aperture.

Cassegrain sub reflectors would have to be situated between the reflector and the cusp and would therefore lie in the region where each point is the intersection of two separate rays (the tangents to the caustic) as at points P in

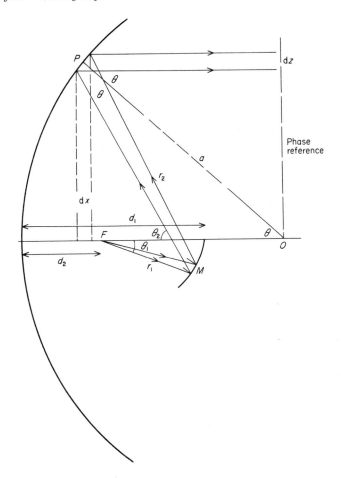

FIG. 1.7(a). Corrector for spherical mirror.

Fig. 1.7(b). Obviously no reflector can exist to focus two separate rays at their point of intersection. If the reflector aperture were limited to isolate only one of the two possible rays, then a Cassegrain geometry becomes possible. The angular aperture and long focal length resulting however make this unattractive in any microwave antenna configuration. This conclusion is borne out by the analysis above, for a Cassegrain geometry the second geometrical relation becomes the negative of that already given, namely

$$r_1 \sin \theta_1 = a \sin \theta - r_2 \sin \theta_2$$

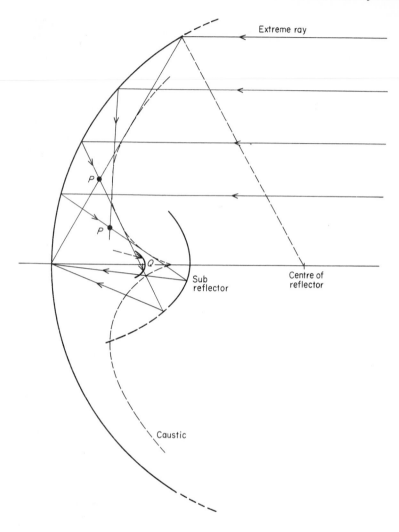

FIG. 1.7(b). *Corrector for spherical mirror. At points P no reflector can be placed to redirect rays to the focus, at Q a symmetrical reflector collects unsymmetrical rays.*

Consequently the solution for r_1, depending as it does on the eventual square of this, gives the identical, concave or Gregorian result.

More general shapes of sub-reflectors can be obtained by including a realistic value of d_2 in the analysis. The design for these given in parametric form can be found in the reference given.

1.3 SINGLE SURFACE REFRACTING LENSES

A lens will be defined to be a single surface lens if the change in ray direction through refraction occurs at only one of the surfaces of the element involved. The remaining surface or surfaces are then orthogonal to all the rays incident

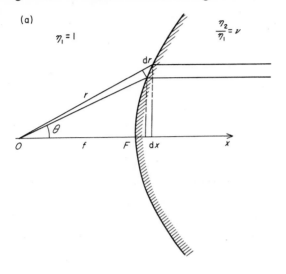

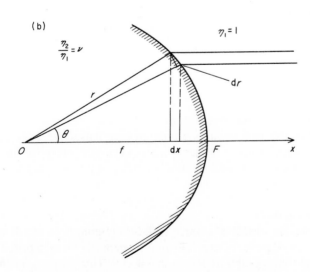

FIG. 1.8. Single surface lenses. (a) Hyperbola; (b) Ellipse.

upon them. The analysis is confined, as in the previous section, to the cross section of an axisymmetrical system or of a two dimensional system.

Consideration is first given to the refraction of a focused pencil of rays going from a less dense medium (air) into a more dense medium of refractive index v as illustrated in Fig. 1.8a. The surface is required to transform the pencil into a parallel system and hence Eqn (1.1) has the form

$$dr = v\,dx, \tag{1.8}$$

where $x = r \cos \theta$, and hence Eqn (1.8) integrates directly to give

$$r = vr \cos \theta + \text{const.}$$

The constant being evaluated by the condition $r = f$ when $\theta = 0$ to give the hyperbolic surface

$$r = \frac{f(v - 1)}{v \cos \theta - 1}. \tag{1.9}$$

When the refraction takes place from a more dense to a less dense medium Eqn (1.8) becomes

$$v\,dr = dx. \tag{1.10}$$

This is identical to the transformation $v \rightarrow 1/v$ in the ensuing analysis and in the solution resulting in the elliptical profile Fig. 1.8(b)

$$r = \frac{f(v - 1)}{v - \cos \theta} \tag{1.11}$$

with pole at the focus most distant from the surface. For a refractive index $v = -1$ the refraction law becomes a law of reflection and both Eqns(1.9) and (1.11) give

$$r = \frac{2f}{1 + \cos \theta}$$

the parabola of the previous section.

These are the two commonest microwave lenses for focusing rays at infinity. An application of the same analysis can be made to the more general problem of refocusing the rays at a second, and possibly more convenient, point.

The Cartesian ovals

The Cartesian oval[6] is a single surface refracting lens which converts a perfectly focused pencil of rays into another perfectly focused pencil of rays at at a general (finite point) as shown in Fig. 1.9. This property of the surface will be termed astigmatic. These, in the terms of reference given, are redis-

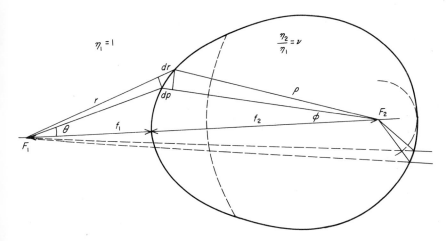

FIG. 1.9(a). *Cartesian oval. The dashed circles centred on* F_2 *form possible second lens surfaces for focal projections.*

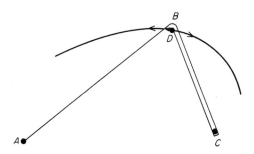

FIG. 1.9(b). *Drawing the cartesian oval ABCD is a taut string.*

covered every ten years or so and have not found any useful application. They do have the capability however of projecting a focal point a considerable distance forward, if an exit surface, in this case purely spherical, is added (Fig. 1.9).

For the real focus the surface is subject to the relation

$$dr = -v\,d\rho, \qquad (1.12)$$

which integrates as usual to give the law of constancy of the optical path

$$r + v\rho = \text{const} = f_1 + vf_2, \qquad (1.13)$$

the constant being evaluated for the ray along the axis.

The geometrical relation required in addition is one of either

$$\rho^2 = r^2 + F^2 - 2rF \cos \theta,$$

or

$$r^2 = \rho^2 + F^2 - 2\rho F \cos \phi; \qquad F = f_1 + f_2.$$

Differentiating these and using Eqn (1.12) gives

$$\frac{dr}{d\theta} = \frac{Fr \sin \theta}{\rho/v - r + F \cos \theta},$$

and

$$\frac{d\rho}{d\phi} = \frac{\rho F \sin \phi}{vr - \rho + F \cos \phi}$$

or

$$(\rho/v + r)\, dr = d(Fr \cos \theta),$$

and

$$(vr + \rho)\, d\rho = d(F\rho \cos \phi).$$

Substituting for r or ρ as appropriate from Eqn (1.13) and integrating gives

$$Fr \cos \theta = \left(\frac{f_1}{v^2} + \frac{f_2}{v}\right) r + \frac{r^2}{2}\left(1 - \frac{1}{v^2}\right) - \left\{ f_1 f_2 \left(\frac{1}{v} - 1\right) + \frac{f_1^2}{2}\left(1 - \frac{v}{v^2}\right)\right\},$$

$$(1.14a)$$

or

$$F\rho \cos \phi = (f_1 v + f_2 v^2)\,\rho + \frac{\rho^2}{2}(1 - v^2) - \left\{ f_1 f_2 (v - 1) + \frac{f_2^2}{2}(1 - v^2)\right\}.$$

$$(1.14b)$$

The final term in each of these relations is the constant of integration determined by the values of r and ρ at the boundary value $\theta = \phi = 0$ that is $r = f_1$; $\rho = f_2$.

Equations (1.14) are the polar equations defining the Cartesian ovals a comparatively simple version when compared with the Cartesian equations themselves.[6] The substitution $v = -1$ for reflecting surfaces in Eqns (1.14) reproduces the conic, elliptical and hyperbolic mirrors (depending upon the relative values of f_1 and f_2).

$$r = \frac{2f_1 f_2}{(f_1 + f_2) \cos \theta - (f_1 - f_2)}; \qquad \rho = \frac{2f_1 f_2}{(f_1 + f_2) \cos \phi - (f_2 - f_1)}.$$

In the limit of either f_1 or f_2 becoming infinite these become the equations for parabolic mirrors.

If the same limit is taken in Eqns (1.14) for general v the elliptical and hyperbolic lenses of Eqns (1.9) and (1.11) result. It is also easily confirmed that Eqns (1.14a) and (1.14b) transform into each other by the transformation

$$v \to \frac{1}{v} \qquad f_1 \leftrightarrow f_2.$$

Equation (1.14) therefore embodies the entire system of single reflecting and refracting surfaces with astigmatic property that have been dealt with so far. A method of drawing the Cartesian ovals, similar to that of the ellipse, with a taut string, looped about the two "foci" is also shown.

The aplanatic points of the sphere

The Cartesian ovals develop in another manner to give a surface configuration with both optical and microwave applications. That is the condition for which the surface of refraction is purely spherical. The situation is illustrated in Fig. 1.10 where a and b are the distances of perfect foci from the centre O of the sphere. We have then from the relation for a virtual image

$$dr = v \, d\rho,$$

$$\rho^2 = d^2 + a^2 + 2ad \cos \psi,$$

$$r^2 = d^2 + b^2 + 2bd \cos \psi,$$

where d is the radius of the sphere.

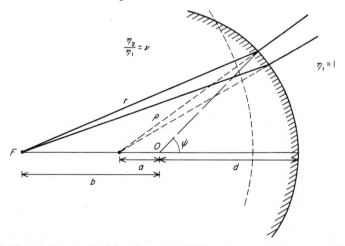

FIG. 1.10. *The aplanatic points of the sphere. The dashed circle centred on F is a possible second surface for a lens.*

Differentiating the last two eliminating ψ gives

$$b\rho \, \mathrm{d}\rho = ar \, \mathrm{d}r$$

and integrating the first gives

$$r = v\rho + \text{const.}$$

We choose this constant to be zero and hence obtain the standard relationship for the aplanatic points of the sphere $b = v\mathrm{d}; a = \mathrm{d}/v$.

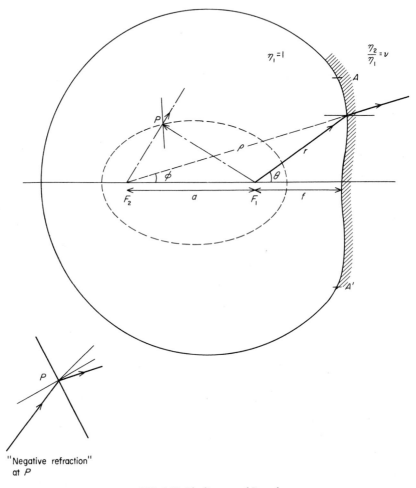

"Negative refraction" at P

FIG. 1.11. The limaçon of Pascal.

The limacon of Pascal

We require finally the convex refracting surface with the property of transforming a point focus into a second point focus. The second focus in this case will be at a fixed distance a from the first as in Fig. (1.11) and will of necessity be a virtual focus. In this case the focus is displaced away from the surface and the action differs from that of the previous section. The two relations required are

$$dr = v\,d\rho$$

$$\rho^2 = a^2 + r^2 + 2ra\cos\theta, \tag{1.15}$$

or quite simply

$$r = v\{r^2 + a^2 + 2ar\cos\theta\}^{\frac{1}{2}} + D, \tag{1.16}$$

where D, the constant of integration is the value of $r = f$ when θ is zero, that is

$$D = f - v(a + f).$$

Re-arranging Eqn (1.16) gives

$$(1 - v^2)r = D + v^2 a\cos\theta + \{(D + v^2 a\cos\theta)^2$$
$$- (1 - v^2)(D^2 - v^2 a^2)\}^{\frac{1}{2}}. \tag{1.17}$$

This curve is a particular form of the Cartesian oval known as the limacon of Pascal as shown in Fig. (1.11). It is more simply given by its direct bipolar form[7] which is the integral of Eqn (1.15)

$$r = v\rho + D. \tag{1.18}$$

The complete limacon includes a second curve shown by the broken line in Fig. (1.11). This derives from taking the negative root in Eqn (1.17). It is the same as the Cartesian oval of the previous section but in the present context arises from taking a negative form of the law of refraction, that is the refracted ray is on the same side of the surface normal as the incident ray as shown in the inset in Fig. (1.11).

The similarity between this curve between points A and A' and the usual surface of a correcting plate for spherical aberration, the Schmidt correcting plate allows us to use the principle of displacing foci by the above means in a more general manner.

1.4 TWO SURFACE LENSES

The single surface lenses are completed by having the second surface orthogonal to the rays of either the incident pencil or the transmitted pencil, as has been illustrated in the foregoing. In microwave practice, since the source of rays is not isotropic and since the finally radiated pencil is affected by this non-isotropy[8] a measure of control is lost by this over-simplified procedure. This can be recovered by the use of an astigmatic surface to "throw" the focus to a more suitable position which in turn would give a more preferred amplitude distribution. If for example we use the aplanatic points of the sphere for this purpose the centre of the spherical surface is moved forward and a thinner lens results as is shown in Fig. 1.12. In the same way the focus can be projected

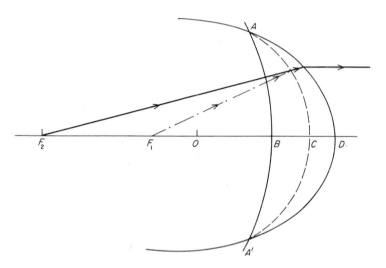

FIG. 1.12. Projection of focus of elliptical lens. F_2 is focus of ellipse ADA'. ABA' is circular centred on F_2 and is "natural" second surface. ACA' is circular centred at O for which F_1 and F_2 are aplanatic points. Source is "projected" from F_2 to F_1 resulting in thinner lens.

backwards by the application of a limacon surface as in Fig. 1.11(b). This then uses the degree of freedom presented by the essential introduction of the second surface in a practical manner. This procedure can be generalized to the situation where the surface designed is not essentially for focusing purposes as is the ellipse just illustrated. A focus is a very particular form of a caustic which has degenerated to a point and[9] a surface can always be designed to create any other given caustic. We illustrate this principle by the following well-known optical device.

The Schmidt corrector plate

The Schmidt corrector plate is dealt with in every textbook on optics, for example Born and Wolf.[10] It is a device for correcting the path length of every ray of a parallel system incidence on the concave surface of a spherical reflector so as to produce a perfect focus at a given point as shown in Fig. 1.13a. The analytical calculation of this surface can only be made by approximation, and, since the displacement from parallelism is small, this is given as

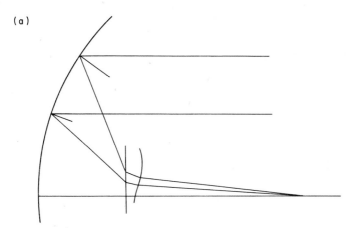

(a)

FIG. 1.13(a). *Schmidt corrector for spherical concave mirror.*

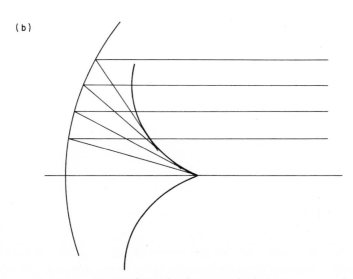

(b)

FIG. 1.13(b). *Nephroid—caustic of a circle.*

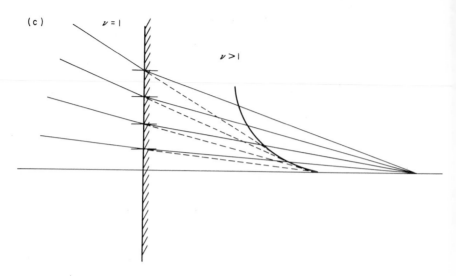

FIG. 1.13(c). Caustic of a plane refraction.

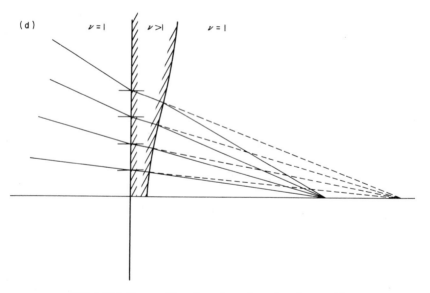

FIG. 1.13(d). Superposition of caustics and creation of external focus.

a series for which only the first few terms are retained. The general (but exaggerated) shape of the profile is shown in the figure. Now it is well known that the uncorrected spherical reflector produces a cusp like caustic when illuminated in this way by a parallel beam. The caustic curve, a cross-section of this surface, is given in Fig. 1.13(b) and is in fact a nephroid (Ref. 7, p. 207 and p. 70).

Now the caustic surface created by refraction in a plane surface Fig. 1.13c has much the same shape (Ref. 6, p. 174). It is probable that in the region of the cusp itself the agreement is to the same or better approximation than that of the usual series design for corrector plates. However, this refraction is created in going from the more dense to the less dense medium and it would appear to require a source to be embedded inside a dielectric volume. This is overcome by creating a virtual point focus through the use of a surface with astigmatic properties. The surface used in the Schmidt corrector plate is the limacon of Pascal illustrated in Fig. 1.11. The action of the corrector plate is therefore the reverse of that usually understood. It is in fact the plane surface that creates the necessary caustic to give the corrective action, the figured surface merely being a device (and by no means a unique one) for creating the necessary virtual source. The successive stages of this process are shown in Figs. 1.13(c) and 1.13(d).

This principle has obvious further application in the correction of caustics to give pure foci. Most such caustics however occur in asymmetrical optical systems and we have so far only been concerned with centred systems. In the more general case the caustic surface becomes highly complex and highly complex structures are necessary to create similar sources.[11] But one further illustration can be tolerated if we confine ourselves to a two dimensional system, or alternatively consider only the plane cross section of the caustic surface of an asymmetrically illuminated paraboloid in the plane of the asymmetry (optically the sagittal plane). In this plane the caustic has the form shown in Fig. 1.14. In the coordinate system symmetrical with the curve Salzer[12] derives the equation of the caustic to be the form

$$3bx''^2 = y''(y'' - b)^2, \qquad (1.19)$$

which we find to be (with a translation of coordinates) a form of sinusoidal spiral known as Tschirnhausen's cubic (Ref. 7, p. 88). It is comparatively simple to reproduce the operative part of this caustic by refraction from the interior of a dielectric medium to free space. This together with the method of establishing external virtual sources, gives a correcting lens for the offset paraboloid. The process could be repeated for the tangential plane.

The theory of Damien

In view of the foregoing a theory that can have great value in the derivation of caustic curves is very valuable. This is a theory given by Damien[13] and is based on essentially the same statements of Fermat's principle as are incorporated in Eqns (1.1), (1.2) and (1.3). In actual fact the theory was also

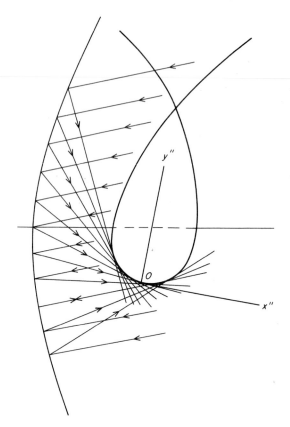

FIG. 1.14. *Caustic of asymmetrically illuminated parabola. Tschirnhausen's cubic.*

stated in much more general terms by Hamilton (Ref. 9, p. 291) as an extension to previous studies on the same subject by Cauchy. In the heading Hamilton terms it "... a new theory of reciprocity giving a new construction for the wave ... connecting the cusps and circles of contact on Fresnel's wave with circles and cusps of the same kind on the 'surface of components'" (adapted).

The connection between them is made through "reciprocal radii" that which we would now call an inversion.

More explicitly enunciated by Damien to whom we attribute the theory it states: (Fig. 1.15)

"If the surface of the wave given by rays issuing from a point source S, after refraction in a surface g separating two media of refractive indices η_1 and η_2, is a surface h; then the surface g' which is the inverse of g with respect to S will be the surface of the wave given by rays issuing from S and refracted by h' which is the inverse of h with respect to S, the magnitudes of the inversions being the same in both cases".

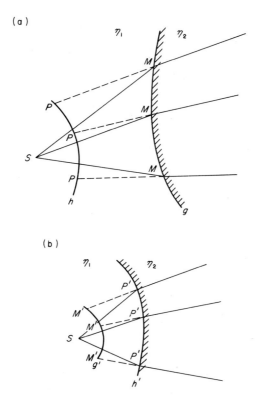

FIG. 1.15. *Damien's theorem.* (a) $\eta_1 SM = -\eta_2 MP$. h is wave surface of zero distance. (b) g' is wave surface of zero distance; the result of inversion in a circle centre S.

Note particularly that refracting surfaces invert into wave fronts and vice versa. The wave surface referred to is that obtained by projecting back along the ray an equal optical length to the distance from the source. This is termed the wave front of zero distance.

Of the many interesting caustics that Damien develops from this theory such as the reflection of rays in the involute of the circle† and in the hyperbolic spiral‡ we give that of the reflection of a point source in a circle, Section 14 of the reference given:

† The wave front is an Archimedean spiral!
‡ The wave front is a spiral tractrix!!

Consider a parabola of focus S and directrix d (Fig. 1.16). Rays from S reflected in the parabola are perpendicular to d which is thus the "wave front of emergence". Then the wave front of emergence for the inverse of d

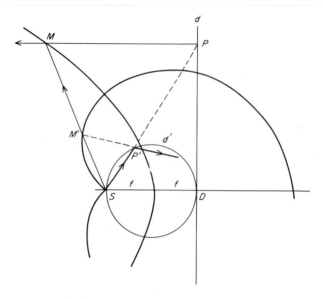

FIG. I.16. *Damien's theorem for the parabola.*

with respect to S, that is a circle passing through S, is the inverse of the parabola in the same circle. This is a cardioid with cusp at S. Letting D be the foot of the perpendicular from S to d and putting $SD = 2f$ the equation of the parabola is

$$r = \frac{2f}{1 + \cos \theta}$$

and the equation of the vertical line is

$$r = \frac{2f}{\cos \theta}.$$

Making an inversion of magnitude $4f$ the line d becomes the circle $r = 2f \cos \theta$ and the parabola the cardioid $r = 2f(1 + \cos \theta)$. The cardioid is then the zero distance wave surface for a source on the surface of a circular reflector.

With regard to refraction Damien uses the theory to convert refraction at a spherical interface into the Cartesian ovals and hence to the aplanatic points of the sphere, with which we are already familiar. The inverse of the conic refracting lens is a limacon of Pascal by which method the curve of Fig. 1.11 could have been derived from the lens of Fig. 1.18(b). Ideally one would like to continue this process for a progression of refractive surfaces but this has not been performed as yet.

The question then arises whether such a transformation of this kind is unique. In general a wave front as it progresses through a uniform medium does not remain form invariant. In every family of progressive wave fronts created by the refraction of rays from a point source however, there exists a unique one, that is the wave front whose optical distance from the source is

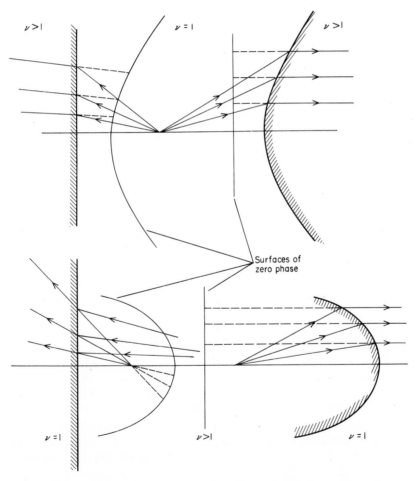

FIG. 1.17. Zero phase surfaces for plane, elliptical and hyperbolic refraction.

zero. This implies simply that this particular wave front is virtual. With this choice we find[14] that refraction at a plane interface creates either a hyperbolic or elliptical zero distant wave-front depending on whether the refraction was from a less dense to a more dense medium or vice versa. We therefore have the situation in which a hyperbolic profile creates a plane wave front and a plane profile a hyperbolic wave front when refraction is from a less dense to a more dense medium, and an elliptical profile has the same reciprocity when the refraction is the other way. We can see therefore that the transformation is that of a simple translation but with the substitution of wave fronts for (essentially) refracting surfaces and the reverse. This is as illustrated in Fig. 1.17.

We shall continue this discussion on the transformations of optical surfaces in a subsequent chapter.

1.5 MICROWAVE LENSES

In addition to the optical lenses which operate at microwave frequencies in the same way as with light, it is possible to use a refractive medium at microwave frequencies based upon the difference in phase velocity that exists between a wave in free space and a wave confined between parallel metal walls. Where the separation of the walls is between a half and one wavelength of the radiation being considered, the dominant TE_{01} mode alone propagates when the polarization is parallel to the walls themselves. In this situation a cylindrical lens fed by a line source is designed by the same principles as for the optical lens with the final result modified only by the fact that the apparent refractive index of the medium is less than unity. This refractive index is given by the relation

$$\eta = \sqrt{\left[1 - \left(\frac{\lambda_0}{2a}\right)^2\right]} \tag{1.20}$$

where a is the spacing between the plates and λ_0 the operating wavelength in free space.† Substitution of a refractive index less than unity directly into the profiles given by Eqns (1.9) and (1.11) convert them from hyperbolic and elliptical respectively into elliptical and hyperbolic (Fig. 1.18). The second surface is then chosen to be normal to the incident or transmitted rays as appropriate. When a point source is used however this situation arises only in one plane of incidence that containing the direction of the polarization

† In practice the range of values for η is between 0·5 and 0·8.

of the source. In the orthogonal plane, the wave is constrained to change its direction to be parallel to the plates. A constraining action also occurs when the polarization is perpendicular to the set of parallel plates. These constraining actions give rise to pseudo refractive indices which are variable with the angle of incidence of the wave.

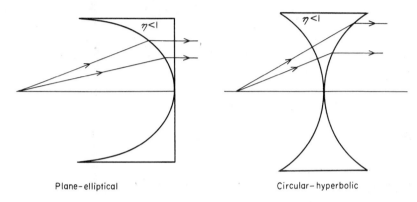

Plane-elliptical   Circular-hyperbolic

FIG. 1.18. Metal plate lenses.

In order to make a lens which will affect all angles of incidence and polarization equally, as would be required by a rotationally symmetric device, a metal plate medium is constructed of two orthogonal sets of metal plates equally spaced (usually) to give a square waveguide medium with refractive index given by Eqn (1.20). Such lenses have to be designed on the basis of optical path length because of the variability of the pseudo-refractive index with angle of incidence.

The principles of these two fundamental designs are well-established over many years.[10, 15]

It should be noted that the frequency dependence in Eqn (1.20) gives a more highly dispersive medium than is found with natural dielectrics. Hence all designs based upon it will have an optimum frequency of operation and an associated band width over which the performance will only be retained to within an acceptable limit. This means that the additional aberration of chromatism has now been added to the one or two normal ones associated with microwave optical devices.

Other forms of refractive index media of the waveguide type can be constructed from other shapes of waveguide notably the hexagonal medium, the rectangular medium (for a single polarized source) and if the 10% loss can be tolerated, a medium of circular waveguides. A further valuable degree of freedom can also be obtained by making the medium variable in refractive index by altering the dimensions of the waveguides in the appropriate areas, subject to practical construction of course. The main distinction between microwave lenses of this type and optical lenses is that the latter have their maximum axial width at the centre. As can be seen from Fig. 1.18, the reverse is the case for refractive index media less than unity, therefore in the following the thickness of the lens at the centre will be taken to be zero.

We now show that, because of the constraining action of the lens, the two main forms with elliptical and hyperbolic profiles respectively do not form two isolated instances as do the corresponding optical lenses but are two types of an infinitely continuous variation in form any of which can be chosen for an advantageous geometry or illumination distribution

1.5.1 Focusing lenses

The general two surface lens with constant refractive index will consist of two curves OPP' and OQQ' (Fig. 1.19) where the front face is a curve

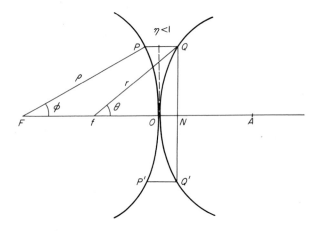

FIG. 1.19. General two surface wave-guide lens.

(ρ, ϕ) described from an origin at the source F. and the rear face a curve (r, θ) described from an origin f. Without loss of generality the lens can have zero thickness at the centre, since in this case the addition of any constant optical length to the rays interior to the lens would produce a curve OQQ' parallel to the one with zero thickness as shown.

Letting for brevity $OF = F$ and $Of = f$ then equality of the optical path to produce a plane wave front QNQ' normal to the axis requires that for any points P and Q on the surfaces

$$FP + \eta PQ = FO + ON, \tag{1.21}$$

or

$$\rho + \eta[F - \rho \cos \phi + r \cos \theta - f] = F + r \cos \theta - f$$

and hence

$$\rho(1 - \eta \cos \phi) - F(1 - \eta) = (1 - \eta)(r \cos \theta - f). \tag{1.22}$$

The principle of constraint states that the rays PQ are all parallel to the axis, that is

$$r \sin \theta = \rho \sin \phi. \tag{1.23}$$

We can thus make a separation of variables between Eqns (1.22) and (1.23) by making the (ρ, ϕ) sides of each functionally dependent and the (r, θ) sides the same functional dependence.

That is the two surfaces can be defined generally to be

$$\rho(1 - \eta \cos \phi) - F(1 - \eta) = \text{any function of } \rho \sin \phi$$

and

$$(1 - \eta)(r \cos \theta - f) = \text{same function of } r \sin \theta. \tag{1.24}$$

The most elementary solution is to make the arbitrary function simply a constant multiplier which we take to be $\alpha(1 - \eta)$; $0 < \alpha < \infty$. This gives

$$\rho(1 - \eta \cos \phi) - F(1 - \eta) = \alpha(1 - \eta)\rho \sin \phi$$

and

$$(1 - \eta)(r \cos \theta - f) = \alpha(1 - \eta) r \sin \theta.$$

The resulting profiles are the ellipses

$$\rho = \frac{(1 - \eta)(F + \alpha \rho \sin \phi)}{1 - \eta \cos \phi},$$

or

$$\rho = \frac{F(1 - \eta)}{1 + \alpha \sin \phi + \eta(\alpha \sin \phi - \cos \phi)}, \tag{1.25}$$

and the hyperbolas

$$r = \frac{f}{\cos \theta - \alpha \sin \theta}. \tag{1.26}$$

When α is zero the elliptical, plane profiles are obtained.

The asymptotes of the hyperbola make angles $\pm \tan^{-1}\alpha$ with the axis. Consequently a continuous variation of the single parameter α changes the shapes of both profiles causing the lens to shear in the axis direction retaining a changing elliptical profile on its front face and a corresponding changing hyperbolic profile on its rear face.

Thus a technique of "lens bending" becomes essentially simple to perform with this type of constrained medium. The analogous technique for "ordinary" optical lenses is currently the subject of much investigation.[16]

Equations (1.24) may also be used to determine the second profile of the lens from any given first profile. Thus to design a lens whose second surface QOQ' is a circle of radius a centred at A we put its equation in the form

$$2a(r \cos \theta - f) - (r \cos \theta - f)^2 = r^2 \sin^2 \theta,$$

which is a functional of the left hand side of the second of Eqns (1.24) in terms of another function of the right hand side.

The required profile for the (ρ, ϕ) face will then be obtained by repeating the functions in terms of the first of Eqns (1.24) that is

$$2a\left[\frac{\rho(1 - \eta \cos \phi)}{1 - \eta} - F\right] - \left[\frac{\rho(1 - \eta \cos \phi)}{1 - \eta} - F\right]^2 = \rho^2 \sin^2 \phi.$$

In general terms the functional relations of Eqns (1.24) do not need to be explicit.

Some generalizations of this basic procedure can now be envisaged. For instance a non-uniform lens can be considered in which η is a function of distance from the axis. This makes it a function of both $r \sin \theta$ and $\rho \sin \phi$ and it can be included in its appropriate form in the respective equation of the two given in Eqns (1.24). Another generalization would be to alter the law of constraint to make say $r \sin \theta = \beta \rho \sin \phi$ (β a second constant parameter) but structural complexity would doubtless outweigh most of the advantages that such designs may induce. On the theoretical side however the existence of these additional degrees of freedom does imply that beside the basic focusing property, consideration can be given to other properties such as amplitude distribution, bandwidth, additional points of focus, or application of the sine condition (Appendix I).

1.5.2 Wide angle lenses

Microwave lenses of the fully constrained type can use the available degree of freedom to provide exact focus at two symmetrically placed points transverse to the axis. The design due to Ruze[17, 18] shows that as performed through the equality of the optical path from the source to each focus in turn, the shape of the front profile is established but only a single relation between the thickness of the lens and the refractive index is given, and thus the shape of the second surface, is not fully determined. A second condition is therefore required to make this relationship determinate and this allows for a third axial point to be made either a point of exact focus or at least of minimum residual aberration. The three points thus defined specify an arc along which the source can travel at each point of which the residual deviation from a·plane wave front can be kept to within specified limits. This arc is usually taken for simplicity to be the circle containing the three points given above and the angle of travel is marginally greater than that subtended by the outer two perfect foci. The design of this lens formed something of a departure from the normal design of optical lenses and its main features will be summarized here for completeness.

The two curved surfaces required are POP' and $QO'Q'$ (Fig. 1.20) where at this stage a finite thickness d_0 is assumed at the centre. F_1 and F_2 are the

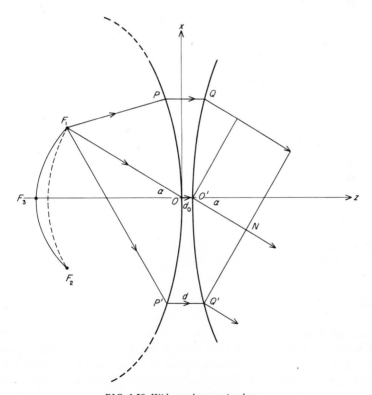

FIG. 1.20. Wide-angle scanning lens.

two chosen perfect foci subtending angles $\pm \alpha$ with the axis at O'. The refractive index η is a function of distance from the axis which we take to be the x coordinate axis. The axis of the lens is the z axis and the value of the refractive index at the centre is η_0

Then since F_1 is a perfect focus

$$F_1P' + \eta P'Q' = F_1O + \eta_0 OO' + O'N. \qquad (1.27)$$

Calling the thickness $P'Q' = d$ and $F_1O = f$, Eqn (1.27) in Cartesian coordinates becomes

$$[(x + f \sin \alpha)^2 + (z + f \cos \alpha)^2]^{\frac{1}{2}}$$
$$= f + \eta_0 d_0 - \eta d + (d - d_0 + z) \cos \alpha + x \sin \alpha. \qquad (1.28)$$

The condition that F_2 be a perfect focus is obtained by replacing α by $-\alpha$ in the above, subtracting, and squaring to remove roots gives firstly

$$[(x + f \sin \alpha)^2 + (z + f \cos \alpha)^2]^{\frac{1}{2}} - [(x - f \sin \alpha)^2 + (z + f \cos \alpha)^2]^{\frac{1}{2}}$$
$$= (x \sin \alpha + f) - (f - x \sin \alpha) = 2x \sin \alpha \qquad (1.29a)$$

and

$$z^2 + 2xf \cos \alpha + x^2 \cos^2 \alpha = 0. \qquad (1.29b)$$

Substitution back then gives[18]

$$\eta_0 d_0 + \eta d + (d - d_0 + z) \cos \alpha = 0. \qquad (1.30)$$

Equation (1.29b) gives the profile of the first surface POP'. This is an ellipse, but contrary to the previous designs has its major axis perpendicular to the lens axis, and with F_1F_2 as its foci.

Equation (1.30) is one relation between η and d and a second arbitrary relation can be chosen. Those selected in the original reference were

(a) a lens of constant thickness
(b) a lens with a plane second surface
(c) a third and arbitrary point of perfect focus on the axis of the lens (the preferred position being on the circle through F_1F_2 centred at O), and
(d) a constant refractive index.

Each of (a), (b) and (d) then gave a preferred point of minimum phase distortion on the axis. The errors in phase from the plane wave front for a general position of the source along the circle through the three points of best focus can now be computed directly from the path lengths along the rays. If a level is then set for permissible deviations from the plane phase front, the scan angle is then determined as that angle subtended by the source at which these levels are attained. The levels are somewhat arbitrary and permit a deterioration up to a degree of the radiated pattern shape but once selected they serve as a basis of comparison for the different possibilities inherent in the design.

It was immediately apparent that even with the three points of focus so defined, the optimum curve for the source to travel, the so-called scanning arc, was not apart from (c) above, the circle through them. When the best scanning arc was achieved by a refocusing procedure for minimum phase error at all points along the path the total scan angle achieved was found to be

$$158\left(\frac{f^2\lambda}{a}\right)^{\frac{1}{3}} \text{ degrees,}$$

where λ is the operating wavelength and a the full aperture of the lens.

The refocusing procedure is essential for any optical system that requires an optimum point of focus over a wide range of angles. It consists of estimating to the first order the change in the optical path obtained by an incremental movement of the source along the radius vector to the centre of the lens. This particular motion is not the only one that could be considered but it is the most obvious practical choice.

Then if the feed is moved a small fraction εf (Fig. 1.21) of the distance f from the centre of the lens the resulting path length change to the point $P(x, z)$ is given by (Ref. 8, p. 78)

$$\Delta = FP - F'P - \varepsilon f \simeq \frac{\varepsilon x^2}{2f}(1 - \theta x). \tag{1.31}$$

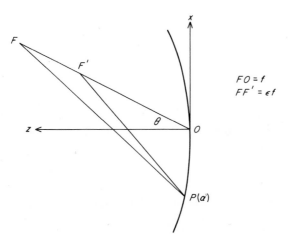

$$FO = f$$
$$FF' = \epsilon f$$

FIG. 1.21. Refocusing.

Such a term can thus be used to reduce both second and third order aberration coefficients, that is spherical aberration and coma. From the point of view of beam shape and beam position, where a choice has to be made it would be preferable to correct the third order term and leave the residual second order.

We note for future reference that the ellipse shape of the front surface of this lens can be given parametrically in the form

$$z = f \cos \alpha(1 - \cos t), \tag{1.32}$$

$$x = f \sin t,$$

where f and α are constants defining the polar coordinates of the foci F_1 and F_2.

Three-ray lens

The second method for designing a wide-angle scanning lens is derived from the above by an application of the principle of duality, a fundamental principle of projective geometry. The resultant method shows greater adaptability for use in the design of natural dielectric lenses with a similar wide-angle property, as well as in the design of lens-reflectors for which the former method is not totally suited.

In the simplified form required for this adaptation the principle of duality states that two dual geometrical propositions can be obtained from each other through the interchange of lines into points and points into lines.[19] For example as two points define a straight line so do two straight lines define a point. A similar duality occurs in high dimensions for example the duality between points and planes in three dimensions.

In the design method above we have arranged that all the rays from three given directions converge individually into three perfect foci. The directions are $\pm\alpha$ and the axial direction and the foci F_1, F_2 and the third axial point F_3 as shown in Fig. 1.20. We form the geometrical dual of this statement as follows; at every point along an arc F_1F_2 three of the rays of the complete pencil are correctly in phase.

These rays may be arbitrarily chosen but from the simplicity of the subsequent analysis we choose

  (i) the main ray to the centre of the lens,
  (ii) the ray parallel to the axis of the lens, and
  (iii) the ray meeting the lens at a point diametrically opposite that of the second ray.

This arbitrary choice receives further justification in its natural occurrence in the design of bi-cylindrical lenses (p. 65)

The effect as the source moves over the arc is to spread the rays further apart making more of the surface of the lens conform at least in part to the required focusing property. The design is performed in exactly the same way,

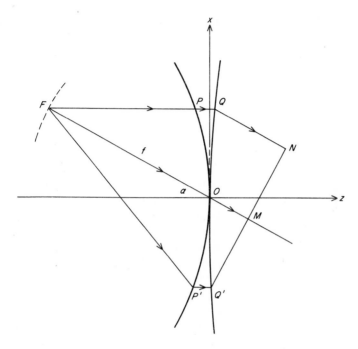

FIG. 1.22. *Three-ray lens. Rays from the variable position of F through variable points P, P' are in phase.*

by assuring constancy of the optical length of the rays chosen. Then from Fig. 1.22 we require

$$FP + \eta PQ + QN = FO + OM = FP' + \eta P'Q' \qquad (1.33)$$

where a zero centre thickness has been assumed. By symmetry we must have $PQ = P'Q'$ hence the outer equation of (1.33) is simply

$$FP + QN = FP'$$

or squaring and rearranging

$$Q'N^2 = 2FP \cdot QN$$

that is $4f^2 \sin^2 \alpha \cos^2 \alpha = 2(f \cos \alpha - z) \cdot 2f \sin^2 \alpha$ and hence in parametric form

$$z = f \cos \alpha (1 - \cos \alpha), \tag{1.34}$$

$$x = f \sin \alpha,$$

where f and α are now essentially variable and f as a function of α defines the scanning arc.

Solving the remaining equation in (1.33) now gives

$$\eta PQ = FT - FP$$

or

$$PQ = \frac{f \cos^2 \alpha (1 - \cos \alpha)}{\cos \alpha - \eta}. \tag{1.35}$$

With f now given as a function of α Eqns (1.34) and (1.35) give parametrically the two surfaces of the lens. (A second condition does not exist here since we have already chosen a constant refractive index.) Totally general lenses can now be designed from the a priori specification of any one of the three curves alone, that is the scanning arc, the front surface, or the rear surface. Variation in refractive index with distance from the axis can be included by making η a function of $\sin \alpha$ and incorporating it directly into Eqn (1.35).

Noting the similarity of description that exists between the parametric descriptions of the front surface of the three ray design and the two focus design Eqns (1.34) and (1.32) respectively, we find that if we were to specify the three rays to be (Fig. 1.23)

(i) the central ray FO, and

(ii), (iii) the rays through two *fixed* points P and P'

equality of the optical path then gives

$$QN + FP = FP'$$

or if P is the point z, x_p where x_p is constant

$$[(z + f \cos \alpha)^2 + (x_p + f \sin \alpha)^2]^{\frac{1}{2}} - [(z + f \cos \alpha)^2 + (x_p - f \sin \alpha)^2]^{\frac{1}{2}}$$

$$= 2x_p \sin \alpha$$

which is identically Eqn (1.29a) with x_p now a constant. The resulting profile in this case is therefore

$$z^2 + 2zf \cos \alpha + x_p^2 \cos^2 \alpha = 0$$

in analogy to Eqn (1.29b), but with f and α variable and x_p constant.

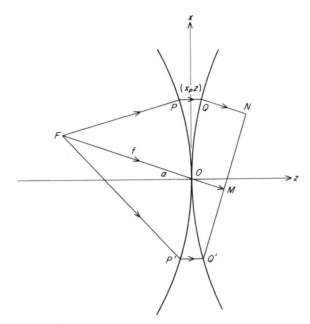

FIG. 1.23. Three-ray lens (alternative design). Rays from the variable position of F through fixed points P, P' are in phase.

1.6 PHASE CORRECTED REFLECTORS

1.6.1 Design procedure

We now consider the two surface system containing one reflecting and one refracting surface. The additional degree of freedom will be used to obtain a high degree of collimation when the source is at a wide angle of displacement from the axis. The first such mirror designed to be free of coma and spherical aberration was described by Mangin in 1876.[20] This consisted of two spherical surfaces the convex one being silvered. The application of such a mirror to microwave antennas has been fully investigated.[21] If it is required to use such a mirror as a mechanical scanner it is an obvious advantage to have the scanning arc on a circle. If further this circle was centred at the centre of the lens, a two to one advantage can be obtained by keeping the source fixed and rotating the mirror. Such a circular scanning arc is not a fundamental property of the original design of Mangin mirror. An improvement can best be sought by using the wide angle property of microwave lenses, and introducing a reflecting surface in the interior of the lens to convert it into a reflector with a correcting layer of refractive material. When this process is performed on the lenses designed by the two focus method, the degree of freedom remaining that allows for a third point on the scanning arc to be determined, is no longer available. This leaves the scanning arc unde-

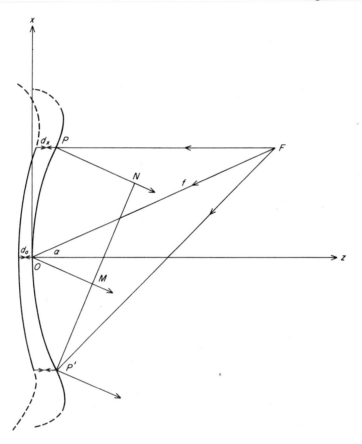

FIG. 1.24. *Phase corrected reflector.*

fined. This is not the case with the three ray method.[22] With the original choice of the "spreading" rays as shown in Fig. (1.24), we require for equi-phased points $P'MN$

$$FP + 2\eta(x)d(x) + PN = FO + 2\eta_0 d_0 + OM = FP' + 2\eta(x)d(x). \tag{1.36}$$

The outer equation gives

$$FP + PN = FP'$$

and the geometry is identical to that of the complete lens resulting in the same parametric form of profile as Eqn (1.34).

$$z = f \cos \alpha (1 - \cos \alpha),$$

$$x = f \sin \alpha.$$

The solution for the thickness $d(x)$ of the lens coating is then

$$2[d(x)\eta(x) - d_0\eta_0] = z \cos \alpha. \tag{1.37}$$

In particular if $\eta(x)$ is made constant for ease of construction Eqn (1.37) becomes

$$2\eta[d(x) - d_0] = z \cos \alpha = f \cos^2\alpha(1 - \cos \alpha). \tag{1.38}$$

A comparison of the curves given by Eqn (1.34) (with f assumed to be constant), Eqn (1.32), and the circle given by

$$z = r(1 - \cos \alpha)$$

$$x = r \sin \alpha$$

shows that the central portion of the former more closely approximates the circle than does the ellipse of Eqn (1.32). Thus the three ray method produces a profile more in agreement with the sine condition for minimum coma than does the two focus profile.

1.6.2 Residual phase distribution

Since only three of the rays from each point are correctly phased, it is important to obtain expressions from the phase distribution for the other rays. Let the feed be in any angular position θ. We now require the difference in path length between the main ray to the centre and any other ray meeting the system at a point of parameter α. With the surfaces of the system given in terms of α by Eqns (1.34) and (1.38), this path-length difference, E_p, is

$$E_p = f(\theta) - f(\alpha) \sin \alpha \sin \theta + f(\alpha) \cos \alpha \cos \theta (1 - \cos \alpha)$$
$$- f(\alpha) \cos^2 \alpha (1 - \cos \alpha) - \{[f(\theta) \cos \theta$$
$$- f(\alpha) \cos \alpha (1 - \cos \alpha)]^2 + [f(\theta) \sin \theta - f(\alpha) \sin \alpha]^2\}^{\frac{1}{2}}, \tag{1.39}$$

where the sign of α is the same as or opposite to that of θ, according as the surface point α is on the side adjacent to, or remote from, the displaced feed at θ. From this it can be seen that $E_p = 0$ when $\alpha = \pm\theta$ and when $\alpha = 0$, in accordance with the choice of the three focused rays.

1.6.3 Scanning properties of particular reflectors

From the parametric expressions of Eqns (1.34) and (1.38) (with constant refractive index) and the phase distribution given by Eqn 1.39, we can analyse systems which are specified in one of three different ways:

(a) definition of f in terms of α. This specifies the scanning arc and hence the two profiles.

(b) Specification of the refracting profile. Comparison of the equation to this curve with the parametric form required provides a definition of f in terms of α and hence the scanning arc and the reflecting surface.

(c) Specification of the reflecting profile. When this profile is given, it is apparent that the shape of the refracting profile is dependent upon the refractive index of the surface coating. Thus since the scanning arc depends on the shape of the refracting profile, it too contains terms involving the refractive index. In this case, if the refractive index is given, the refracting profile can be obtained and the procedure is then the same as in (b).

Specification of scanning arc

Two scanning arcs are of interest, the circle with centre at the reflector centre, and the straight line perpendicular to the axis. The former provides a scan motion which is mechanically simple to perform either for movement of the feed or for tilt of the reflector. The latter focuses plane waves arriving at an angle to the axis into points of a plane, which, in pure optics, is a requirement for astronomical photographic processes.

Circular scanning arc. Until a later investigation revealed other refocused systems with a circular scanning arc, the first experiments were carried out with a reflector with $f = \text{constant} = c$.

This gives

$$\left.\begin{array}{l} z = c\cos\alpha(1 - \cos\alpha), \\[4pt] x = c\sin\alpha, \\[4pt] d = \dfrac{x\cos\alpha}{2\eta}, \end{array}\right\} \tag{1.40}$$

the centre thickness, d_0, being zero. The residual phase distribution is shown in Fig. 1.25.

Straight line scanning arc. To obtain a straight scanning arc at right angles to the axis, put $f = c/\cos\alpha$. The system is thus defined by

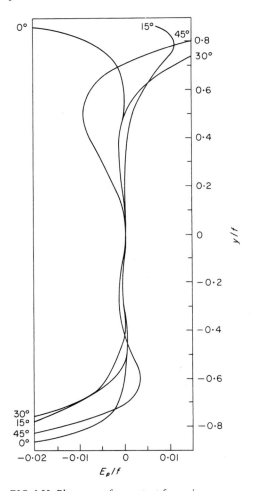

FIG. 1.25. Phase error for constant-f scanning arc.

$$z = c(1 - \cos \alpha),$$
$$x = c \tan \alpha,$$
$$d = \frac{x \cos \alpha}{2\eta},$$

$$(1.41)$$

the centre thickness being zero.

The residual phase distribution is given in Fig. 1.26. The high degree of asymmetry greatly limits the achievable angular aperture.

Specification of the refracting profile

Let $x^2 = F(x)$ be the equation of the refracting profile. If this profile is

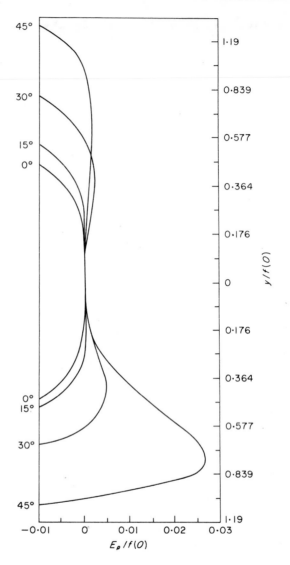

FIG. 1.26. *Phase error for straight-line scanning arc.*

also to be given by the parametric form of Eqn (1.34),

$$f^2 \sin^2 \alpha = F[f \cos \alpha (1 - \cos \alpha)]$$

and thus f can be obtained as a function of α, defining the required scanning arc.

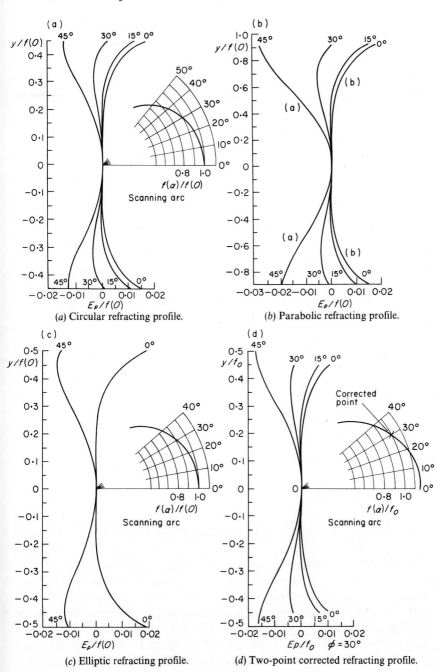

(a) Circular refracting profile.

(b) Parabolic refracting profile.

(c) Elliptic refracting profile.

(d) Two-point corrected refracting profile.

FIG. 1.27.

C

For example, if the refracting profile is a circle of radius r defined by the equation $z^2 + x^2 - 2zr = 0$, substituting for z and x from Eqn (1.34) gives

$$f^2 \cos^2 \alpha(1 - \cos \alpha)^2 + f^2 \sin^2 \alpha - 2rf \cos \alpha(1 - \cos \alpha) = 0,$$

i.e.

$$f = \frac{2r \cos \alpha}{\cos^2 \alpha(1 - \cos \alpha) + (1 + \cos \alpha)}.$$

This equation defines the scanning arc for a circular cylindrical refracting surface.

The results of this operation for several refracting profiles are given in Table 1.1. The phase distributions given in Fig. 1.27 have a high degree of symmetry about the main ray for offset angles up to 45°. The similarity between these curves and the profiles of Schmidt correcting plates for spherical mirrors is marked since the main residual phase error is spherical aberration. Such systems may be refocused with advantage, and this will be dealt with in a later section.

The scanning arc for the 2-point corrected profile passes through the two correcting points as expected (Fig. 1.27(d)). It is not, however, the anticipated circle centred on the reflector centre as can be seen.

It may be noted that attempts to design a reflector with a tapered or plane refracting surface prove to be impossible. If the apex is at the origin no solution for f in terms of α can be obtained. If the apex is behind the origin the distance f becomes infinite.

Specification of the reflecting profile

We assume the systems to have zero thickness at the centre and constant refractive index. From Eqns (1.34) and (1.38) the reflecting surface has equations

$$x_1 = x = f \sin \alpha$$

$$z_1 = z - d = z\left(1 - \frac{\cos \alpha}{2\eta}\right) = f \cos \alpha(1 - \cos \alpha)\left(1 - \frac{\cos \alpha}{2\eta}\right). \quad (1.42)$$

When this is expressed as $x_1^2 = F(z_1)$, the same analysis can be made as in the previous section. This method has been applied to the design of the corrected parabola and corrected circle, with the results as shown in Table 2 and Fig. 1.28. In both cases the axial focal length $f(0)$ contains a factor $2 - 1/\eta$ which gives a lower limit of 0·5 to the possible range of refractive indices that can be used for the correcting layers. Furthermore, the axial focal length of a corrected parabola of focal length a is double that of a corrected circle of radius a, as is the case for uncorrected reflectors. In both

TABLE 1.1

Refracting profile	Equation of profile	Polar equation of scanning arc	Axial focal length $f(0)$	Phase distribution and scanning arc reference
Circle (radius r)	$z^2 + x^2 - 2rz = 0$	$f = \dfrac{2r\cos\alpha}{\cos^2\alpha(1-\cos\alpha)+(1+\cos\alpha)}$	r	Fig. 1.27(a)
Ellipse (semi-axes a, b)	$z^2 + \dfrac{a^2}{b^2}x^2 - 2az = 0$	$f = \dfrac{2a\cos\alpha}{\cos^2\alpha(1-\cos\alpha)+a^2/b^2(1+\cos\alpha)}$	$\dfrac{b^2}{a}$	Fig. 1.27(b)
Parabola (focal length a)	$x^2 = 4az$	$f = \dfrac{4a\cos\alpha}{1+\cos\alpha}$	$2a$	Fig. 1.27(c)
Two-point correction. Ellipse [foci $(f_0, \pm\phi)$]	$z^2 + x^2\cos^2\phi - {}$ $-\,2f_0 z\cos\phi = 0$	$f = \dfrac{2f_0\cos\phi\cos\alpha}{\cos^2\alpha(1-\cos\alpha)+\cos^2\phi(1+\cos\alpha)}$	$\dfrac{f_0}{\cos\phi}$	Fig. 1.27(d) $\phi = 30°$

TABLE 1.2

Reflecting profile	Equation of profile	Polar equation of scanning arc	Axial focus length $f(0)$	Phase distribution and scanning arc reference
Parabola	$x_1^2 = 4az_1$	$f = \dfrac{4a \cos \alpha \left(1 - \dfrac{\cos \alpha}{2\eta}\right)}{1 + \cos \alpha}$	$a\left(2 - \dfrac{1}{\eta}\right)$	$\eta = 1{\cdot}6$ Fig. 1.28(a)
Circle	$z_1^2 + x_1^2 - 2az_1 = 0$	$f = \dfrac{2a \cos \alpha \left(1 - \dfrac{\cos \alpha}{2\eta}\right)}{1 + \cos \alpha + \cos^2 \alpha \left(1 - \dfrac{\cos \alpha}{2\eta}\right)^2 (1 - \cos \alpha)}$	$\dfrac{a}{2}\left(2 - \dfrac{1}{\eta}\right)$	$\eta = 1{\cdot}6$ Fig. 1.28(b)

cases a valid solution is obtained for $\eta = 1$. This means that the constraint alone enables phase correction to be obtained.

The angular aperture

From the diagrams of the residual phase error it is possible to determine either the maximum angular displacement that a system with a given numerical aperture can permit or, for a maximum required angular displacement, the numerical aperture that will allow it. This can be done by defining the maximum permissible residual phase error and offsetting the feed along the scanning arc to the angle at which this maximum occurs; or conversely, by offsetting the feed to the maximum angle required and limiting the aperture to the point at which the maximum permissible error is obtained.

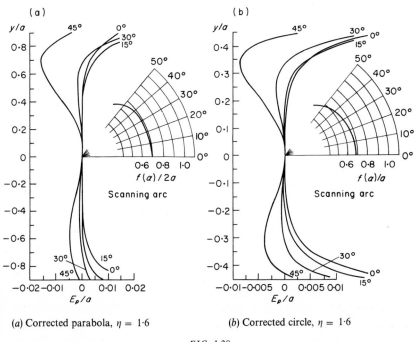

(a) Corrected parabola, $\eta = 1\cdot6$ (b) Corrected circle, $\eta = 1\cdot6$

FIG. 1.28.

1.6.4 Refocusing

Those systems exhibiting large symmetrical or nearly symmetrical aberrations about the main ray can be refocused to a certain extent in off-axis positions by a radial movement of the source.

The effect of refocusing is to introduce a quadratic phase error which can be made equal and opposite to the quadratic component of the symmetrical phase distribution. The phase-distribution diagrams for the conic-section refracting profiles (Fig. 1.27) show that the central portions of the symmetrical phase-error curves can be adequately approximated by a quadratic term.

If the feed is at a position F (Fig. 1.21) on the scanning arc of one of these reflectors, a displacement inward along FO by an amount $\varepsilon f(\theta)$ where $FO = f(\theta)$ results in a change in the path length relative to the central ray given by Eqn (1.31).

$$\Delta \simeq - \frac{\varepsilon}{2} \left[\frac{x^2}{f(\theta)} - \frac{\theta x^3}{f(\theta)^2} \right],$$

which is correct to the first order in ε and θ and to the third order in $x/f(\theta)$.

If Δ is then taken as the value of the quadratic approximation to the phase error at any intermediate value of α, the refocusing increment is given by

$$\Delta = \frac{-\varepsilon}{2f(\theta)} \left[f(\alpha) \sin \alpha \right]^2 (1 - \theta \sin \alpha). \qquad (1.43)$$

This can be calculated for several values of θ, with α and E_p given by the phase-error curve corresponding to the particular value of θ. With appropriate sign, this adjusts the scanning arc to a best-focal-position arc.

The cubic term introduces some asymmetry, particularly at the larger values of θ. The refocusing increments, calculated for the reflector with a circular refracting profile, are shown in Fig. 1.29. After removing the quadratic component a small residual phase error remains and a slight asymmetry, consisting mainly of the cubic component, is observed. The refocusing results in a best-focus arc which is circular with radius 0·94 of the original axial focal length.

When this method is applied to the 2-point corrected profile and scanning arc as in Fig. 1.30 the scanning arc once more becomes circular but with radius $1·09 f_0$, and hence it no longer passes through the two correction points.

From the similarity of the parametric description of the scanning arcs and the phase error distribution curves between these two systems and the other systems given in Table 1.1, it can be expected that refocusing these also produces circular scanning arcs. This has been shown to be the case but a complete analysis has not been made. These systems have in common the fact that the refractive profile is given by a curve that is a conic section. Hence these refracting–reflecting combinations constitute a class whose best-focus scanning arc is the circle with centre at the centre of the system. Although this refocusing procedure has destroyed the *a priori* relationship between the refracting profile and scanning arc, the discovery of this class of related curves

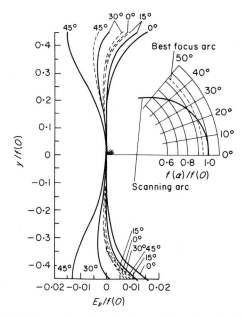

FIG. 1.29. Effect of refocusing on reflector with circular refracting profile.

provides a basis for the iterative design of similar coated reflectors with a natural dielectric correcting layer.

1.6.5 Phase correction with a natural dielectric medium

It may be noted that the shape of the reflectors so far considered, for which the correcting layer is an array of metal wave-guides, is, in general, the same as that of the original negative meniscus lens of Mangin, i.e. concavo–convex. This fact is contrary to the usual experience when comparing microwave with optical lenses. It means that, for these reflectors at least, the shape with a constrained refractive medium is in itself a satisfactory first approximation to a natural dielectric unconstrained correcting layer. This is particularly so in fully stepped systems, in which the thickness of the correcting layer need never be greater than half the wavelength in the medium used. The approximation involved in assuming that the rays are still parallel to the axis when they are within the correcting medium causes the quadratic error of the system to increase slightly. This can be cancelled, therefore, by further focusing.

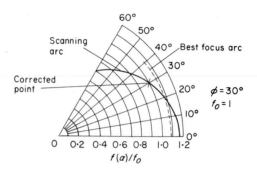

FIG. 1.30. *Refocusing scanning arc of 2-point corrected refracting profile.*

A better approximation for the natural dielectric medium can be made by calculating the refracting profile from the original formula (Eqn 1.34) and then calculating the reflecting profile on the assumption that the rays in the correcting layer travel along the normals to the refracting profile. In this case, the path length along each normal has to be the value of d calculated from Eqn (1.38). This approximation is particularly appropriate to the circular refracting profile, where, in the axial position, the rays do, in fact, travel along these normals. The approximation here also results in an increase in the residual quadratic error and may similarly be corrected.

However, in all natural dielectric systems a further parameter is available in the thickness of the corrected reflector at its centre. It is found that variation of this parameter too gives rise mainly to a quadratic component of phase error which could be used for the cancellation of the above effects.

Finally, the 3-ray principle itself can be used directly for the natural dielectric case. Apart from the necessity of refocusing which has been demonstrated, the method by which the movement of the feed over a given arc describes both this arc and the refracting profile is a stepwise procedure suitable for programming a computer. To do this, one establishes the phase equality of the three rays concerned by considering the localities at which the outer two rays meet the reflector to be sections of thin reflecting prisms, similar in shape and symmetrical with respect to the axis. The 3-ray principle, however, does not give a defined axial focal position. The scanning arc limits towards a point on the axis at which, of course, the three rays coincide. This point cannot therefore be taken as a starting point for the step-wise procedure. This means that a guess has to be made at the thickness at the outer edge, the guess being acceptable if, on reaching the axial position, a zero thickness has not been passed. The procedure is similar to that to be described in Section 1.14.

1.6.6 Experimental results

The analysis presented in the previous Sections has dealt with 2-dimensional or cylindrically symmetrical reflectors. The profiles concerned could, however, also be considered as cross-sections of a rotationally symmetrical reflector. Where such a reflector is required it is constructed by the rotation of the profiles about its axis. This, of course, introduces the further aberration of astigmatism which limits the angle of scan of rotationally symmetrical reflectors to a much lower value than cylindrically symmetrical ones. With the type of construction possible at microwave frequencies, namely a 2-dimensional array of wave-guides, systems with rectangular symmetry can be considered. In these the surfaces would be formed by the translation across a profile of a similar profile at right angles to it. The astigmatism of such systems, however, is still to be investigated.

Experimental work so far has been concerned with the construction of systems to test the elementary theory and to assess the limitations imposed on the scanning properties by the astigmatism of the rotationally symmetrical reflector and the dielectric approximation of the previous paragraph.

To test the basic theory a linear reflector (cylindrical system) was constructed for which the correcting layer was a parallel array of metal plates with effective refractive index of 0·6. The profiles were determined from Eqns (1.34) and (1.38) with f constant: this gives the uncorrected scanning

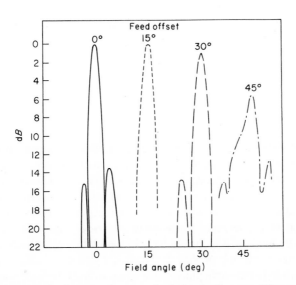

FIG. 1.31. Polar diagrams of constant-f cylindrical reflector at different offset angles.

arc as the circle $f = c$. The reflector is 0·92 m long and has a focal length of
0·77 m operating at a frequency of 9·375 GHz. The f/D ratio is 0·83. From the
residual phase-error curves it is to be expected that coma effects will become
noticeable beyond 30° of angular displacement, because of the asymmetry in
the phase-error curve. This is found to be the case, as shown by the scan patterns of Fig. 1.31. A beam displacement of 30° (approximately 15 beam widths)
is achieved without any observable deterioration in the beam shape, and is
followed quite suddenly by a decrease in gain and a large increase in the
coma lobe.

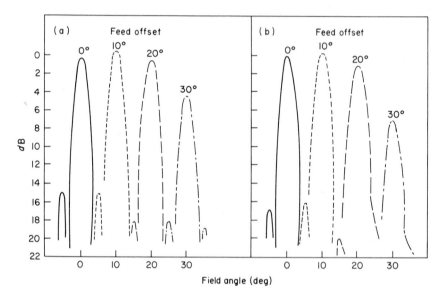

*FIG. 1.32. Polar diagrams of constant-f spherical reflector at different offset angles. (a) H-plane
scan; (b) E-plane scan.*

A rotationally symmetrical system constructed from square-section metal
tubes has scan patterns shown in Fig. 1.32 demonstrating that the gain and
beam shape deteriorate steadily on displacing the feed source along the
scanning arc. Very similar results are achieved with a dielectric-coated reflector designed in accordance with the approximation given in Section 1.6.5.

Examination of the residual phase-error curves of Fig. 1.29 shows that
an expected scan angle of 45° should be achieved with a cylindrical reflector
having a circular refracting profile. A system for which $r = 20\lambda$ is possible
with the maximum permissible phase error equal to $\lambda/10$ and with a refocused
scanning arc which is circular.

A spherical natural dielectric system with the same aperture thus has an expected useful scan approaching 30°. A reflector of this type has been constructed. The scan patterns achieved give a useful scan of 30°, at which point the gain has decreased by 2 dB without any great deterioration in the main beam shape and before the coma lobe level has become pronounced.

Thus to summarize, the design procedure whereby three rays from every point of a specified scanning arc are kept equi-phased, gives rise to a class of phase-corrected reflectors with the following properties:

(a) The refracting profile is a conic section and can, for simplicity, be circular.

(b) The residual phase errors at angles of scan up to 45° are largely symmetrical about the main ray.

(c) Refocusing the system at angles off-axis makes the scanning arc circular.

With such systems used as beam-scanning devices, i.e. with stationary feed and mechanical scan of the reflector, linear scans approaching ±90° can be achieved with cylindrical reflectors, and a cone of semi-angle 60° can be scanned with spherical reflectors.

1.7 TRANSLATION REFLECTORS

A class of reflectors exists, usually designed by empirical methods which consists of the surface obtained by the movement of a given curve in a parallel fashion across a second orthogonal curve (Fig. 1.33). Apart from the particular case of the paraboloid itself, which is the translation of a parabola in this manner around an *identical* parabola, these reflectors are not pure focusing devices. They have thus to be designed from principles somewhat different from the focusing rules of Eqns (1.1) and (1.3). This procedure will be gone into in a later section (see Section 1.13.3). A common form of such a reflector is that of the translation about a parabola of that curve which in a two-dimensional-line source situation would have created a specific shaped radiation pattern.[22] The translation method thus converts what would have been a line source requirement into a point source one. To the much poorer degree of approximation to a specified shape of pattern that is usually allowed, this procedure is permissible.

Other reflectors of this type involve the more simple procedure of rotating a curve of the required kind about a vertical axis. As is well known the parabola itself does not differ greatly from an optimum circle[23] and so the

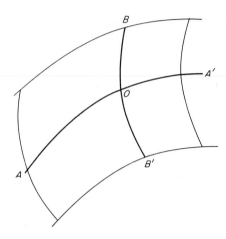

FIG. 1.33. Translation reflector. The reflector is formed by the translation of the generating curve BB' across the curve AA' maintaining a fixed angle AOB.

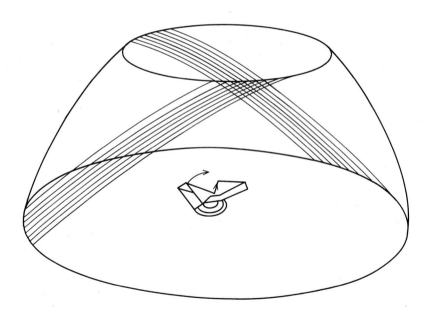

FIG. 1.34. Barrel reflector. The surface is created by rotating a sector of a parabola about a vertical axis and constructed of parallel wires orientated at 45° to the vertical. This enables the polarized feeds to "see through" the opposite surface after reflection. With a switching arrangement the two feeds rotate giving a complete 360° scan with the possibility of elevation of the beam at each rotation.

translation of a parabola about a circle gives a good approximation to the paraboloid, with the added attraction of circular symmetry in one plane. Using the polarization properties of the field it is possible by this means to create a complete azimuth scanning beam[24] as illustrated in Fig. 1.34. Several picturesque names have been attributed to this design. It can also be improved in pattern quality by using a combination of feed sources in the horizontal plane more applicable to the circular form of the cross section in that plane.

If the axis of rotation for the parabola is taken to be on the convex side the reflector known as the Wullenweber results. In the horizontal plane a combination of feed point sources is essential to the achievement of a relatively narrow beam shape.

Apart from the design method to be given in Section 1.13.2, little can be done to codify such reflectors. However, they do establish a more general principle, that of replacing a two dimensional general curved reflector surface with two single dimensional surfaces, that is with two cylindrical surfaces.

1.7.1 Cylindrical reflectors

The simplest translation of this kind is the parallel motion of a given curve along a straight line, to give a right cylindrical reflector. If the equation to the curve is taken to be $f(x, y) = 0$, and the translation made parallel to the z axis, then a source in the plane $z = 0$ gives rise to a virtual line source in the same plane, a focal line. This is obtained by tracing the rays after reflection back to their intersection with this plane and this curve is *independent* of the height of the point of reflection. The equation to this focal line can be obtained by comparatively simple means, but it has a great value in the ensuing theory and is applicable to reflectors in general.

With the equation of the curve $f(x, y) = 0$ and a source of rays in the plane $z = 0$ at the point (d, h) (Fig. 1.35) rays after reflection from the point (a, b, c) on the reflector will intersect the plane $z = 0$ in the point, independent of c, given by

$$x = d + 2\frac{\partial f}{\partial x}\left\{(a - d)\frac{\partial f}{\partial x} + (b - h)\frac{\partial f}{\partial y}\right\}\bigg/\left\{\left(\frac{\partial f}{\partial x}\right)^2 + \left(\frac{\partial f}{\partial y}\right)^2\right\},$$

$$y = h + 2\frac{\partial f}{\partial y}\left\{(a - d)\frac{\partial f}{\partial x} + (b - h)\frac{\partial f}{\partial y}\right\}\bigg/\left\{\left(\frac{\partial f}{\partial x}\right)^2 + \left(\frac{\partial f}{\partial y}\right)^2\right\},$$

(1.44)

where the differentials are evaluated at the point (a, b). Since (a, b) obeys the same relation $f(a, b) = 0$ it is possible to eliminate one or both from these equations to obtain the cartesian equation to the focal line. This curve has

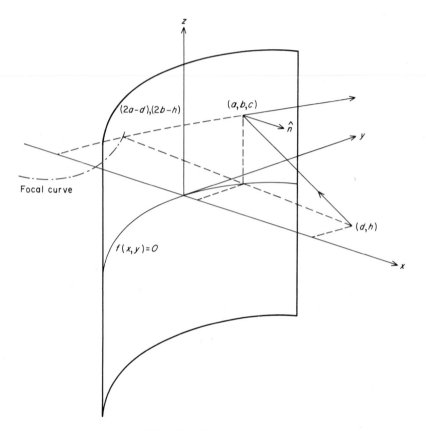

FIG. 1.35. Cylindrical reflector.

more than one interpretation. In a practical sense it is the virtual line source
created by the reflector and point source combination and as such it can
be used to feed a second appropriately shaped translation reflector. By its
very construction however it is the wave front of zero distance for the reflector
cross-section. In those cases where symmetry is preserved this cross section
can be rotated about the axis of symmetry ($y = 0$) to give the complete wave
front. In cases of asymmetry it applies only to the two dimensional problem
or to the single plane of the three dimensional one. These are the wave
fronts to which the theory of Damien applies and thus by inversion can be
used to derive other reflector and wave front configurations. Finally, Eqn
(1.44) gives by far the simplest means of obtaining the wave fronts in cases of
aberration occurring through feed offset from true focal positions. The exact

formulae for such wave fronts will be a necessary requirement for the evaluation of actual radiation patterns by scalar diffraction theory. The following results, which illustrate these aspects can be derived directly from Eqn (1.44).

(a) Parabolic cylinder, focal length t, source at focus

$$f(x, y) = y^2 - 4t^2 + 4tx$$

hence, $x = 2t$; $y = b$ and the curve is the directrix of the parabola in the plane $z = 0$. This is therefore a straight line virtual source, with practical applications that will be dealt with in a later section (Section 1.7.3).

(b) Circular cylinder, radius R, source at centre

$$f(x, y) = x^2 + y^2 - R^2 \text{ giving } x = 2a, y = 2b \text{ and hence } x^2 + y^2 = 4R^2$$

that is a circular source with the same centre radius $2R$.

(c) Circular cylinder, radius R, source at $d = R/2$. In this case $x = a(2R - a)/R + R/2$; $y = b(2R - a)/R$, which upon elimination of a and b results in

$$\left(x - \frac{R}{2}\right)^2 + y^2 = \left\{R - \left(\frac{3R^2}{2} - Rx\right)^{\frac{1}{2}}\right\}^2.$$

This curve shows the resultant spherical aberration from feeding a spherical (by rotation) reflector from its nominal focal point. One could, by varying the source about this position, obtain a result for which the curvature of this focal line was minimum. This would then be the "best" focal point for a spherical reflector.

(d) Elliptical cylinder, source at focus

$$f(x, y) = \frac{x^2}{l^2} + \frac{y^2}{m^2} - 1, \qquad d = (l^2 - m^2)^{\frac{1}{2}}.$$

After some arithmetical complexity this gives the known result, namely a circle of radius $2l$ centred upon the other focus. A similar result is obtained for hyperbolic cylinders.

(e) Axially defocused parabola. With the source position $x = d$, $y = 0$ the result of example (a) becomes on solution

$$(x^2 + y^2 - d^2)(x - 2t - d) = 4td(d - x).$$

This result is subject to rotation to give the wavefront of the axially defocused paraboloid.

(f) Transversely defocused parabola. With the source position $x = 2t$, $y = h$ we obtain the curve

$$\{y(x - 2t) - 2ht\}^2 = x^2\{h^2 - (x - 2t)^2\}.$$

This applies to the two dimensional situation only or in the case of the paraboloid to the cross-section of the wave front by the plane containing the offset feed.

The resulting curve given in (e) closely approximates to a circle and upon rotation then gives the exact equation to the wave front associated with spherical aberration. The curve given in example (f) closely approximates to a cubic and is the exact form of wave front associated with primary coma.

1.7.2 Conical reflectors

The rotation of a general straight line about an axis which intersects it generates a right circular cone for which we can determine the optical properties by the same methods. With the apex of the cone at the origin and axis the z axis the equation of the surface is given by

$$x^2 + y^2 = m^2 z^2,$$

where $m = \tan \theta$ and θ is the cone semi angle (Fig. 1.36).

A source is positioned at a distance h from the apex along the axis then at a general point (a, b, c) on the cone the normal is given by

$$\hat{\mathbf{n}} = \frac{a\hat{\mathbf{i}} + b\hat{\mathbf{j}} - m^2 c\hat{\mathbf{k}}}{(a^2 + b^2 + m^4 c^2)^{\frac{1}{2}}}$$

and an incident ray by

$$S_i = \frac{a\hat{\mathbf{i}} + b\hat{\mathbf{j}} + (c - h)\hat{\mathbf{k}}}{(a^2 + b^2 + (c - h)^2}.$$

Then from formula (A.I.3) of Appendix I the reflected ray will have direction cosines proportional to

$$S_{r,x} = a(a^2 + b^2 + m^4 c^2) - 2a(a^2 + b^2 - m^2 c(c - h))$$
$$S_{r,y} = b(a^2 + b^2 + m^4 c^2) - 2b(a^2 + b^2 - m^2 c(c - h)),$$
$$S_{r,z} = (c - h)(a^2 + b^2 + m^4 c^2) + 2m^2 c(a^2 + b^2 - m^2 c(c - h)).$$

The equation to the reflected ray is therefore

$$\frac{x - a}{S_{r,x}} = \frac{y - b}{S_{r,y}} = \frac{z - c}{S_{r,z}}.$$

Anticipating that the image will lie in the same plane as the image of the source in a tangent plane, we find that the intersection of this ray with the plane

$$z = \frac{h(1 - m^2)}{1 + m^2}$$

is given by

$$x = \frac{2ah}{c(1 + m^2)}; \qquad y = \frac{2bh}{c(1 + m^2)}$$

and thus

$$x^2 + y^2 = \frac{4m^2h^2}{(1 + m^2)^2} = h^2 \sin^2 2\theta.$$

That is the virtual image in this plane is a circle of radius $h \sin 2\theta$.

Of particular interest is the situation where the cone has a half angle of 45°. The plane containing all virtual images of axial sources is then the plane

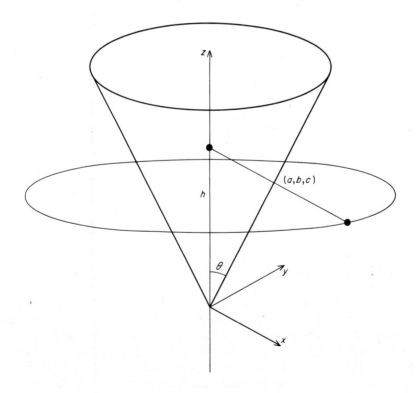

FIG. 1.36. Conical reflector.

containing the apex. The virtual images of an array of sources along the axis is then an array of circular rings and with each ring associated with its own source a considerable degree of control over phase and amplitude can be exercised. The pattern from a distribution of this type will be dealt with in the appropriate place, but this effect has already gained some attention in the field of optics.[25] It can also be noted that the resultant virtual aperture is double the radius of the real aperture of the cone and hence gives a gain for the antenna over and above the real aperture gain. The additional effect is due to the array factor of the sources themselves which for efficient illumination would have an end-fire effect. Polarization effects too have to be taken into consideration, but one antenna based on this concept can be visualized (Fig. 1.37).

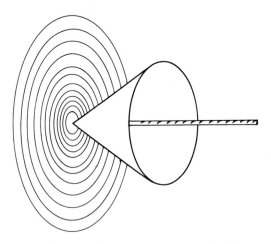

FIG. 1.37. *Conical antenna. The array of axial sources is converted by the conical reflector into an array of ring sources.*

1.7.3 Bi-cylindrical reflectors

The derivation of the virtual line source by a parabolic cylinder with a point source on the focal line as shown above, makes it the obvious feeding system for a second parabolic cylinder at right angles to the first. This principle has been used both in optics[26] and for microwaves.[27] In the former case focusing has not necessarily been confined to parallel rays at infinity and elliptical and circular cylinders have been used. In particular Ref. 26b indicates that the system, at least for rays of small obliquity (paraxial rays) can be orientated so as to deflect the beam in a periscope fashion over a considerable portion of the sphere. In more practical microwave terms the use of singly curved reflectors must always be advantageous in the constructional sense and most certainly in its effect upon the polarization of the resultant radiation when compared with the equivalent asymmetric doubly curved reflectors. It is apparent that much of the application found so useful in the field of optics has yet to be adapted to microwave purposes.

1.8 BI-CYLINDRICAL LENSES

1.8.1 Focusing lenses

In the same way that the reflective caustics of two cylindrical reflectors can be matched to create a bi-cylindrical reflector, two refractive caustics

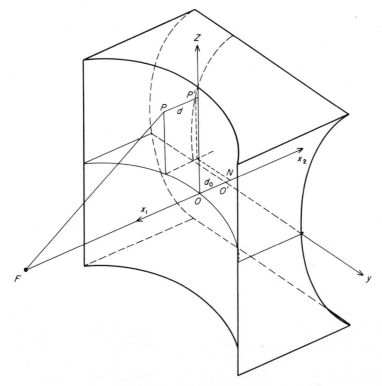

FIG. 1.38. Bi-cylindrical focusing lens.

can be matched to create a bi-cylindrical lens. This principle is also well-established in optical practice and has been used in the design of spectacle lenses. In microwave antenna design, we investigate the same concept involving the use of microwave lens like media, that is the fully constraining wave-guide media of the lenses of Section 1.5. As always the design for this medium has to be done by equalizing the optical path length from source to required wave-front, and since rotational symmetry no longer applies the bi-polar coordinate system cannot be used. We therefore use Cartesian coordinates.[28]

We have as shown in Fig. 1.38, with a source at the origin, a lens whose first surface is a cylinder with independence of the z coordinate, whose profile we assume to have the form $y = f(x_1)$. The second surface will therefore be independent of the y coordinate and thus have a profile given by $z = g(x_2)$ and the x axis is the axis of the lens.

Then if P is a general point on the first surface, we require for parallel rays the equality of the optical path lengths FPP' and $FOO'N$

$$FP + \eta(y, z)\, d(y, z) = OF + \eta_0 d_0 + (x_2 - OF - d_0) \qquad (1.45)$$

where η the refractive index is a function of position and d the thickness of the lens at position (y, z).

Rearranging Eqn (1.45) gives

$$FP = [\eta_0 d_0 - \eta(y, z)\, d(y, z)] + (x_2 - d_0).$$

The first factor occurs frequently in lens design (cf. Section 1.5.2) and so for simplicity let

$$D(y, z) \equiv \eta_0 d_0 - \eta(y, z)\, d(y, z). \qquad (1.46)$$

Squaring then gives

$$x_1^2 + y^2 + z^2 = \{D + x_2 - d_0\}^2, \qquad (1.47)$$

but x_1 is independent of z and x_2 is independent of y. Therefore Eqn (1.47) separates into two equations

$$\begin{aligned} x_1^2 + y^2 &= \text{const} = A^2 \\ z^2 + A^2 &= (D + x_2 - d_0)^2, \end{aligned} \qquad (1.48)$$

provided that $D(y, z)$ can be made to be a function of z only.

Solutions to Eqn (1.48) can be freely chosen and depend upon the (almost) arbitrary choices of A and $D(z)$. The limiting consideration is not in the design mathematics but in the feasibility of the final construction, that is upon the limited range that exists for the refractive index in this type of medium.

The simplest solution is to make A the focal length OF and $D(z)$ equal to zero. The resulting profiles are then

$$x_1^2 + y^2 = F^2,$$

and

$$z^2 + F^2 = (x_2 - d_0)^2. \tag{1.49}$$

The first of these is the circle centred at F and the second a hyperbola passing through O'.

There now arises the rather enigmatical consequence. The choice of $D(z)$ to be zero implies $\eta_0 d_0 - \eta(y, z)d(y, z) = 0$ for all positions of the surface point P. Hence the optical path length of each parallel ray element PP' is the same everywhere in the lens. Since it is the same as the value at the centre by this reasoning, the lens cannot have a zero thickness there as do the other lenses so far designed. This at first sight seems strange when one is accustomed to using the different optical widths of a lens to apply the required phase correction to the rays. However, considering a lens the first face of which is purely spherical and the second face totally plane, with a fully constrained waveguide medium connecting them, the plausibility of equal phase lengths for all rays internal to the lens establishes itself. The bi-cylindrical lens is in fact a symmetrically sheared version of such a lens in the manner of the lens bending process of Section 1.5.1.

The result given by Eqn (1.49) further establishes the following statement: if for any lens of this type, which focuses the rays from a point source to infinity, the first surface has a circular cross-section, then the second surface will have a hyperbolic one and vice versa. In a practical design study it is found that the difference in profile between the circular and hyperbolic is very small and within the usual tolerance of $\lambda/16$ for phase differential. Thus there is no great error in making both profiles the same, say circular, and if necessary adjusting even this small error by a modification of the medium refractive index. We then have a lens which focuses exactly at infinity from a source on either side at the same focal length.

1.8.2 Scanning bi-cylindrical lenses

In accordance with the principle that the two surfaces can be used to a second effect beside that of ordinary focusing we establish a design for a feed point displaced from the point F, to a point with polar coordinates (ρ, α). That this has to take place in the x, y plane is apparent if one visualizes the cross section of the centre of the lens in the other plane. This will be straight on the incident side and curved on the transmitted side. From Fig. 1.39 it can be seen that for a displacement of the source through an angle α

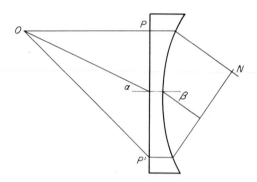

FIG. 1.39. Scan in vertical plane.

the beam position will be given by an angle β where

$$\tan \alpha = \tan \beta \sec \beta.$$

Although this effect is small for small angles of scan and in practice when the remainder of the surface is brought into action hardly observable, we will nonetheless confine the scanning action to the plane specified above. Then as shown in Fig. 1.40 we require

$$SP + \eta(y, o)PP' + P'N = SO + \eta_0 OO' + O'M = SQ + \eta(y, o)QQ' \qquad (1.50)$$

where S is the displaced source at the point (ρ, α) with respect to the origin at O.

From the outer equation in Eqn (1.50) we obtain the parametric description of the curve PQ. In terms of coordinates (x_1, y, z) at O this is identically that derived for the three ray lens given by Eqn (1.34) namely

$$\left.\begin{array}{l} x_1 = \rho(\alpha) \cos \alpha (1 - \cos \alpha), \\ y = \rho(\alpha) \sin \alpha, \end{array}\right\} \qquad (1.51)$$

where ρ as a function of α defines the scanning arc travelled by the source S.

Solving the remaining equation in (1.50) for the optical thickness $\eta(y, o)$ $d(y, o)$ we obtain

$$\eta(y, o)d(y, o) - \eta_0 d_0 = 0,$$

a condition in this case and not the free choice made in the previous section. Note that it only applies as yet to the section of the lens given by $z = 0$.

Thus we find that the bi-cylindrical lens in its capacity as a scanning lens automatically gives the principle of equal path length for each element of lens from the three-ray principle of lens design.

The second surface is obtained by considering (Fig. 1.40) a general point

R with coordinates referred to F given by

$$\{F - \rho(\alpha) \cos \alpha(1 - \cos \alpha), \rho(\alpha) \sin \alpha, z\} \qquad (F = OF).$$

We require for a plane wave-front

$$FR + \eta(y, z)RR' = FO + \eta_0 d_0 + O'V \qquad (1.52)$$

that is

$$FR = x_2 - d_0 + D$$

where

$$D = \eta_0 d_0 - \eta(y, z)d(y, z)$$

as before. Squaring Eqn (1.52) and separating for x_2 and D dependent on z only we obtain

$$\{F - \rho(\alpha) \cos \alpha(1 - \cos \alpha)\}^2 + (\rho(\alpha))^2 \sin^2 \alpha = \text{const} = A^2$$

$$(x_2 - d_0 + D)^2 - z^2 = A^2. \qquad (1.53)$$

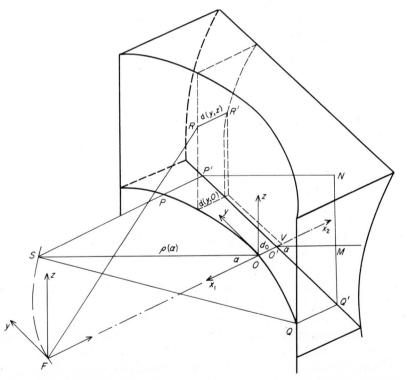

FIG. 1.40. *Scanning bi-cylindrical lens. Scanning motion is in the horizontal plane SPOQN.*

The first of these establishes the scanning arc by solving for ρ as a function of α. The second is the profile of the second face of the lens.

For example taking $A = F$ and $D = O$ we recover the hyperbolic surface and the equality of optical length for each element of the entire lens as was obtained for the simple focusing lens. This means that the profile derived from Eqn (1.53) with this function of ρ inserted *must* be circular.

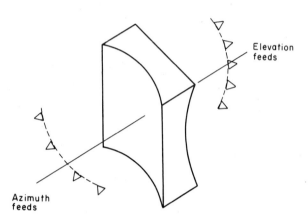

FIG. 1.41. *Multi-element lens.*

Thus from the first of Eqn (1.53) with this choice we find

$$\rho = 2F \cos \alpha / (1 + \cos \alpha + \cos^2 \alpha - \cos^3 \alpha) \qquad (1.54)$$

and hence from equation (1.51)

$$x_1 = 2F \cos^2 \alpha (1 - \cos \alpha)/(1 + \cos \alpha + \cos^2 \alpha - \cos^3 \alpha),$$

$$y = 2F \sin \alpha \cos \alpha / (1 + \cos \alpha + \cos^2 \alpha - \cos^3 \alpha).$$

That this is a circular arc with centre at F is readily confirmed since

$$(F - x_1)^2 + y^2 = F^2,$$

but it is not a description of a complete circle. As the parameter α (*not* polar coordinate here) is varied this expression describes a limited circular arc traversing it once in each direction. This delimits the aperture of the lens and the scanning arc simultaneously. Being circular the profile also satisfies the requirements of the sine condition for this plane of scan.

Generalizations of this lens are possible through the various choices of the ·constant A and the function of thickness $D(z)$. Variation of A alters the hyperbolic profile of the second surface, hence the radius of the circular first surface and thus the scanning arc.

Variations in $D(z)$ would, as for the ordinary focusing lens, allow the scanning lens to have the same circular profiles on both faces. The symmetrical lens would then have the ability to focus rays from oblique angles within the range of the scanning arc, at individual points on either side of the plane of symmetry. This situation is indicated in figure 1.41.

1.8.3 Line source lenses

The bi-cylindrical structure can, with even greater simplicity be designed to produce a line source. From Fig. 1.42, a line source will be created at a distance a from the point source $(a \neq 0)$ if the lens thickness is given by

$$\eta(y, z)d(y, z) = TP' - OP$$
$$= \{(x_2 - a)^2 + z^2\}^{\frac{1}{2}} - (x_1^2 + y^2 + z^2)^{\frac{1}{2}}. \quad (1.55)$$

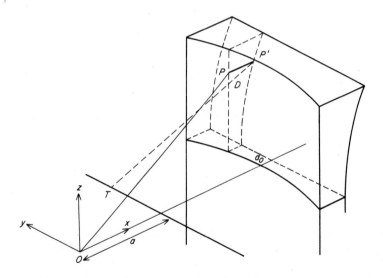

FIG. 1.42. Line source lens. The source at O has a line image at right angle to the lens axis.

The refractive index at the centre is given by

$$\eta_0 d_0 = d_0 - a$$

and can be arbitrarily chosen so that the range of $\eta(y, z)$ in Eqn (1.55) falls within the limits set by the nature of this medium, say $0.5 < \eta < 0.8$. One surface of the lens can also be arbitrarily chosen for example if $x_2 = $ constant $= 1.2F$ the second surface is plane $d_0 = 0.2F$ and $\eta(y, z)$ lies in the range $(0.4, 0.75)$. The object of such a design would be to provide a scanning *line* source or virtual line focus through the movement of a single point source.

1.8.4 Experimental results

A scanning lens designed on these principles has been constructed using waveguides of square cross-section and obtaining the variation in refractive index by filling each element with a dielectric of appropriate refractive index. It is found with the lossless dielectric foam materials that the refractive index is a linear function of the density and hence the required relation

$$\eta(x, y) = \left[\kappa_e(x, y) - \lambda^2/4a^2\right]^{\frac{1}{2}}, \qquad (1.56)$$

which gives the refractive index as a function of position in the lens can be converted to a relation between the density of the expanded dielectric medium and position of the waveguide element. The range of refractive index values that can be achieved by this method lies between that of the lightest manufacturable foam and the solid dielectric (the unfilled wave-guide being too fragile for the manufacturing process). The requirement that each element has the same *electrical* length places a limit on the maximum width of the lens at any point, unless wavelength stepping is introduced. This also defines the final aperture that can be achieved. The design, at a wavelength of 3 cm, of an unstepped lens was between 5 and 10 cm in thickness and gave an aperture (near circular) of 60 cm.

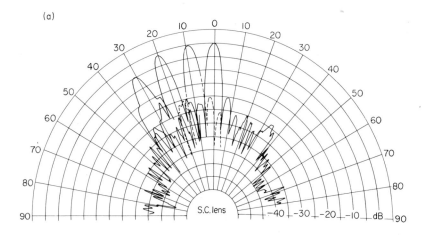

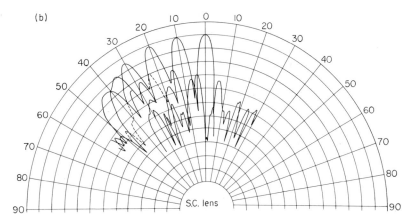

FIG. 1.43. Scanning bi-cylindrical lens patterns. (a) horizontal plane–E plane scan, (b) vertical plane–H plane scan.

Radiation patterns of this lens when scanned in both E and H planes are shown in Fig. 1.43. The E plane scan is the plane of offset as designed, the H-plane is that facing the linear central section and thus subject to the possibility of the squint as described earlier. In practice both planes give a performance over $\pm 30°$ much in line with the scanning patterns of spherical lenses designed by the alternative methods given previously. Thus the squint effect predicted for a central cross-section is largely overcome when the entire aperture is brought into operation.

1.9 THE GENERAL CONSTRAINED LENS

The fully general constrained lens is a term applied to a two surface lens in which the first surface is constructed an array of receiving elements and the second transmitting surface a similar array of elements connected one to one with the elements of the first surface by a length of transmission line. In microwave technique these arrays could be receiving dipoles over reflecting surfaces connected by wire or coaxial transmission lines. In optical technique the elements could be the shaped ends of optical fibres which themselves form the connecting lines. Waveguide connections and elements could similarly be used. No further principle of design is required to derive the action of such a lens, called a "boot-lace lens" by its designers,[29] other than the direct measurement, or calculation, of the optical path lengths from the source to the required phase front under the various conditions to be considered.

In principle the two surfaces can be totally dissociated and the connecting transmission lines give a one-to-one mapping of points on one surface with the points on the other.

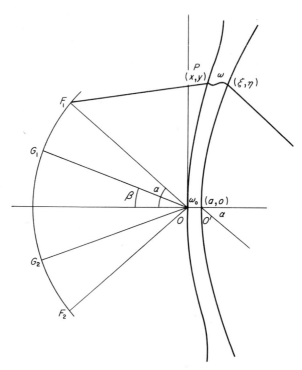

FIG. 1.44. *Boot-lace lens.*

Before a concrete design can be entered into assumptions have to be made as to the eventual nature of the lens. These so far have had to assume axial symmetry both of the surfaces and their method of inter-connection. Even with these limitations the additional degree of freedom gives rise to a wide choice of lens surfaces or scanning arcs. It becomes possible for example to double the number of *exact* focal points as in the design of the constrained wide angle lens of Ruze, Ref. 17.

Referring to Fig. 1.44, if the points

$$
\left.
\begin{aligned}
F_1 &= (-f\cos\alpha,\ f\sin\alpha),\\
F_2 &= (-f\cos\alpha,\ -f\sin\alpha),\\
G_1 &= (-g\cos\beta,\ g\sin\beta),\\
G_2 &= (-g\cos\beta,\ -g\sin\beta),
\end{aligned}
\right\} \tag{1.57}
$$

are symmetrically placed (in pairs) points of exact foci which respectively give rise to plane phase fronts with normals at angles $\pm\alpha$ and $\pm\beta$ to the axis of symmetry then the optical path length conditions are

$$
\left.
\begin{aligned}
F_1 P + \omega - \xi\cos\alpha + \eta\sin\alpha &= f + \omega_0 - a\cos\alpha,\\
F_2 P + \omega - \xi\cos\alpha - \eta\sin\alpha &= f + \omega_0 - a\cos\alpha,\\
G_1 P + \omega - \xi\cos\beta + \eta\sin\beta &= g + \omega_0 - a\cos\beta,\\
G_2 P + \omega - \xi\cos\beta - \eta\sin\beta &= g + \omega_0 - a\cos\beta,
\end{aligned}
\right\} \tag{1.58}
$$

in which (ξ, η) is the point on the second surface connected with $P(x, y)$ on the first surface through the transmission line of optical length $\omega(x, y)$ and $\omega_0 = \omega(0, O')$.

Subtracting the first pair and the second pair of these relations and dividing gives

$$
\frac{F_2 P - F_1 P}{2\sin\alpha} = \frac{G_2 P - G_1 P}{2\sin\beta}, \tag{1.59}
$$

which is a single relation between x and y and thus describes the contour of the first surface. The second surface is then obtained from the solution of the remaining equations giving

$$
\begin{aligned}
\xi &= \frac{a + (G_2 P + G_1 P - 2g) - (F_2 P + F_1 P - 2f)}{2(\cos\beta - \cos\alpha)},\\[1mm]
\eta &= \frac{F_2 P - F_1 P}{2\sin\alpha},\\[1mm]
\omega(x, y) &= \frac{\omega_0 + (G_2 P + G_1 P - 2g)\cos\alpha - (F_2 P + F_1 P - 2f)\cos\beta}{2(\cos\beta - \cos\alpha)}. \tag{1.60}
\end{aligned}
$$

Various forms of lens can now be derived from conditions on the position of the points F_1, F_2, G_1 and G_2 or on $\omega(x, y)$ for example $\omega(x, y) = $ constant.

When the four focal points lie on a circle particularly simple forms of surface result. Thus with F_1, F_2, G_2 and G_2 on a circle through the point

$(-h, 0)$ with radius $h/(1 - 2\mu)$ the front surface contour becomes the circle

$$x^2 + y^2 - xh = 0$$

of radius $h/2$ and centre $(-h/2, 0)$, and the second surface the ellipse

$$\xi = a + 2\mu[h - (h^2 - \eta^2)^{\frac{1}{2}}.$$

which has its centre at $(a + 2\mu h, 0)$ and axes $(2\mu h, h)$. The required path length delay between points (x, y) on the front surface and (ξ, η) on the second surface is given by

$$\omega = \omega_0 + (2\mu + 1)[h - (h^2 - \eta^2)^{\frac{1}{2}}].$$

Thus when $\mu = -\frac{1}{2}$ the focal arc is of radius $h/2$ and the connecting transmis-

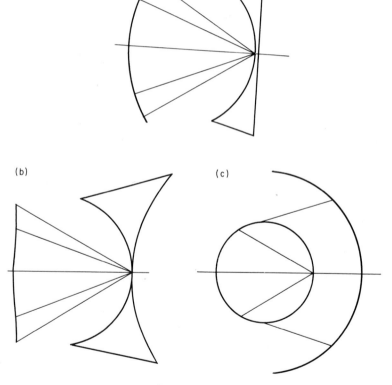

FIG. 1.45. *Boot-lace lens designs (from Ref. 30).* (a) *Lens with plane outer face,* (b) *Lens with near-linear scanning arc,* (c) *Frequency independent lens with circular scanning arc.*

sion lines are all of the same electrical length $\omega = \omega_0$. A judicious choice of this length could be made to cancel the transmission line mismatch reflections. For a final assessment of the lens the phase front for the source at intermediate points of the focal arc require to be calculated and the errors minimized if necessary be refocusing. Other designs[30] (and Gent, 1957) can be made based upon the minimization of aberrations and on adaptations of the 3-ray method. In the reference given three particular cases are discussed

(i) a lens with a plane second surface with an almost circular scanning arc and low aberrations

(ii) a lens with a near linear scanning arc

(iii) a frequency independent lens with zero geometrical aberrations on a circular scanning arc.

Some of the designs are illustrated in Fig. 1.45.

The practical difficulties to be overcome are those facing the designers of most planar arrays. That is the consistency of the receiving and transmitting arrays of elements under conditions of varying wide angle incidence of plane waves, due to element interactions and impedance variations. Care has also to be taken that grating lobes due to element spacing do not occur in the field of operation.

Further generalizations can be considered such as the inclusion in the interior transmission lines of electronically controlled phase shifters or even of active elements amplifiers and frequency changers, but such a lens becomes an entire system in its own right and ceases to be simply an optically designed device.

The inclusion of short-circuits alone in the transmission lines creates a phase corrected reflector, but the limitation that only one surface is to be used both for incidence and transmission gives it no apparent advantage over the previous design. The reason for this is that with short-circuits the one-to-one correspondence between elements is the identity relation. If completely dissociated surfaces were to be considered with a general one-to-one correspondence between the elements on each then it is possible to imagine the second surface folded back over itself and into coincidence with the first surface. The array will then consist of a series of receiving and transmitting elements over a single reflecting surface connected in pairs by the bent over transmission lines. This forms the fundamental reflect-array.[31] Applications of symmetry conditions can make this a monostatic reflector, for example if each element were connected to a diametrically opposite element (with respect to the centre of the array) by transmission lines all of equal length. Variations of the line lengths, the introduction of active elements and so on produce generalizations of this design. Fibre optical and acoustical versions based on the same principle are also known.

1.10 GRATING-REFLECTORS

The reflecting and refracting surfaces that have gone to make up the optical systems that have been discussed so far in this chapter have been considered to be smooth continuous surfaces. Consequently, these act as designed by ray optics over the entire frequency range from the infinite, where the geometrical optics design is exact, down to the longest wavelength where the approximation is at a limit which is decided upon by the system designer. The results are usually confirmed experimentally and the degree of agreement and modifications required are determined in that way.

Hence there is a hidden degree of freedom available in this large range of frequencies which, if sacrificed, could be used for some other system criterion. The resultant device would then only have a narrow and defined frequency range for its optimum operation.

This principle can be applied to the extension of the basic paraboloid reflector and hyperbolic refractor surfaces to give, in the example chosen, a considerably extended range of off-axis operation. The design, optically, requires to conform to the Abbe sine principle and be coma free. As a microwave antenna this implies a larger scanning capability. Such a design was originally proposed by de Coligny[32] and based on the following considerations.

If one wished to use a paraboloid to scan a beam through a given angle without distortion, the ideal method would be to rotate the entire reflector about a fixed axis such as I in Fig. 1.46. At the intersection of the paraboloid in its original position with the paraboloid in its rotated position there is a region, which, within a given tolerance has remained static, shown as the region $H'H''$ in the figure. The extent of this region depends upon the choice of the permitted tolerance of the phase front. This is normally taken to be about $\lambda/16$ so that the excursion of the end points of section $H'H''$ could be within $\lambda/32$ of the original paraboloid surface without creating undue phase front distortions. We note that if the centre of rotation had been taken to be the original focus of the paraboloid the region $H'H''$ would have been the apex of the paraboloid only.

The position of H is thus a function of the angle rotated through, the geometry of the paraboloid and the position of the axis of rotation.

The procedure is now applied to an ensemble of paraboloids all with the same axis and focal point but with focal lengths increasing by multiples of $\lambda/2$. The points H for each of the paraboloids will then lie on a curve which de Coligny calls the "support curve", illustrated in Fig. 1.47. Each element $H_i'H_i''$ is essentially a section of a paraboloid with focal length $p\lambda/2$ from the original, p being an integer, and the resultant phase front

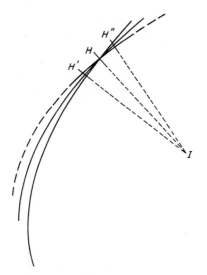

FIG. 1.46. Static region of rotated parabola.

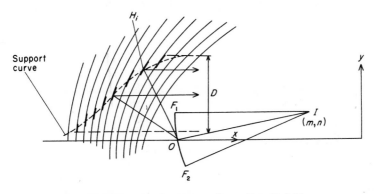

FIG. 1.47. Generation of the grating reflector (from Ref. 32).

for a source at the focus is plane, incorporating as it does "steps" of magnitude $2p\pi$. Along the axis direction however, the distances between the edges of the successive elements are greater than $\lambda/2$ in general and thus the system is a true diffraction grating and the higher order lobes have to be taken into consideration. The reference quotes experimental evidence which shows that the usual limitation of about 5 beam-widths of scan for the ordinary paraboloid can be extended by this method to between 40 and 50 beam-widths.

D

The form of the support curve is derived in the following manner.

The basic equation of the parabolic cross-section referred to the focus as origin, and with focal length f is

$$y^2 - 4f(x + f) = 0.$$

If the centre of rotation is the point (m, n) (Fig. 1.47) then referred to this point as origin the equation becomes

$$(y + n)^2 - 4f(x + m + f) = 0.$$

Then the set of confocal parabolas with focal lengths increasing by multiples of $\lambda/2$ will have the same equation with f replaced by $F = f + p\lambda/2$ (p, integer) that is

$$(y + n)^2 - 4F(x + m + F) = 0. \tag{1.61}$$

On rotation through the *fixed* angle θ equation 1.61 becomes

$$S \equiv (x \sin \theta + y \cos \theta + n)^2 - 4F(x \cos \theta - y \sin \theta + m) - 4F^2 = 0. \tag{1.62}$$

A "characteristic curve" is given by $\left| \partial S / \partial \theta \right|_{\theta = 0} = 0$ and the "support curve" is given by the curve containing the intersection of each parabola with its corresponding characteristic curve.

From Eqn (1.62) the characteristic curves are the hyperbolas (with parameter p)

$$x(y + n) + 2Fy = 0, \tag{1.63}$$

and hence the support curve, obtained by the elimination of F between Eqns (1.61) and (1.63) is

$$x^2(y - n) + y^2(x + n) + 2mxy = 0.$$

This is referred to the origin at the point (m, n) and thus reverting to the original focus as origin the support curve has the equation

$$(x^2 + y^2)(y - 2n) + 2mnx + (n^2 - m^2)y = 0. \tag{1.64}$$

The form of this curve is a strophoid as illustrated in the Fig. 1.47.

The procedure can be generalized, as is also stated in the references, by using a more general form of displacement of the paraboloid. This, in addition to the required rotation, could include a translation and a p-dependent, θ-dependent change in the focal lengths from parabola to parabola. The analysis using characteristic curves remains the same for the more complex forms of S (Eqn 1.62) that arise. The plane phase front criterion for the central position has to be retained of course, but this generalization allows the development of more advantageous grating support curves as the references illustrate.

The special form of grating reflector that arises when the centre of rotation is on the axis of the original parabolic system, is recognizable as the zoned mirror or diffraction reflectors of Ronchi and Toraldo di Francia.[33] These were designed from the principles of parageometrical optics.

We can obtain the result directly by putting $n = 0$ in Eqns (1.63) and (1.64). The hyperbolas of Eqn (1.58) become straight lines parallel to the y axis, spaced by $\lambda/2$ and the grating support curve the circle

$$x^2 + y^2 = m^2.$$

The obvious choice for m, and thus incidentally the circular scanning arc for the source, is the focal length of the most distant of the confocal parabolas to be used in the design. This results in the well known construction shown in Fig. 1.48.

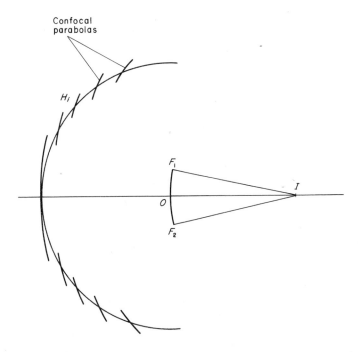

FIG. 1.48. *Special case of grating reflector. Centre of rotation I is on axis, support curve becomes circle centre O.*

As shown by de Coligny (Ref. 32, p. 693) the principle applies equally well to the case of refraction. This produces a wide angle version of the hyperbolic lens illustrated in Fig. 1.49. For a centre of rotation on the hyperbola axis the support curve is again circular, as shown and the result is a form of scanning Fresnel lens.

This principle by which *any* system of continuous reflecting and refracting surfaces can be replaced by a grating could then be applied absolutely generally to *all* the previously designed optical devices. The sacrifice of the infinitely high pass band of the system for a narrower frequency range is no loss in microwave terms when measured against the possible degree of flexibility that can be achieved by this means. Since the system is ultimately a diffraction grating, different focal properties would be possible at different frequencies. The well known duality between angular effects and frequency effects† can thus be utilized as an additional design criterion.

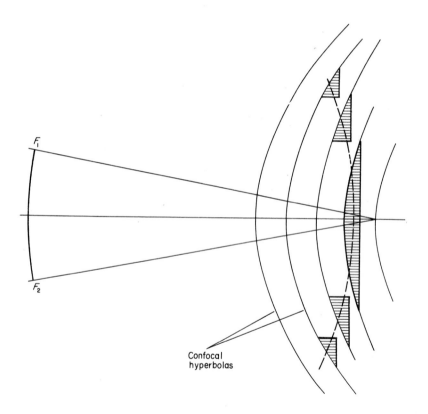

Confocal
hyperbolas

FIG. 1.49. Grating lens.

† See for example the multilayer theory of Chapter 2.

1.11 DIFFRACTING–FOCUSING ARRAYS OF APERTURES

One of the major differences between optical and microwave systems is the increased importance of diffraction effects in the latter.

Many devices are operational in optical practice where these effects have been put to system design purposes. Using the principle that in microwave practice a bad effect is only a good effect out-of-phase, it can be seen that many of the optical devices have direct microwave analogues. Following from the discussion of the previous section, zoned diffracting systems with reflecting or refracting elements have a practical microwave antenna application. There remains the optical systems which rely upon diffracting effects solely. The well known Fresnel zone plate is such a device and the theory of its operation can be found in Chapter 4 where the zoned circular aperture is fully discussed.

In optics the very common aperture stop is a vital element which has as

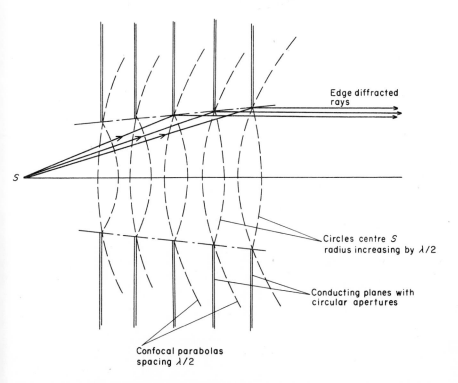

FIG. 1.50. Cascaded aperture system (Ref. 34 adapted). The adaptation is to account for a focus at infinity thus parabolas are used instead of the ellipses in the reference.

yet not been adapted to microwave use through the complications that may be introduced by its diffractive effects.

The concepts of the Fresnel zone plate and aperture stop have been carried a stage further in recent optical studies by a fundamental change of emphasis. That is, instead of having diffracting edges all arranged in a plane transverse to the optical axis as in the Fresnel zone plate, the edges are spaced along the axis of the system. The concept has been studied by Lit and Tremblay[34] and is illustrated in Fig. 1.50.

Similar near field properties to that of the Fresnel zone plate have been achieved and there would appear to be no reason why the operation should not be as successful in the microwave region. It would in any event provide a field day for the proponents of the geometrical theory of diffraction.

An interesting development results from this design in the microwave sense. As shown in the figure the peripheries of the apertures basically lie upon a set of confocal ellipses when focusing is arranged for a local point. These can be adjusted to lie on a cylinder and presumably, with suitable spacing upon other curves of choice. The apertures themselves are considered

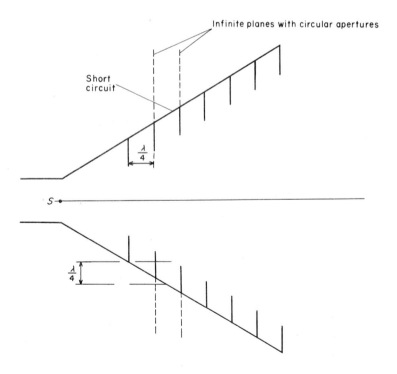

FIG. 1.51. Development of corrugated horn from cascaded aperture lens.

to be made in an infinite conducting screen. If we choose the curve of the aperture edges to be a cone, then with closely spaced planes (approximately $\lambda/4$) the infinite region between then will only support the TEM mode of propagation. The open circuit presented by each pair of planes at the edges of the apertures can be induced by placing short circuits at a distance of $\lambda/4$ from the edges. This progression is illustrated in Fig. 1.51 from which it can be seen that the optical system proposed by Lit and Tremblay develops in this way into the corrugated horn as used in many microwave antenna feed systems.[35]

1.12 DOUBLE-REFLECTOR SCANNING ANTENNAS

One of the major applications for the additional degree of freedom available in two surface reflector antennas is in the wide-angle scanning of the beam. The design and theoretical performance of these antennas remain standard exercises in geometrical optics, on the lines already discussed, but the range of alternatives is not always apparent and we can illustrate this by discussing in general terms the design of three such antennas.

1.12.1 The Rakovin antenna

Variations in the geometry of the two reflector Cassegrain or Gregorian antenna, in which the source is no longer constrained to lie on the axis of symmetry of the main paraboloid reflector, can be made in an obvious manner. All that is basically required is that the source and the focus of the paraboloid are conjugate foci of the intermediate (auxiliary or sub-) reflector. Some of the variations that are possible are illustrated in Fig. 1.52 and the constructions are such that shadowing of the main reflector by the sub-reflector can be avoided or minimized. A known drawback to this design is the effect of cross-polarized components of the radiated field created by the asymmetric illumination of the paraboloid[37] but from the geometrical optics of the system the only requirement is that the sub-reflector has rotational symmetry about the line joining the source to the focus of the paraboloid.

The process can be adapted to antennas that do not have a point focus, as does the paraboloid, but a caustic, and in particular, because of its advantages of symmetry to the spherical reflector. Such an antenna is called by its designer (Ref. 36) a spherical Rakovin antenna and is used as a means of

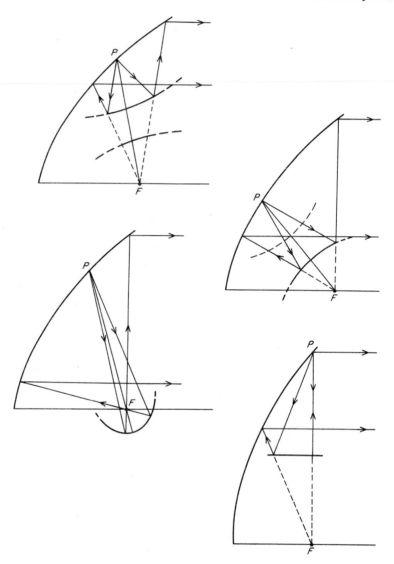

FIG. 1.52. Sub-reflector geometries for offset feed in paraboloid. Sub-reflectors have rotational symmetry about line PF joining source point to focus.

scanning the beam for very large (>100 metres) radio astronomy antennas by the movement of the smaller reflector and the feed system only.

The principle is illustrated in Fig. 1.53 in which the curve M is a cross-section in the diametral plane of the drawing, of the corrector for the spherical reflector fed by a source at *A*. This profile can be obtained by the method given in Section 1.2 for the corrector for a spherical mirror with only minor changes of sign to account for the new geometry of the ray path.

In this case however, rotational symmetry about the central ray from *A* cannot be applied and the skew rays have to be investigated to obtain the

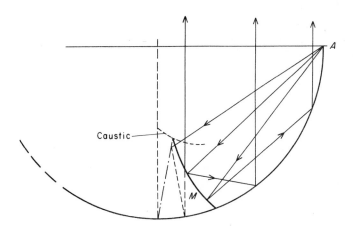

FIG. 1.53. *Sub-reflector for offset feed in a sphere.*

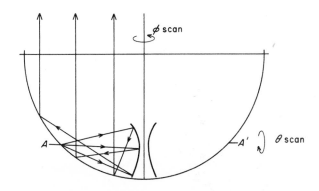

FIG. 1.54. *Scanning with static spherical reflector (from Ref. 36).*

full three dimensional surface required as is done in the reference given above. To use the spherical surface more completely a second in-phase source at *A'* together with a mirror image of the sub-reflector can be used. In the final system the sub-reflector is rotated about an axis through *AA'* which is horizontal and a second rotation of *AA'* about a vertical axis produces a

(θ, ϕ) scan as shown in Fig. 1.54. In a static form other sources on the periphery could be added each with its appropriate sub-reflector.

1.12.2 Scanning Cassegrain antennas

Once the collimation of a beam has been achieved its deflection into any other direction can be simply performed by a plane mirror at a suitable orientation. Normally the directions into which such a beam is turned would have to avoid the original collimating device and the shadowing that this would create. This shortcoming can be overcome by using the polarization properties of the field in an ingenious manner. The original collimating device can be made polarization sensitive by using for its reflecting surface a grating of closely spaced conductors, fine wires for example, which then appears as a continuous conducting surface to a wave polarized parallel to the direction of the grating elements. The plane mirror is then surfaced by a layer which besides reflecting the wave, rotates its polarization through 90°, a so-called twist-reflector. The design of such elements is given in Chapter 5, Section 5.9. Then, Fig. 1.55, the reflected wave is polarized perpendicularly

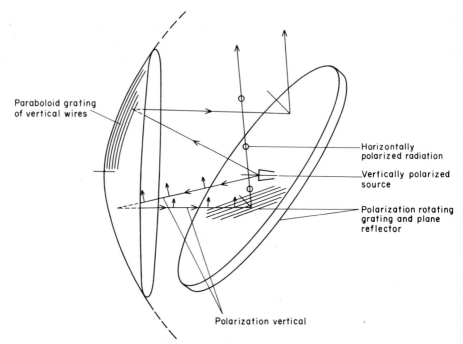

FIG. 1.55. Hemispherical scanning cassegrain antenna (after Ref. 38).

to the grating of the original reflector and so is transmitted totally. Deflection of the beam can then be achieved by orientation of the plane twist-reflector with a magnitude of twice the angle of the rotation of the normal to its surface. Scan angles covering the entire hemisphere can be achieved in this way, since by intersecting a plane wave-front with a plane surface no further aberrations can be introduced.

In certain situations where such a comprehensive angular coverage is not essential, other reflecting profiles can be considered which use the same polarization properties and scanning motion, but with possibly some advantage in geometry or in amplitude distribution over the aperture.

In such cases, the reflectors can be designed entirely by the basic theory. For example a spherical first reflector will have a limacon profiled second twist-reflector as shown in Fig. 1.56. The scan angle then permissible becomes

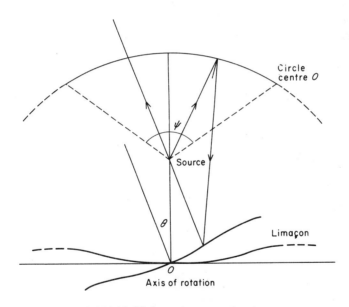

FIG. 1.56. *Modified scanning cassegrain antenna.*

limited by the onset of aberration effects, particularly coma, but since the two-to-one beam scan angle advantage still remains, most practical system requirements can still be met. Computed phase error curves in the plane of the beam offset for an antenna of this type are shown in Fig. 1.57. With this result a beam scan of θ up to $\pm 90°$ can be anticipated, with an angular aperture ψ of 70°.

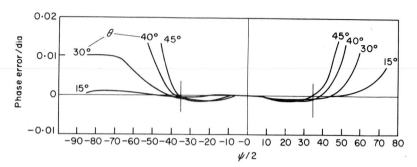

FIG. 1.57. Phase error curves for modified cassegrain antenna.

1.12.3 Double cylinder scanning antennas[39]

Similar considerations apply to the periscopic antenna created by two cylindrical reflectors, Fig. 1.58. The curves derived above for fully rotational systems would be the same for the orthogonal cross-section of such cylindrical reflectors. The use of two mirrors of equal curvature has already been proposed for this purpose in the field of optics and the reference given determines the degree to which a beam can be scanned by this method.

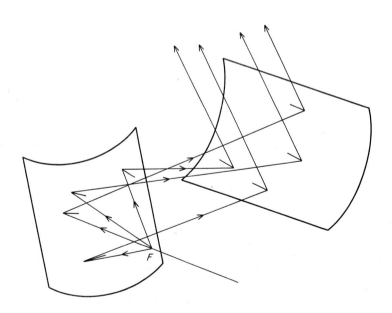

FIG. 1.58. Double cylindrical scanning reflector (after Ref. 39).

1.13 SHAPED BEAM ANTENNAS AND APERTURE AMPLITUDE DISTRIBUTIONS

Two important situations arise in the operation of optical antenna systems which require a different method of design from that of the optics of focusing rays as given in the previous sections of this chapter. One of these arises naturally, in that the source of rays, considered to be totally isotropic in the geometrical optics design, is in fact, in a practical microwave antenna, a directive source with a known angular distribution of the field. The optical design is still required for the condition that rays from such a non-isotropic source are finally in the correct required phase, and the variety of designs that have been illustrated is such that the non-isotropic field of the source can be converted by the optical device into a variety of aperture amplitude distributions among the rays of the final system. This is one of the major criteria for which the additional degree of freedom is used in two surface antennas.

The way in which such amplitude and phase distributions affect the final radiated pattern can be assessed by the method of the scalar Kirchhoff integral, to be given for the circular aperture in Chapter 3. The approximation of this method gives a result sufficiently close to experimentally verified patterns to allow its use in relating the properties of the various optical designs to the expected resulting radiation patterns. The link required is a method, with the same degree of approximation, that will give the expected variation of amplitude over the final aperture from the known angular distribution of the source of rays.

In the case of a symmetrical circular or cylindrical reflector this can be simplified to a standard procedure. The principle that is used is that of the conservation of the energy flux in a tube of rays as that tube is traced through its various reflecting or refracting processes. Being a scalar formulation, dealing as it does with energy density, effects such as interference between adjacent or intersecting tubes of rays are ignored. The scalar diffraction integral applied to the resultant distribution applies a corrective effect making up in part for this deficiency. This procedure can thus be applied to the analysis of radiation patterns from an aperture in which by geometrical optics the phase distribution has been determined and by conservation methods the field distribution is derived. As will be shown in Chapter 4, the reverse procedure whereby an aperture distribution is to be determined from the requirements of a specified radiation pattern is far more complex and in most cases insoluble.

In such a case the use of the conservation principle alone will enable the derivation of the reflector profiles from a knowledge of the pattern characteristics of the source and the prescribed radiation pattern. The approximation that results is much poorer and usually the amplitude and phase distributions resulting are too complex for the simplified Kirchhoff theory to provide a more accurate result. In most of the systems to which this method has to be applied the results are usually accepted to this degree of approximation.

The same conservation principle is therefore applicable to the main problems of

(i) deriving the aperture or near field distribution of symmetrical reflectors or lenses

(ii) deriving a profile cross section of a reflector to give a required far-field radiation distribution in one plane

(iii) the generalization of (i) that is the scattering of incident and general radiation from a known source by a general surface (without edges)

(iv) the generalization of (ii) that of deriving the complete reflector surface from the total spatial distribution required and source characteristic.

1.13.1 Aperture amplitude distribution

The first of these is applicable to nearly all of the focusing reflectors and lenses discussed previously. In the geometrical optics approximation Fig. 1.59, we have that the energy confined to the tube of rays represented by $d\Omega'$ is after reflection or refraction transmitted as a tube of rays confined to the angle $d\Omega$.

In the case of an axially symmetric system with near field collimation of the rays this gives (Ref. 15, p. 391)

$$d\Omega' = \sin\theta \, d\theta \, d\phi, \text{(Fig. 1.59(a))}$$
$$d\Omega = \rho \, d\rho \, d\phi,$$

and hence the final radial power distribution is given by

$$P_1(\rho) = \frac{P_2(\theta)\sin\theta \, d\theta}{\rho \, d\rho}, \tag{1.65}$$

where $P_2(\theta)$ is the source distribution. Because of this factor $P_2(\theta)$ it is impossible to express $P_1(\rho)$ explicitly as a function of ρ. If the surface in which the transmitted rays meet their corresponding incident rays, that is the reflector surface in a single reflector system, has a cross section given by $r = f(\theta)$, then since $\rho = r\sin\theta$ Eqn (1.65) can be rewritten

$$P_1(\rho) = P_2(\theta)/\{(f(\theta))^2 \cos\theta + f(\theta) \, f'(\theta)\sin\theta\}. \tag{1.66}$$

Where the system obeys the sine condition (q.v.) $f(\theta)$ is constant $= a$ and the

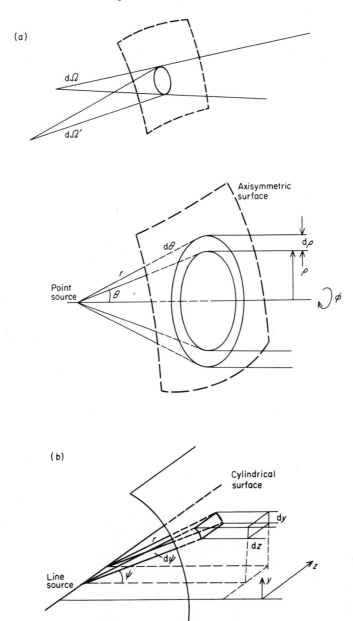

FIG. 1.59. *Conservation of flux. (a) for spherical systems, (b) for cylindrical systems.*

second term in the denominator is zero leaving

$$P_1(\rho) = P_2(\theta)/(a^2 \cos \theta).$$

In the case of a cylindrical system we consider an element dz of the entire reflector and its line source then (Fig. 1.59b)

$$d\Omega' = d\psi\, dz \quad \text{and} \quad d\Omega = dy\, dz.$$

Hence the power per unit length of the system in the plane aperture is

$$P_1(y) = \frac{d\psi}{dy} P_2(\psi), \tag{1.67}$$

where $P_2(\psi)$ is the power distribution per unit length of the line source. If the incident and refracted or reflected rays intersect on a surface given by $r = f(\psi)$ then since $y = r \sin \psi$

$$P_1(y) = P_2(\psi)/\{f(\psi) \cos \psi + f'(\psi) \sin \psi\}. \tag{1.68}$$

For a system obeying the sine condition this becomes

$$P_1(y) = P_2(\psi)/(a \cos \psi).$$

Relation 1.66 is that required for the eventual determination of the far field radiated pattern by the methods of Chapter 3. It incorporates the fact that for a paraboloid $r = 2f/(1 + \cos \theta)$ uniform illumination $P_1(\rho) =$ constant is obtained when $P(\theta) \propto (1 + \cos \theta)^{-1}$ that is when the *amplitude* distribution $|P(\theta)|^{\frac{1}{2}} \propto \sec^2(\theta/2)$.

1.13.2 Shaped beam reflectors

The standard method for obtaining the profile of a cylindrical reflector giving a specified shaped radiation pattern from a line source with a known characteristic has not been improved upon and is included here, yet again, for completeness (Ref. 41, p. 124; Ref. 15, p. 498).

If the line source has a radiation pattern per unit length given by $P_1(\psi)$ and the required secondary pattern is to be $P_2(\theta)$ then the assumption made is that all the energy within any angular range contained within the source distribution will be contained in a corresponding angular range within the radiated distribution. This expresses itself as

$$\frac{\int_{\psi_0}^{\psi} P_1(\psi)\, d\psi}{\int_{\psi_0}^{\psi_1} P_1(\psi)\, d\psi} = \frac{\int_{\theta_0}^{\theta} P_2(\theta)\, d\theta}{\int_{\theta_0}^{\theta_1} P_2(\theta)\, d\theta}, \tag{1.69}$$

where ψ and θ are any values in the range $[\psi_0, \psi_1]$ and $[\theta_0\, \theta_1]$ respectively, which define the limits of the incident radiation and those of the radiated pattern. By graphical means (Ref. 41), Eqn (1.69) can be used to derive a relationship between θ and ψ, say $\theta = g(\psi)$.

If the shape of the reflector with respect to the origin of ψ is $\rho = f(\psi)$ then the reflector geometry (Appendix I) gives

$$\mathrm{d}\rho/\rho = \tan\left(\frac{\psi - \theta}{2}\right)\mathrm{d}\psi,$$

or

$$\log \rho/\rho_0 = \int_{\psi_0}^{\psi} \tan \tfrac{1}{2}[\psi - g(\psi)]\,\mathrm{d}\psi, \tag{1.70}$$

where ρ_0 corresponds to ψ_0 and θ_0 and is the "starting" position for the reflector profile.

In general Eqn (1.70), due to the procedure for deriving the function $g(\psi)$ will require numerical integration.

To convert the reflector from a cylindrical system with associated line source to a reflector requiring only a point source more complex analysis is required.[41] A most elementary method is to convert the reflector to a translation reflector by moving the profile obtained above in a parallel fashion about a second curve. For simplicity this could be a circle, more naturally it would be a parabola. A closer approximation would be to construct the reflector from a series of parabolas each with focus at the feed source and with apex lying on the cross-sectional profile. In all such instances the degree of finesse has to be weighed against the essential approximation built into the design through the use of the conservation principle.

1.13.3 The general scattering of waves by surfaces

The totally general case has been studied recently in both the optical and microwave antenna contexts. The connecting mathematical process is the use by both of the differential geometry of surfaces to connect the incident tube of rays or its intersection with a reflecting or refracting element with the radiated tube of rays or its intersection with a receiving surface.

A full survey of the principles of differential geometry involved is beyond the scope of this volume. The results of these studies will therefore be summarized so that their application to microwave antennas can be appreciated. The conclusions can be found in Section 10 of Appendix I.

Burkhard and Shealy[43] give the conservation principle in the form

$$F(\mathrm{d}S_1 \to \mathrm{d}S_2) = \rho\sigma \cos\phi_i\,\mathrm{d}S_1/\mathrm{d}S_2 \tag{1.71}$$

in which $F(\mathrm{d}S_1 \rightarrow \mathrm{d}S_2)$ is the flux per unit area on the receiving surface $\mathrm{d}S_2$ which corresponds to the reflector or refractor surface element $\mathrm{d}S_1$ (Fig. 1.60), ρ is the reflectance or transmittance of the element (and could be a function of ϕ_i if required) and σ is the flux density of the incident beam. The directions of the tubes are obtained by the laws of reflection or refraction.

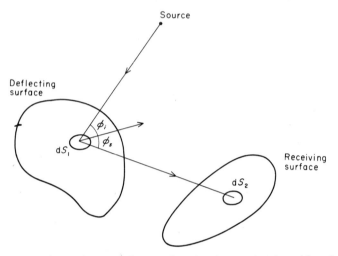

FIG. 1.60. Geometry of general source deflector and receiver in scattering theory (from Ref. 43).

The application of differential geometry then provides the method of obtaining the ratio $\mathrm{d}S_1/\mathrm{d}S_2$ required in Eqn (1.71) and the results of this analysis are presented in the appendix.

It has two main applications to the requirements of microwave antenna design. Firstly if the "receiving surface" be the plane exit aperture of the optical system and σ the flux density of a non-isotopic source, aperture distributions appropriate to the scalar method could be obtained. Alternatively if the "receiving surface" be a surface through or near to the expected geometrical optics focus, the caustic of an incoming plane wave could be derived. A "shaped beam" can be constructed from a variety of such incoming plane waves and a composite feed system with sources at the caustics would then create a specific radiation pattern. The theory is in fact tailored to obtain these caustics as is shown in the appendix.

The microwave study by Westcott and Norris is a direct attack on the reflector synthesis problem and is concerned essentially with far-field radiation distributions. We have as in Fig. 1.61 the conservation law in the form

$$D = \frac{\text{reflected power density}}{\text{incident power density}} = \left| \frac{\mathrm{d}\Omega'}{\mathrm{d}\Omega} \right|, \qquad (1.72)$$

which was used in Section 1.13.1. The law of reflection at P is (in terms of unit vectors p. 355)

$$\hat{\mathbf{y}} = \hat{\mathbf{r}} - 2\hat{\mathbf{n}}(\hat{\mathbf{r}} \cdot \hat{\mathbf{n}}),$$

or

$$\mathbf{N} = r\hat{\mathbf{y}} - \mathbf{r}, \qquad (1.73)$$

where $\mathbf{r}$ is the position vector of the point P on the reflector with respect to the origin of rays at 0, and $\mathbf{N}$ is a non-unit vector in the direction of the normal at P.

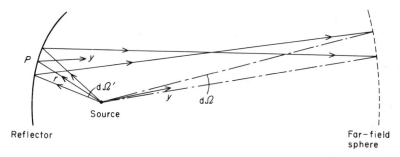

FIG. 1.61. Geometry for far field of a reflector (from Ref. 44).

If the surface is given parametrically by

$$\mathbf{r} = x(u, v)\hat{\mathbf{i}} + y(u, v)\hat{\mathbf{j}} + z(u, v)\hat{\mathbf{k}},$$

and the direction of $\hat{\mathbf{y}}$ parametrized by the *same* coordinates (say $u = \theta$ $v = \phi$ in a spherical polar system) then

$$d\Omega' = \frac{-1}{r^3}\mathbf{r} \cdot \frac{\partial \mathbf{r}}{\partial u} \times \frac{\partial \mathbf{r}}{\partial v},$$

and

$$d\Omega = \hat{\mathbf{y}} \; \frac{\partial \hat{\mathbf{y}}}{\partial u} \times \frac{\partial \hat{\mathbf{y}}}{\partial v}$$

and hence

$$D(u, v) = \left| \frac{-\mathbf{r} \cdot \dfrac{\partial \mathbf{r}}{\partial y} \times \dfrac{\partial \mathbf{r}}{\partial v}}{r^3 \, \hat{\mathbf{y}} \cdot \dfrac{\partial \hat{\mathbf{y}}}{\partial u} \times \dfrac{\partial \hat{\mathbf{y}}}{\partial v}} \right|. \qquad (1.74)$$

The problem of synthesis as in the reference is to find the surface $\mathbf{r}(u, v)$ satisfying both Eqns (1.73) and (1.74). The theory is obviously adaptable to analysis where $\mathbf{r}(u, v)$ would be known in advance.

1.14 THE GRAPHICAL DESIGN OF TWO-SURFACE SYSTEMS

Approximate methods for the design of two surface reflectors or lenses depend for their success upon the ease by which the criterion concerned with the available degree of freedom can be used to create an iterative step-wise procedure along the curves of the system profiles (assuming an axisymmetric or cylindrical system). For sufficiently closely spaced points, the piecewise linear solution can be made as close to the exact curve as the usual $\lambda/16$ total phase discrepancy requires. In the case of lenses this is a differential phase shift between the ray in the precisely shaped lens and in the graphical approximation, and usually permits a greater physical surface tolerance than in the case of reflectors. In the reflector case a summation of the free space tolerances must be kept within the minimum permitted phase error and this has to be halved at each reflection. Thus a closer step procedure is desirable for reflectors, but with computing techniques this is easily met to within the tolerances required by engineering templates.

It is of interest to see how an otherwise indeterminate problem, the general two surface optical device, becomes determined by the inclusion of a design criterion particularly for the case of the Abbe sine condition as applied to collimating systems. This requirement (Appendix I, Section 9) is for every transmitted ray to intersect its corresponding incident ray at points which lie on a circle centred at the source of the rays. Using this criterion one can establish a connection between succeeding linear elements and their orientations for both surfaces. We give here the method of Ponomarev[45] for this design procedure.

The basis is the circle of intersection of the rays which is given (Fig. 1.62). This is divided into small equal angular increments of a size determined by the eventual phase discrepancy. For the reflecting system shown a starting point A on the outer ray of the system is chosen. The horizontal ray through the intersection of the outer ray with the circle has then got to be the reflected ray from the second surface. The ultimate form of the system depends upon whether A is taken inside the circle or outside, and upon the initial slope of the increment at A. The choice of a tangential element at A then defines the starting point on the second reflector at B, and the tangential increment there. The three entities, the position of B and the incremental slopes at A and B are all determined once any one of them has been specified. The second ray and its horizontal counterpart then intersect the incremental elements at A

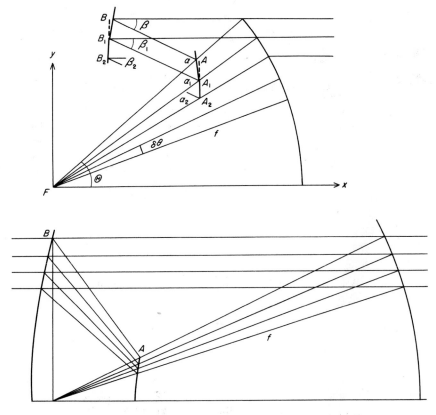

FIG. 1.62. Graphical construction of two-surface reflector antennas obeying the sine condition (*from Ref. 45*).

and *B* respectively in points A_1 and B_1. Joining $A_1 B_1$ new incremental elements and their orientations are obtained from which the intersections A_2 and B_2 can be derived and so on.

Two of the many diverse reflectors that can be obtained in this way are illustrated in the figure. Ponomarev gives the following recurrence formulae for deriving the coordinates of the points A_k and B_k from the preceding points $A_{k-1} B_{k-1}$ using this procedure

$$x_k = x_{k-1} \frac{\tan\{\theta - (k-1)\delta\theta\} + \cot(\{\theta - (k-1)\delta\theta + \beta_{k-1}\}/2)}{\tan\{\theta - k\,\delta\theta\} + \cot(\{\theta - (k-1)\delta\theta + \beta_{k-1}\}/2)},$$

$$y_k = x_k \tan(\theta - k\,\delta\theta),$$

for *A* coordinates,

and

$$x_k = x_{k-1} + 2f \tan(\beta_{k-1}/2) \sin(\delta\theta/2) \cos[\theta - (2k - 1)\delta\theta/2],$$
$$y_k = f \sin(\theta - k\,\delta\theta),$$

for B coordinates,

where

$$\tan\beta_{k-1} = \frac{y_B - y_A}{x_B - x_A} \text{ at } k - 1 \text{ points.} \qquad (1.75)$$

This process has assumed the division of the Abbe circle into equal elements $\delta\theta$. Other similar procedures could be devised with a weighted division or by a ray to ray process.

The design of aplanatic lenses can be achieved in much the same way by using the law of refraction at A and B successively instead of the law of reflection, as shown in Fig. 1.63.

The recurrence formulae in this case are for A coordinates

$$x_k = x_{k-1} \frac{\tan[\theta - (k - 1)\delta\theta] + \cot[\alpha_{k-1} - \beta'_{k-1}]}{\tan[\theta - k\,\delta\theta] + \cot[\alpha_{k-1} - \beta'_{k-1}]},$$

$$y_k = x_k \tan(\theta - k\,\delta\theta),$$

and for B coordinates

$$x_k = x_{k-1} + 2f \tan\beta_{k-1} \sin(\delta\theta/2) \cos[\theta - (2k - 1)\delta\theta/2],$$

$$y_k = f \sin(\theta - k\,\delta\theta). \qquad (1.76)$$

Where from the law of refraction

$$\tan\beta_{k-1} \text{ at } B_{k-1} = \frac{\eta \sin\alpha_{k-1}}{\eta \cos\alpha_{k-1} - 1},$$

$$\tan\beta'_{k-1} \text{ at } A_{k-1} = \frac{\sin([\theta - (k - 1)\delta\theta - \alpha_{k-1}]}{\eta - \cos[\theta - (k - 1)\delta\theta - \alpha_{k-1}]}$$

and

$$\tan\alpha_{k-1} = \frac{y_B - y_A}{x_B - x_A} \text{ at } k - 1 \text{ points.} \qquad (1.77)$$

Different criteria naturally result in different iterative procedures. The three ray principle (Section 1.6) for phase corrected reflectors gives rise to the following method. We consider the outer horizontal ray of the system to intersect a small element of a reflecting prism but for which the incident ray

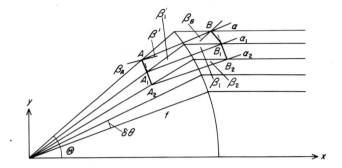

FIG. 1.63. *Graphical construction of two surface lens obeying the sine condition.*

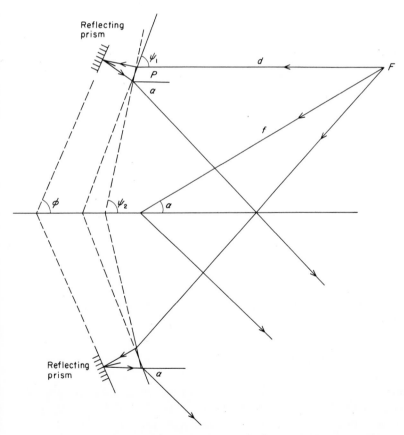

FIG. 1.64. *Graphical design of a phase corrected reflector using 3-ray principle.*

is refracted at a surface making an angle ψ_1 with the horizontal and after reflection the ray intersects a surface making an angle ψ_2 with the horizontal ψ_1 and ψ_2 being incrementally different (Fig. 1.64). The finally transmitted ray makes an angle α with the horizontal and the source is therefore at a position (f, α).

From the three ray principle an identical symmetrically placed prism exists which likewise reflects the incident ray making an angle θ with the horizontal into a ray making α with the horizontal. By simple laws of refraction we have for the upper prism

$$-\cos(\psi_1 + \alpha) = \eta \sin\{\psi_1 + \psi_2 - 2\phi + \sin^{-1}((1/\eta)\cos\psi_1)\}$$

and for the lower prism

$$\cos(\psi_1 - \alpha) = \eta \sin\{2\phi - \psi_1 - \psi_2 - \sin^{-1}((1/\eta)\cos(\theta + \psi_2))\},$$

$$\text{(1.78)}$$

where ϕ is the angle with the horizontal of the reflecting surface and θ is given by

$$\tan \theta = (2f \sin \alpha)/d$$

where d is the horizontal distance between the source and the point of refraction at P.

As shown in Section 1.6, one can specify one of the three curves, the reflecting profile, the refracting profile or the scanning arc. The first two give incremental values of either ϕ or ψ from which incremental values of θ and hence d can be derived by simultaneous solution of Eqns (1.78). Given a prescribed scanning arc and starting values of d and either ϕ or ψ further steps can be taken. In this case the resultant curves may intersect and the procedure must be restarted with new values or a greater thickness at the edge.

The approximate small angle (paraxial) lens can be derived simply from the Abbe condition. A considerable amount of the lens design is specified by its diameter, focal length and centre thickness. Thus as in Fig. 1.65 if OA, d, Θ and t are given the point P is defined from the relation

$$PQ = F + \eta t - d/\sin \Theta \qquad (1.79)$$

d being the semi diameter of the lens and t the central thickness.

The Abbe circle is then the circle centre O and radius OP. Taking a single intermediate ray as for example OMN then the tangents to surface profiles through M and N cannot intersect the lines PA or PB within their end points, in order to be realistic lens surfaces. Making an arbitrary choice of position of M on the incident ray, the refracted ray and thus the element MN and the

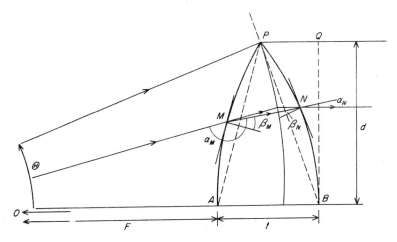

FIG. 1.65. *Approximate design of small angle lens.*

surface tangent at N are all derived from the refractive relations

$$\beta_M = \tan^{-1} \frac{-\eta \sin \alpha_M}{1 + \eta \cos \alpha_M},$$

$$\beta_N = \tan^{-1} \frac{\eta \sin \alpha_N}{\eta \cos \alpha_N - 1}, \qquad (1.80)$$

where α_M is the larger angle between OM and MN and α_N the smaller angle between MN and the horizontal.

Provided a practicable solution for the tangent at N is obtained in this way, the curve with three points and a tangent at the central one can be approximated in many ways. These approximate profiles for small angle lenses can be expected to be well within the wide tolerance that applies to microwave lenses of this kind.

REFERENCES

1. J. L. Synge. "Geometrical Optics", Cambridge University Press, 1937, p. 1.
2. A. W. Conway and J. L. Synge (Eds). "The collected papers of W. R. Hamilton Vol. I Geometrical Optics", Cambridge University Press, 1931, p. 10.
3. H. P. Brueggemann. "Conic Mirrors", The Focal Press, 1968.
4. S. H. Moss. Private communication.
5. D. W. Fry and F. K. Goward. "Aerials for Centimetre Wavelengths", Cambridge University Press, 1950.

6. O. N. Stavroudis. "The Optics of Rays, Wavefronts and Caustics", Academic Press, New York and London, 1972, p. 97.

7. J. D. Lawrence. ."A Catalogue of Special Plane Curves", Dover Paperback No. 0-486-60288-5, 1972, p. 115.

8. J. Brown. "Microwave Lenses", Methuen, 1953, p. 33.

9. A. Sommerfeld. "Lectures on Theoretical Physics, Vol IV, Optics", Academic Press, London and New York, 1964, p. 318.

10. M. Born and E. Wolf. "The Principles of Optics", Pergamon Press, 1953, p. 246.

11. C. J. Sletten, R. B. Mack, W. G. Mavorides and H. M. Johanson. Corrective line sources for paraboloids, *Trans. I.E.E.E.* **AP6** (1958), 239.

12. H. J. Salzer, Jnr. Comment on the caustic of a parabola, *Applied Optics* **4** (9), (1965), 1205.

13. R. Damien. "Théorème sur les Surfaces d'Onde en Optique Geometrique", Gauthier-Villars, Paris, 1955.

14. O. N. Stavroudis. Refraction of wavefronts: a special case, *Jour. Opt. Soc. Amer.* **59** (1969), 114.

15. S. Silver. "Microwave Antenna Theory and Design", MIT Radiation Laboratory Series Vol. 12, McGraw Hill, 1949, p. 402.

16. T. P. Vogl and A. Lavi. Signal transfer functions of optical components, *Jour. Opt. Soc. Amer.* **60** (1970), 813.
   F. J. Lopez-Lopez. Generalized bending and thickening of lenses (abstract), *Jour. Opt. Soc. Amer.* **62** (1972), 712.

17. J. Ruze. Wide-angle metal plate optics, *Proc. I.R.E.* **38**, (1950), 53.

18. R. E. Collin and F. J. Zucker. "Antenna Theory Part II", McGraw Hill, 1969, p. 123.
   F. G. Friedlander. A dielectric lens aerial for wide angle beam scanning, *Jour. I.E.E.* **93**, part IIIA (1946), 658.
   Jour. I.E.E. **93**, part IIIA (1946), 658.

19. H. F. Baker. "Principles of Geometry Vol. I", Cambridge University Press, 1928, p. 5.

20. L. C. Martin. "Technical Optics Vol. II", Pitman, 1954, p. 253.

21. R. C. Gunter, Jnr., F. S. Holt and C. F. Winter. "The Mangin Mirror", AFCRC Report No. TR-54-111, April 1955.

22. S. Cornbleet. A new design method for phase-corrected reflectors at microwave frequencies, *Proc. I.E.E.* **107**C (1960), 179.

23. D. W. Fry. Some recent developments in the design of centimetre aerial systems, *Jour. I.E.E.* **93** part IIIA (1946), 658.
   J. Ashmead and A. B. Pippard. The use of spherical reflectors as microwave scanning antennas, *Jour. I.E.E.E.* **93** part IIIA (1946), 627.

24. J. D. Barab, J. G. Marangoni and W. G. Scott. "The Parabolic Dome Antenna", IRE Wescon Convention Record 1958, Vol. 2 part 1, p. 272.

25. J. W. Y. Lit and E. Brannen. Optical properties of a reflecting cone, *Jour. Opt. Soc. Amer.* **60** (1970), 370.
   R. E. Ward, Jnr. Optical properties and uses of the conical mirror, *Applied Optics,* **4** (2), (1965), 201.

26. J. P. C. Southall. "Mirrors, Prisms and Lenses", McMillan, New York, 1933, Chapter IX.

27. R. C. Spencer, F. S. Holt, H. M. Johanson and J. Sampson. Double parabolic cylinder pencil beam antennas, *Trans. I.R.E.* **3** (1955), 4.

28. S. Cornbleet. Bi-cylindrical lenses, Proceedings of European Microwave Conference, IEE Publication No. 58, 8–12 September 1969, p. 353.
29. H. Gent. "Lectures on Developments in Microwave Optics", Royal Radar Establishment U.K. 1955 (unpublished).
    H. Gent. The bootlace arial, *R.R.E. Journal* (1957).
    S. S. D. Jones, H. Gent and A. A. L. Browne, British Patent No. 860826 February 8, 1961.
30. C. Pomot, P. Sermet, J. Munier and J. Benoit. Lentilles et reflecteurs bidimensionelles a grand changes angulaire. "Electromagnetic Wave Theory" (J. Brown, Ed.), Part 2. Proceedings of Symposium, Delft 1965, Pergamon 1967, p. 685.
31. D. G. Berry, R. G. Malech and W. A. Kennedy. The reflectarray antenna, *Trans. I.E.E.E.* **APII** (1963), 645.
    N. Amitay, V. Galindo and G. P. Wu. "Theory and Analysis of Phased Array Antennas", Wiley-Interscience, 1972.
32. G. de Coligny and A. Fournier. Physique et technique des systems focalisants a reseau stationnaire, *L'Onde Electrique*. Special Supplement on the International Congress on Ultra-high frequency circuits and Antennas, Paris, 21–26 October, 1957; Vol. II. August, 1958, p. 690.
    G. de Coligny, V. A. Altovsky, A. Fournier and A. Depauw. Construction of a three-dimensional radar aerial provided with a fixed grating focusing device, *loc. cit.* p. 770.
    J. Deschamps and J. Combelles. A new zoned toric reflector providing a large amplitude-volume sweep, *loc. cit.* p. 694.
33. L. Ronchi and G. Toraldo di Francia. An application of parageometrical optics to the design of a microwave mirror, *Trans. I.R.E.* **AP6** (1958), 129.
    J. H. Provencher. Experimental study of a diffraction reflector. *Trans. I.R.E.* **AP6** (1960), 331.
    L. Ronchi, V. Russo and G. Toraldo di Francia. Stepped cylindrical antennas for radio-astronomy, *Trans. I.R.E.* **AP9** (1961), 68.
34. J. W. Y. Lit and R. Tremblay. Boundary-diffraction-wave theory of cascaded apertures, *Jour. Opt. Soc. Amer.* **69** (5), (1969), 559.
    J. W. Y. Lit. Focusing capabilities of cascaded apertures. *Jour. Opt. Soc. Amer.* **62** (4), (1972), p. 491.
    J. W. Y. Lit, R. Boulay and R. Tremblay. Diffraction fields of a sequence of equal radii circular apertures, *Optics Communications*, **1** (6), (1970), 280.
35. A. F. Kay. "The Scalar Feed", AFCRC TRG Report 19(604)-8057, 30 March, 1964.
    R. E. Lawrie and L. Peters, Jnr. Modifications of horn antennas for low side-lobe levels, *Trans. I.E.E.E.* **AP14** (5), (1966), 605.
36. V. I. Prishlin. Some possible modifications of large radio astronomical antennas, *Rad. Eng and Electron Phys.* **14** (12), (1969), 1817.
37. A. W. Rudge, T. Pratt, M. Shirazi. Radiation fields from offset reflector antennas. "Proc. European Microwave Conference, Brussels, 1973", C4.3, p. 1.
    P. J. Wood. Depolarization with Cassegrain and front fed reflectors, *Elect. Lett.* **9** (1973), 181.
38. C. A. Cochrane. High frequency radio aerials. British Patent No. 700 868, February, 1952; December, 1953.
    P. F. Mariner and C. A. Cochrane. British Patent No. 716939, August, 1953; October, 1954.

P. F. Mariner. "Microwave aerials with full hemispherical scanning", L'Onde Electrique Special Supplement (as Ref. 32), p. 767.

39. J. A. Arnaud and J. T. Ruscio. Focusing and deflection of optical beams by cylindrical mirrors, *Applied Optics* **9** (10), (1970), p. 2377.
W. D. White and L. K. de Size. Scanning characteristics of two-reflector antenna systems, "Convention Record of the I.R.E.", 1962, part 1, p. 44.

40. J. B. Keller. The inverse scattering problem in geometrical optics and the design of reflectors, *Trans. I.R.E.* **AP7** (1959), 146.

41. R. C. Hansen (Ed.). "Microwave Scanning Antennas", Chapter 2, Reflecting Systems. L. K. De Size and J. F. Ramsay. Academic Press, London and New York, 1964, p. 124–127 and p. 161.

42. A. S. Dunbar. Calculations of doubly curved reflectors for shaped beams, *Proc. I.R.E.* **36** (1948), 1289.
L. Thourel. Calculation and construction of doubly curved reflectors, *L'Onde Electrique*, **35**, 1955, p. 1153.

43. D. G. Burkhard and D. L. Shealy. Flux density for ray propagation in geometrical optics, *Jour. Opt. Soc. Amer.* **63** (3) (1973), 299.
D. L. Shealy and D. G. Burkhard. Caustic surfaces and irradiance for reflection and refraction from an ellipsoid, elliptic paraboloid and elliptic cone, *Applied Optics,* **12** (12) (1973), 2955.

44. B. S. Westcott and A. P. Norris. Reflector synthesis for generalized far fields, *Jour. Phys. A Math Gen.* **8** (4), (1975), 521.
F. Brickell and B. S. Westcott. Reflector design for two-variable beam shaping in the hyperbolic case. *J. Phys. A. Math. Gen.* **9** (1), (1976), 113.

45. N. G. Ponomarev. Graphical method for the design of profiles of aplanatic antennas, *Radio Eng. and Electronics* **6** (2), (1961), 42.

BIBLIOGRAPHY

LENSES AND LENS DESIGN

J. Brown. Microwave Optics, *Adv. Electron. Electron Phys.* **10** (1958), 107.

R. M. Brown. Dielectric Bifocal Lenses. "Convention Record I.R.E." pt 1, 1956, 180.

J. E. Eaton. Zero phase fronts in microwave optics, *Trans. I.R.E.* **AP1** (1952), 38.

C. Goatley and C. F. Parker. "Symmetrical Microwave Lenses" Convention Record of the I.R.E., pt. 1, 1955, p. 13.

E. K. Proctor and M. M. Rees. Scanning lens design for minimum mean square phase error, *Trans. I.R.E.* **AP15** (1967), 348.

S. Rosin and M. Amon. Colour corrected mangin mirror, *Applied Optics* **6** (5), (1967), 897.

W. Rotman. Wide angle scanning with microwave double-layer pill boxes, *Trans. I.R.E.* **AP6** (1958), 96.

D. H. Shinn. The design of a zoned dielectric lens for wide angle scanning, *Marconi Review* **18** (1955), 37.

R. L. Sternberg. Successive approximations and expansion methods in the numerical design of microwave dielectric lenses. *Jour. Maths Phys.* **34** (1956), 209.

G. Svennérus. Microwave antennas with optical analogues, *Tekn Tid.* **86** (1956), 619.

R. Tremblay and A. Boivin. Concepts and techniques of microwave optics, *Applied Optics* **5** (2), (1966), 249.   (Including 535 refs).

GEOMETRICAL OPTICS

S. W. Lee. Electromagnetic reflection from a conducting surface: Geometrical optics solution, *Trans. I.E.E.E.* **AP23** (2), (1975), 184.

M. M. Sussmann. Maxwell's ovals and the refraction of light, *American Jour. of Physics* **34** (1966), 416.

REFLECTORS

A. Brunner. Possibilities of dimensioning doubly curved reflectors for azimuth-search radar antennas, *Trans. I.E.E.E.* **AP19** (1), (1971), 52.

T. F. Carberry. Analysis theory for the shaped-beam doubly curved reflector antenna, *Trans. I.E.E.E.* **AP17** (2) (1969), 131.

V. Galindo. Design of dual reflector antennas with arbitrary phase and amplitude distributions, *Trans. I.E.E.E.* **AP12** (1964), 403.

K. A. Green. Modified Cassegrain antenna for arbitrary aperture illumination, *Trans. I.E.E.E.* **AP11** (1963), 589.

P. W. Hannan. Microwave antennas derived from the Cassegrain telescope, *Trans. I.R.E.* **AP9** (2), (1961), 140.

F. S. Holt and E. L. Bouche. A Gregorian corrector for spherical reflectors, *Trans. I.E.E.E.* **AP12** (1964), 44.

B. Y. Kimber. On two reflector antennas, *Radio and Electron Phys.* **7** (1962), 914.

T. Li. A study of spherical reflectors as wide-angle scanning antennas, *Trans. I.R.E.* **AP7** (1959), 223.

A. W. Love. Spherical reflecting antennas with corrected line sources, *Trans. I.R.E.* **AP10** (1962), 529.

W. Magnus. Theory of cylindrical parabolic reflector, *Z. Phys.* **118** (1941), 343.

S. P. Morgan. Some examples of generalized Cassegrain and Gregorian antennas, *Trans. I.E.E.E.* **AP12** (1964), 685.

P. J. B. Clarricoats and C. J. E. Phillips. Optimum design of Gregorian corrected spherical reflector antenna, *Proc. I.E.E.* **117** (1970), 718.

P. D. Potter. Aperture illumination and gain of a Cassegrain system, *Trans. I.E.E.E.* **AP11** (1963), 373.

J. F. Ramsay and J. A. C. Jackson. Wide angle reflectors: Scanning performance of mirror aerials, *Marconi Review* **19** (1956), 116.

A. W. Rudge. Multiple beam antennas: Offset-reflectors with offset feeds. *Trans. I.E.E.E.* **AP23** (3), (1975), 317.

W. V. T. Rusch. Phase error and associated cross-polarization effects in Cassegrain fed microwave antenna, *Trans. I.E.E.E.* **AP14** (3), (1966), 266.

R. E. Ward Jnr. Optical properties and uses of the conical mirror. *Applied Optics* **4** (2), (1965), 201.

W. F. Williams. High efficiency antenna reflector, *Microwave Jour.* No. 8 (1965), 79.

W. D. White and L. K. De Size. Scanning characteristics of two reflector antenna systems, "Convention Record of the I.R.E." Part 1, 1962, p. 44.

2

Non-uniform Media

PART 1
NON-UNIFORM MICROWAVE LENSES

The same two fundamental approaches to the subject of the propagation of electromagnetic waves in an isotropic non-homogeneous medium exists as was discussed in the introduction to the first chapter, that is the field approach and that of geometrical optics. Which of these has to be used depends upon the nature of the problem. Where the precise nature of the medium is known or can be conjectured in advance, the field theory has to be used to determine reflection and transmission effects and vector properties such as polarization and field intensity. From these results ray paths and wave fronts could, if required, be derived.

The situation is different if it is intended to create a medium with specified optical properties, usually the focusing of a parallel system of rays to a point.

In this case, the geometrical optics of rays and their behaviour in non-uniform media is the only method that gives a sufficiently close approximation to the required design. Hence this geometry is the first essential for any experimentation in the design of such lenses.

The field method has been most often used as the basis for the study of reflection and propagation of waves in such non-uniform media as the atmosphere and ionosphere, since the earliest days of radio communications. The general theory is highly complex even for the simplest forms of this inhomogeneity. In one or two cases the field problem is relevant to microwave antennas, for example where the field is propagated through transparent windows such as radomes. The simplest form of this problem, that of plane waves incident on a plane infinite layered medium, is the usual approach. The essential extension to other geometries such as cylinders or cones has not as yet proved capable of a simplified analysis that could be used for design purposes.

The first part of the chapter thus deals with the geometry of rays in non-uniform media based upon a natural extension of the laws used in the first chapter for systems of only a few surfaces. The application of this geometry to the design of focusing devices is known for only a few instances and these are included even though the general principles of the designs are well-established. This is not intended as a simple reiteration of this material since the fundamental background to the problem is dealt with, in the hope that this increased basic understanding will permit a greater variation in design. Some of the possibilities are conjectured. Among these is the possibility of applying the theory to media whose description relies on different coordinate systems. It does tend to show for example, that concentration upon spherical media may be an undue limitation upon the design of optical devices in the same way as concentration upon the paraboloid has been for reflecting systems. It also enables a deeper investigation to be made into the basic effect of such media and to illustrate the analogy with other branches of theoretical physics.

The second part then deals with the field problem of reflection and transmission through a plane stratified medium and introduces what is thought to be a new method for investigating the properties of a finite layer such as a radome wall. The method is applicable to the converse problem of the design of such a layer with prescribed properties.

2.1 RAY OPTICS IN A NON-UNIFORM MEDIUM

In much the same way as Snell's laws or Fermat's principle have many diverse formulations, as shown in Appendix I, which range from the simple trigonometrical relations to the dyadic form, the starting point for the generalization of these laws for rays propagating in a non-uniform medium can be as diverse and complex.

The most useful of these relations as derived in the appendix, is the form

$$\frac{d}{ds}\left(\eta\frac{d\mathbf{r}}{ds}\right) = \nabla\eta, \tag{2.1}$$

in which η, the refractive index, is now a scalar function of position specified by the radius vector $\mathbf{r}$ from any convenient fixed origin. The increment ds is measured along the path of the ray itself. Integration of Eqn (2.1) leads to the design of delimited regions of a non-uniform medium with specified ray properties. The integration is carried out in a similar way to the piecewise integration carried out in Chapter 1, the object being to obtain classes of lenses with focusing properties applicable to antenna design.

We first examine Eqn (2.1) for media which stratify along the coordinate surfaces of the three fundamental coordinate systems.

2.2 LINEAR HORIZONTALLY STRATIFIED MEDIUM

In a Cartesian coordinate system for which η is a function of a single variable say $\eta = \eta(x)$ and in which the rays lie in planes parallel to the (x, z) plane, then Eqn (2.1) is

$$\frac{d}{ds}\left\{\eta\frac{d}{ds}(x\hat{\mathbf{i}} + z\hat{\mathbf{k}})\right\} = \nabla\eta = \frac{d\eta}{dx}\hat{\mathbf{i}}, \tag{2.2}$$

with $\hat{\mathbf{i}}, \hat{\mathbf{j}}$ and $\hat{\mathbf{k}}$ the basis vectors of the coordinate system.

Equating $\hat{\mathbf{i}}$ and $\hat{\mathbf{k}}$ components gives

$$\frac{d}{ds}\left(\eta\frac{dz}{ds}\right) = 0 \text{ and } \frac{d}{ds}\left(\eta\frac{dx}{ds}\right) = \frac{d\eta}{ds} \tag{2.3}$$

Both equations lead to the same result†

$$\eta\frac{dz}{ds} = \text{constant} = A \text{ along the path of the ray.}$$

---

† From Figure 2.1 the second of these is

$$\cos\psi\frac{d}{ds}(\eta\cos\psi) = \frac{d\eta}{ds}, \text{ whence } -\eta\cos\psi\sin\psi\frac{d\psi}{ds} = \frac{d\eta}{ds}\sin^2\psi$$

or

$$\eta\sin\psi = \text{constant}$$

From Fig. 2.1 this is

$$\eta \sin \psi = A \qquad (2.4)$$

where ψ is the angle between the ray and the normal to the stratification of the medium and A is necessarily a different constant for each ray.

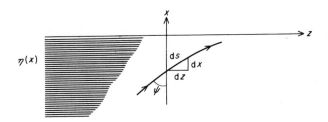

FIG. 2.1. Ray in a linear stratified medium.

Substituting $ds = (dx^2 + dz^2)^{\frac{1}{2}}$ the equation for ray paths becomes

$$dz = \frac{A \, dx}{(\eta^2 - A^2)^{\frac{1}{2}}} \qquad (2.5)$$

One or two particular solutions of Eqn (2.5) may be noted at this stage as they will be required for reference later.

Consider a refractive index law given by

$$\eta(x) = \frac{c}{x}$$

for a ray originating at the point $(x = x_0; z = 0)$ at which point the refractive index will be $\eta_0 = c/x_0$. Let the ray at that point be making an angle α with the vertical. Equation (2.5) gives

$$dz = \frac{x \, dx}{(c^2 - \eta_0^2 \, x^2 \sin^2 \alpha)}$$

or

$$(z\eta_0^2 \sin^2 \alpha + b)^2 = c^2 - \eta_0^2 \, x^2 \sin^2 \alpha \qquad (2.6)$$

where b is the arbitrary constant of integration. For simplicity let $b = c$ then Eqn (2.6) can be seen to be that of ellipses in general. For the particular ray for

E

which

$$\eta_0 \sin \alpha = \frac{c \sin \alpha}{x_0} = 1$$

the ellipse becomes circular. The point to be made is that, allowing the fiction of the infinite refractive index when x tends to zero and the negative refractive index when x is negative, the mathematical solution gives rays that are continuous closed loops. We shall see later that this is one essential condition for completely continuous media. All the rays passing through the point $(x_0, 0)$ intersect again at the point which is at the opposite end of the diameter of the circular ray and hence are exactly refocused. The medium thus has the astigmatic property shown in Fig. 2.2.

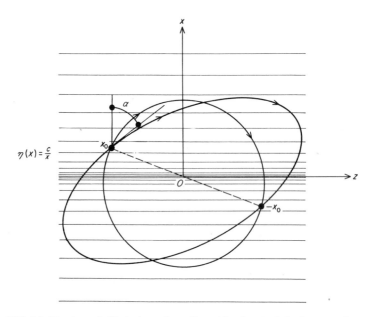

FIG. 2.2. *Circular and elliptical rays in medium with refractive index law* $\eta = c/x$.

To obtain a lens using a medium of this kind it is necessary to find a surface within the physically realizable region of refractive indices, at which the refraction into free space will turn the rays from their elliptical paths into the required focusing (or other) directions. In this instance, a two dimensional case, the source would be a line source but the principle is the same for three-dimensional problems and point sources.

The second example is the refractive index law given by

$$\eta(x) = \eta_0 \operatorname{sech}(ax) \tag{2.7}$$

giving

$$dz = \frac{A\,dx}{(\eta_0^2 \operatorname{sech}^2(ax) - A^2)^{\frac{1}{2}}}$$

where, for rays originating from the origin making an angle α with the vertical

$$A = \eta_0 \sin \alpha.$$

Using the trigonometric and hyperbolic identities converts this equation to

$$dz = \frac{\sin \alpha \cosh(ax)\,dx}{(\cos^2 \alpha - \sin^2 \alpha \sinh^2(ax))}, \tag{2.8}$$

which is directly integrable to give

$$\sin(az) = \tan \alpha \sinh(ax), \tag{2.9}$$

the constant of integration being made zero by a choice of the position of the origin $z = 0$.

The rays in this case are continuous and repeatedly focus at points along the z axis with separation $2f = \pi/a$ (Fig. 2.3).

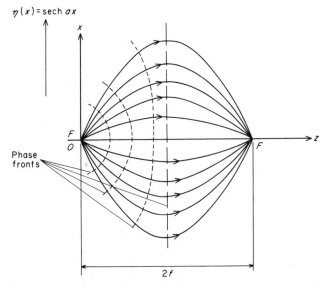

FIG. 2.3. Ray paths and phase fronts in a medium with sech law of refractive index (from Ref. 1).

Hence with the origin of rays at the origin of coordinates, by symmetry the plane midway between two foci will cut each ray orthogonally and hence provides the required surface for focusing the rays, at infinity, without additional refraction. This design of lens is given in Brown[1] under the title of the "short focus horn", and the equation to the wave fronts given as

$$\cos az = p \cosh (ax),$$

p being a variable parameter for the different wave fronts.

The illustration here is of a second type of focusing which is not of the closed loop type. In this case it is periodic and at an infinite number of points along a straight line. If the origin of rays were to be displaced from the axis of symmetry two lines of images are formed as shown in Fig. 2.4 on opposite sides of the axis.

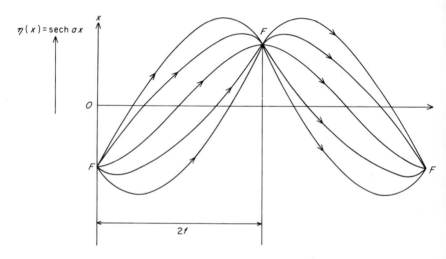

FIG. 2.4. *Ray paths in a medium with sech law of refractive index focus displaced from axis (from Ref. 2).*

2.3 CYLINDRICAL POLAR COORDINATES

In a coordinate system with cylindrical symmetry about the z axis Eqn (2.1) becomes

$$\frac{\mathrm{d}}{\mathrm{d}s}\left\{\eta \frac{\mathrm{d}}{\mathrm{d}s}(\rho\hat{\boldsymbol{\rho}} + z\hat{\mathbf{k}})\right\} = \frac{\partial\eta}{\partial\rho}\hat{\boldsymbol{\rho}} + \frac{\partial\eta}{\partial z}\hat{\mathbf{k}}, \qquad (2.10)$$

since by assumption $\partial\eta/\partial\phi = 0$. $\hat{\rho}$, $\hat{\phi}$ and $\hat{k}$ are the basis vectors of the cylindrical geometry. The complete expansion of the left hand side requires the relations

$$\frac{d\hat{\rho}}{ds} = \frac{d\phi}{ds}\hat{\phi} \quad \text{and} \quad \frac{d\hat{\phi}}{ds} = -\frac{d\phi}{ds}\hat{\rho},$$

then

$$\frac{d}{ds}\left(\eta\frac{dz}{ds}\right) = \frac{\partial\eta}{\partial z}, \tag{2.11a}$$

$$\frac{d}{ds}\left(\eta\frac{d\rho}{ds}\right) - \eta\rho\left(\frac{d\phi}{ds}\right)^2 = \frac{\partial\eta}{\partial\rho}, \tag{2.11b}$$

and

$$\frac{d}{ds}\left(\eta\rho\frac{d\phi}{ds}\right) + \eta\frac{d\rho}{ds}\frac{d\phi}{ds} = 0.$$

The last is identical to

$$\frac{1}{\rho}\frac{d}{ds}\left(\eta\rho^2\frac{d\phi}{ds}\right) = 0 \tag{2.11c}$$

These are of course far too general for a complete solution. If the nature of the medium is specified, that is if $\eta(\rho, z)$ is defined, ray paths can be obtained but the possibility of a practicable focusing device resulting is small. We will discuss therefore those situations in which the rays are confined to the coordinate surfaces of the medium.

(a) Meridional planes

For planes containing the axis of symmetry we put $d\phi$ zero in Eqn (2.11). Equations (2.11a) and (2.11b) then become interchangeable resulting in the complete two dimensional problem of the Cartesian geometry. This can be made one dimensional by the further assumption that there is no variation of the refractive index with respect to one of the variables. In this case, this is obviously to make $\partial\eta/\partial z$ zero, in which case the problem and its solution for meridional planes are identically those of the Cartesian geometry, with, of course, the replacement of the coordinate x by the radial coordinate ρ.

In particular the sech law of refractive index, Eqn (2.7), can be used for a rotationally symmetrical system to focus rays from a point source into repeated point foci along the axis. A similar double line of repeated foci occurs when the source is moved to a point away from the axis of symmetry. The foci then form on each side of the axis alternately in the meridional plane containing the source.[2]

In the symmetrical case with the source on the axis, a "short focus horn" is obtained by terminating the lens at a plane midway between two foci at which plane all the rays, by symmetry alone, must be parallel to the axis and hence collimated at infinity. It has been shown[3] that a lens so formed can tolerate a fair amount of axial defocusing, particularly when only the paraxial rays are considered. This is easily appreciated in microwave terms when it is realised that such an axial defocusing creates only a small *differential* phase shift between the axial ray and any other ray. This turns out to be a particular quality of all of the non-uniform lenses.

If in the limited length "short focus horn" the feed is moved off the axis in a plane containing the original focus the situation is far more complex. In the meridional plane containing the displaced source the rays retain some measure of coherent focusing. Some optical designs of lenses of variable magnification (with varying length of the cylindrical medium) are based on this property.[4] But in other planes, where the skew rays are taken into account, the solution is not at all certain and it remains one of the unsolved problems given in Ref. 3.

As presented, the short focus horn is thus a medium of rotational symmetry in which the index of refraction obeys the law

$$\eta(\rho) = \eta_0 \operatorname{sech}\left(\frac{\pi\rho}{2f}\right)$$

and which has a plane surface distance f from the focus as exit aperture. Other forms of cylindrical refractive index laws can have the same focusing properties if the exit surface is correctly shaped, so that the refraction that will then occur will direct all rays parallel to the axis as required.

The derivation of this surface is possible by geometrical optics for all laws of refractive index and of the resulting class of lenses one may be chosen which more closely obeys the sine condition than the remainder. Such a lens would therefore be better capable of use with an offset feed than the simple design with a plane face. One could include in the design the possibility of z variation of refractive index, arbitrarily chosen so far to be zero, and, with the additional degree of freedom, it may be possible to separate the angle of offset of the feed from the angular position of the beam. We thus arrive at the possibility of a magnifying lens using a cylindrical refractive medium which could scan a beam over a greater angle than the feed is offset.

(b) *The transverse plane*

For rays which are confined to planes perpendicular to the axis we consider the two dimensional problem, with z independence, of a line source parallel to the axis of the cylindrical medium.

We then have from Eqn (2.11c)

$$\eta\rho^2 \frac{d\phi}{ds} = \text{constant} = A. \tag{2.12}$$

For a transverse plane $ds^2 = d\rho^2 + \rho^2 d\phi^2$ which substituted into Eqn (2.12) results in

$$\phi - \phi_0 = \pm \int_{\rho_0}^{\rho} \frac{A \, d\rho}{\rho(\eta^2\rho^2 - A^2)^{\frac{1}{2}}}, \tag{2.13}$$

as the equation for a ray through the point (ρ_0, ϕ_0). In this result the upper sign corresponds to the original direction of the ray, that is the direction towards the axis of symmetry of the medium and the lower sign when directed away from the axis.

Equation (2.12) also states, since $\rho(d\phi/ds) = \sin \alpha$, where α is the angle between the radius vector and ds the direction of the ray itself, that $\eta\rho \sin \alpha = A$ along the path of the ray. The identical situation arises in the following section dealing with the spherically symmetric medium and discussion of this case will be fully covered therein.

(c) Rays confined to a cylinder

It is quite obvious that in order to have rays which are confined to the surface of a circular cylinder coaxial with the axis of symmetry of the medium, variation of refractive index in the axial z direction cannot be permitted. (Consideration of a meridional plane alone would indicate this.) For such a ray variation of ρ and hence of η with the path increment ds must also be zero reducing Eqn (2.11) to the following

$$\eta \frac{dz}{ds} = \text{constant}; \qquad \eta \frac{d\phi}{ds} = \text{constant} = A$$

(confined to a surface $\rho = \text{constant}$ and thus $\eta = \text{constant}$), and

$$-\eta\rho \left(\frac{d\phi}{ds}\right)^2 = \frac{d\eta}{d\rho} \tag{2.14}$$

The first two of these equations show that the path of the ray is helical and the ratio of the two constants concerned will give the helix pitch.

Substitution of the second into the third gives

$$\frac{-\rho}{\eta} A^2 = \frac{d\eta}{d\rho},$$

or

$$\eta^2 = c^2 - A^2\rho^2. \tag{2.15}$$

This is the well-known "parabolic gradient' for the refractive index[5] for which all rays are helical and of the same pitch. Thus coherence is retained and the medium can be used for a fibre optical light guide.

We note however that this result is dependent only upon the final two equations of the set, whereas the pitch of the helix is determined from the first two. It may be surmised therefore, that a variation (for example linear) of refractive index in the z direction, in addition to the radial parabolic gradient, will result in an expanding or contracting helical ray of (linearly) varying pitch. The result is closely analogous to the trajectory of a charged particle in a magnetic "bottle".

(d) Skew rays

Referring to Eqn (2.11c) which applied to the completely general circularly symmetric medium, that is a medium with refractive index $\eta = \eta\,(\rho, z)$, we find a second constant for the ray path given by

$$\eta\rho^2 \frac{d\phi}{ds} = h. \tag{2.16}$$

The implication is that when h is zero the rays are once again confined to meridional planes and thus Eqn (2.16) contains a measure of the skew behaviour of these rays. This is borne out by the analysis given by Luneburg (Ref. 2, p. 188). Recasting Eqn (2.16) as a dependent function of the axial variable z, i.e. substituting $ds^2 = d\rho^2 + \rho^2\,d\phi^2 + dz^2$ one obtains Eqn (30.15) of the reference

$$\frac{\eta\rho^2 \dfrac{d\phi}{dz}}{\left(1 + \left(\dfrac{d\rho}{dz}\right)^2 + \rho^2\left(\dfrac{d\phi}{dz}\right)^2\right)^{\frac{1}{2}}} = h.$$

Solving for $d\phi/dz$ and integrating gives (Ref. 2, Eqn (30.281))

$$\phi_1 - \phi_0 = h \int_{z_1}^{z_0} \frac{\{1 + (d\rho/dz)^2\}^{\frac{1}{2}}}{\rho\{\eta^2\rho^2 - h^2\}^{\frac{1}{2}}}\, dz, \tag{2.17}$$

where (ϕ_0, z_0) and $(\phi_1 z_1)$ are the *angular* positions at the end points of a ray.

We now separate the optical path of a ray from a point in the z_0 plane to a point in the z_1 plane, into an equivalent "radial" path and an "angular" optical path, the latter given by $h(d\phi/dz)$. Then the "radial" path will be given by the total path minus the "angular" path, that is

$$\int_{z_0}^{z_1} \left[\eta\left\{1 + \left(\frac{d\rho}{dz}\right)^2 + \rho^2\left(\frac{d\phi}{dz}\right)^2\right\}^{\frac{1}{2}} - h\left(\frac{d\phi}{dz}\right) \right] dz, \tag{2.18}$$

which equals, after substituting for $d\phi/dz$,

$$\int_{z_0}^{z_1} \left(\eta^2 - \frac{h^2}{\rho^2}\right)^{\frac{1}{2}} \left\{1 + \left(\frac{d\rho}{dz}\right)^2\right\}^{\frac{1}{2}} dz, \tag{2.19}$$

which is (ρ, z) dependent only, hence the definition of "radial" optical path.

This is equivalent to the path of a ray in a *meridional* plane in a medium with refractive index

$$\mu(\rho, z) = \left(\eta^2 - \frac{h^2}{\rho^2}\right)^{\frac{1}{2}} \text{(Ref. 2, Eqn 30.29).} \tag{2.20}$$

Thus for a skew ray we obtain the radial movement, with respect to distance along the axis, from Eqn (2.19) as for a Cartesian geometry but using the refractive index $\mu(\rho, z)$, and the angular motion from Eqn (2.17).

2.4 SPHERICAL POLAR COORDINATES

The expansion of the fundamental Eqn (2.1) in a spherical system can be greatly simplified by *a priori* assumptions regarding the nature of the medium intended. It is customary but by no means obligatory to consider the refractive index to be solely a function of the radial variable r in the coordinate system (r, θ, ϕ) with basis vectors $\hat{\mathbf{e}}_r$, $\hat{\mathbf{e}}_\theta$ and $\hat{\mathbf{e}}_\phi$.

This reduces the three dimensional problem to a single dimensional one and the rays are all confined to diametral planes. The solution then becomes the rotational version of the one dimensional problem noted in the previous Section, part (*b*).

However, against the day when the investigation of such media will incorporate at least a two-dimensional variation of refractive index say $\eta \equiv \eta(r, \theta)$, we will completely expand Eqn (2.1) in this coordinate geometry.

We will note the following relations among the derivatives of the basis vectors

$$d\hat{\mathbf{e}}_r = d\theta \hat{\mathbf{e}}_\theta + \sin\theta \, d\phi \, \hat{\mathbf{e}}_\phi,$$

$$d\hat{\mathbf{e}}_\theta = -d\theta \, \hat{\mathbf{e}}_r + \cos\theta \, d\phi \, \hat{\mathbf{e}}_\phi,$$

$$d\hat{\mathbf{e}}_\phi = -\sin\theta \, d\phi \, \hat{\mathbf{e}}_r - \cos\theta \, d\phi \, \hat{\mathbf{e}}_\theta.$$

Then (2.1) in component form becomes

$$\frac{d}{ds}\left(\eta \frac{dr}{ds}\right) - \eta r \sin^2\theta \left(\frac{d\phi}{ds}\right)^2 - \eta r \left(\frac{d\theta}{ds}\right)^2 = \frac{\partial \eta}{\partial r}, \tag{2.21a}$$

$$\frac{d}{ds}\left(\eta r\,\frac{d\theta}{ds}\right) - \eta r \sin\theta \cos\theta \left(\frac{d\phi}{ds}\right)^2 + \eta\,\frac{dr}{ds}\,\frac{d\theta}{ds} = \frac{1}{r}\,\frac{\partial\eta}{\partial\theta}, \qquad (2.21b)$$

$$\frac{d}{ds}\left(\eta r \sin\theta\,\frac{d\phi}{ds}\right) + \eta r \cos\theta\,\frac{d\theta}{ds}\,\frac{d\phi}{ds} + \eta\,\sin\theta\,\frac{dr}{ds}\,\frac{d\phi}{ds} = \frac{1}{r\sin\theta}\,\frac{\partial\eta}{\partial\phi}, \ (2.21c)$$

When the variation in refractive index is purely spherical any individual ray will be confined to a plane through the centre of symmetry, which will be taken to be the origin of the coordinate system. Then with no loss of generality this can be taken to be a plane of constant ϕ (In analogy with the theory of central force orbits Eqn (2.21b) with $d\phi/ds = 0$ gives the same resultant equation as does Eqn (2.21c) with $\theta = $ constant $= \pi/2$). Thus with

$$\frac{\partial\eta}{\partial\theta} = \frac{\partial\eta}{\partial\phi} = \frac{d\phi}{ds} = 0,$$

Eqn (2.21b) becomes

$$\frac{d}{ds}\left(\eta r^2\,\frac{d\theta}{ds}\right) = 0,$$

and hence

$$\eta r^2\,\frac{d\theta}{ds} = \text{constant} = A. \qquad (2.22)$$

For a ray confined to the plane $\phi = $ constant, $ds^2 = dr^2 + r^2\,d\theta^2$ and thus (2.22) becomes

$$\eta r \sin\psi = A \qquad (2.23)$$

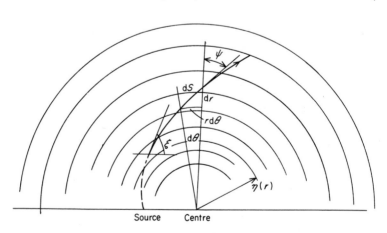

FIG. 2.5. *Ray path in a spherical medium of varying refractive index. Bouguer's theorem.*

where ψ is the angle between the tangent to the ray and the radius vector from the origin (Fig. 2.5).

Solving (2.22) for θ then gives

$$d\theta = \frac{A\,dr}{r(\eta^2 r^2 - A^2)^{\frac{1}{2}}} \tag{2.24}$$

Equation (2.23) is a generalization of Snell's law applicable to spherically symmetrical media, known as Bouguer's theorem.[6]

An interesting transformation can be effected if, as in Ref. 3, we take the angular variable to be the angle between the tangent to the ray and a line parallel to the central diameter containing the source, angle ξ in Fig. 2.5.

Then using the relation (Ref. 3)

$$d\xi = \frac{\tan \psi}{\eta}\,d\eta,$$

and Eqn (2.23), we obtain

$$d\xi = \frac{A\,d\eta}{\eta(\eta^2 r^2 - A^2)^{\frac{1}{2}}}, \tag{2.25}$$

This can be obtained from (2.24) by the direct transformation

$$r \leftrightarrow \eta.$$

The integration of Eqn (2.24) takes on two forms depending upon whether the refractive index $\eta(r)$ is specified in advance or whether the refractive index has to be derived from the required geometry of the rays as given by their (θ, r) variation. In the latter case a standard procedure exists in which the integration is performed through the use of Abel's theorem.[7] For completeness this method is reproduced in Appendix II. It is also given in slightly different form in Ref. 3.

From this theory however only one or two practical antennas have been created. Their design has been obtained by the method indicated above but the results are sufficiently familiar for a summary only of their properties to be included here, in order to extend the discussion subsequently.

(a) Maxwell's fish-eye

The refractive index, law given by

$$\eta(r) = \frac{2}{1 + r^2}, \tag{2.26}$$

defines a medium with focusing properties known as Maxwell's fish eye.[8] As always it can be derived by specifying the required focal property and obtaining the law of the medium through the integration of Eqn (2.24). There are however other methods of deriving this relation either from the focal property or by transformation from other refractive index laws, and these are of fundamental interest.

Considering first the entire space for which the law of Eqn (2.26) applies, that is extending it throughout the region in which the refractive index has a fictitious value less than unity.

The ray paths are then completely circular and all the rays through a given fixed point, the source, intersect again at a second common point the focus. The focusing is thus of the continuous closed loop kind. If the source is at a point on the unit circle where $\eta = 1$, the focus is at the diametrically

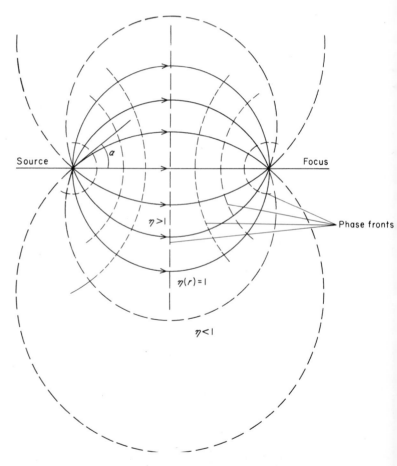

FIG. 2.6(a). Ray paths and phase fronts for a spherical medium with $\eta(r) = 2/(1 + r^2)$. Maxwell's fish eye.

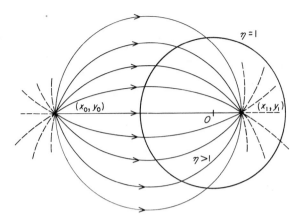

FIG. 2.6(b). *Maxwell fish-eye with displaced source.*

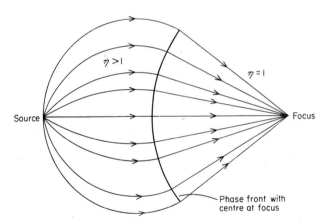

FIG. 2.6(c). *Fish-eye with external focus. For parallel rays the focus goes to infinity and the exit-surface becomes plane.*

opposite point on the same circle. The rays then form a system of circles (Fig. 2.6(a)) with centres on the diameter intersecting the source–focus diameter at right angles. If α is the angle made by an individual ray with this diameter the equations to these circles are given parametrically by (Ref. 1, p. 84)

$$x^2 + y^2 + 2yR \cot \alpha = R^2. \tag{2.27}$$

If the source is at any other point the focusing is of a similar nature, and the focus lies on the straight line through the source and the centre of symmetry of the medium. The relative positions of source and focus are that of a negative inversion in the unit circle, that is (Ref. 2, p. 175) all the rays through a point (x_0, y_0) intersect again at $(x_1 y_1)$ (Fig. 2.6b) where

$$x_1 = \frac{-x_0}{r_0^2}; \qquad y = \frac{-y_0}{r_0^2}$$

where $r_0^2 = x_0^2 + y_0^2$, and if $r_1^2 = x_1^2 + y_1^2$,

$$r_0 r_1 = 1. \tag{2.28}$$

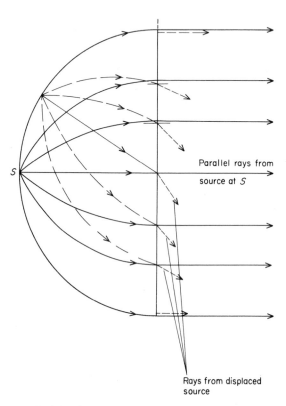

FIG. 2.6(d). *Maxwell's half fish-eye. Rays at plane surface are refracted at different angles due to variation of refractive index when source is rotated about centre.*

In both cases the wave fronts are the orthogonal system of circles. Lenses with external foci may be obtained simply by creating an exit surface at one of the wave fronts, in which case the rays will continue along the normals to this surface, which, being spherical, will have a focus at its centre, (Fig. 2.6c).

As a microwave lens, this wave front could be taken as the plane central wave front creating a system of parallel rays. It is not known to what extent the half fish-eye is capable of beam scanning as shown in Fig. 2.6d before the onset of deterioration through aberrations, but quite obviously this can be calculated in phase error terms from the ray equations above.

The ray paths given in Eqn (2.27) are well known to be the stereographic projection of great circles on a unit sphere onto the equatorial plane. The

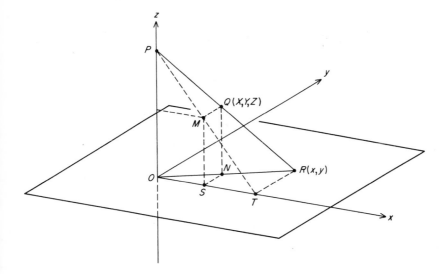

FIG. 2.7. Stereographic projection.

means whereby this fact is converted into the derivation of a refractive index is of great interest as it involves the concepts of the ray as a geodesic path in space and its transformation into a geodesic path in the plane. That is a transformation from the geodesics of one space into the geodesics of the space of one lesser dimension. The stereographic projection is a conformal mapping of the two spaces, and this conformal mapping between spaces will be the subject of a later chapter (Chapter 6).

The basis for the stereographic projection from the pole of the unit sphere is the simple geometric relations governing the similar triangles POR and QNR and OPT and SMT in Figure 2.7.

These are

$$\frac{x}{X} = \frac{y}{Y} = \frac{1}{1 - Z} = \frac{r}{(X^2 + Y^2)^{\frac{1}{2}}} = \frac{(r^2 + 1)^{\frac{1}{2}}}{(X^2 + Y^2 + (1 - Z^2))^{\frac{1}{2}}} \quad (2.29)$$

where $r^2 = OR^2 = x^2 + y^2$ and $P \equiv (0, 0, 1)$.

The point (X, Y, Z) can now be made to conform to any curve in space and its projection onto the (x, y) plane obtained from Eqn (2.29).

For great circles on the unit sphere $X^2 + Y^2 + Z^2 = 1$ and hence

$$r^2 = \frac{1 + Z}{1 - Z},$$

giving

$$X = \frac{2x}{1 + r^2}; \qquad Y = \frac{2y}{1 + r^2}; \qquad Z = \frac{r^2 - 1}{r^2 + 1}. \qquad (2.30)$$

The increment of the circle on the sphere is

$$dS^2 = dX^2 + dY^2 + dZ^2,$$

from Eqn (2.30)

$$dX^2 = \left\{ \frac{2dx}{1 + r^2} - \frac{4rx\, dr}{(1 + r^2)^2} \right\}^2,$$

and on summation the terms in (dr^2) cancel to give

$$dS^2 = \frac{4}{(1 + r^2)^2} (dx^2 + dy^2). \qquad (2.31)$$

If these are the geodesics of rays in a plane then it is to be given by

$$ds^2 = \eta^2(x, y)\,(dx^2 + dy^2),$$

where η is the refractive index variation causing the curvature of the rays. Hence the apparent refractive index is

$$\eta^2 = \frac{4}{(1 + r^2)^2},$$

or

$$\eta = \frac{2}{1 + r^2},$$

which is the Maxwell fish-eye condition.

This procedure can now be applied in the converse manner. By deriving the law of refractive index for a required focusing problem using the integration method, stereographic projection can be used to obtain a surface in space with unity refractive index upon which the rays project as geodesic curves. Such a procedure is possible after one or two other forms of transformation notably conformal mappings. The resulting lenses, named geodesic lenses have been the subject of much investigation (Ref. 1, p. 101) as shown in the associated bibliography to this chapter. It is not intended to enter into the complete design theory of such lenses. For the form of lens which is a two dimensional parallel plate version of a cross section of the complete lens, a description will be given in a section dealing with practical designs of these lenses and with the results of known geodesic designs.

(b) The Luneburg lenses

The original problem dealt with by Luneburg (Ref. 2, p. 182) was to determine a law for a spherical region ($r < 1$) which imaged a source on a diameter onto a focus on the same diameter with both points external to the medium. By excluding any refractive effect at the surface of the spherical medium it was implied that the refractive index law was continuous there and so for a lens in free space the refractive index has to have the value unity at the surface $r = 1$.

The method used is that given in the appendix as the fundamental illustration of the integration method to be applied to this kind of problem.

When the source is placed at the surface of the unit sphere and the focus at infinity the resultant collimating lens has the refractive index law

$$\eta(r) = (2 - r^2)^{\frac{1}{2}}. \tag{2.32}$$

If we now consider this law to apply to the entire (real) space, then the rays cannot extend beyond the boundary $r = \sqrt{2}$ since we may allow fictitious media but not imaginary ones.

The rays in this extended medium are complete ellipses whose equations are given by (Ref. 1, p. 87)

$$x^2 + y^2(1 + 2\cot^2 \alpha) - 2xy \cot \alpha = 1 \tag{2.33}$$

the parameter α being the angle between the ray at the source and the diameter of the sphere.

Differentiation with respect to the parameter and then elimination of α between the resulting equation and Eqn (2.33) shows that the envelope of this system of ellipses is another ellipse

$$x^2/2 + y^2 = 1.$$

The focusing in the extended medium is thus also of closed loops (Fig. 2.8) in this case ellipses and the exit surface normally used is that of the unit sphere itself. It can be shown that the tangents to the ellipses are all parallel to the x axis at the points where they intersect the exit surface and, since the refractive index is unity at that surface, no further refraction occurs, and the entire ray system is then parallel to the axis as required.

Extensions of the Luneburg lens

There are two fundamental extensions to Luneburg's analysis that have been made, both with the practical application of the concept to microwave antennas in mind.

The first of these is by Morgan[9] who continues the original system analysed by Luneburg, that of the spherical region with external source and focus, to that where the refractive index law is piecewise continuous. That is the sphere

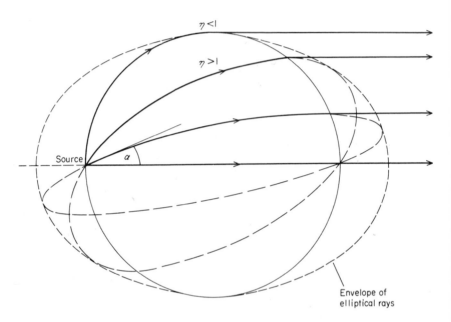

FIG. 2.8(a). Ray paths in a medium with refractive index law $\eta(r) = \sqrt{(2 - r^2)}$. The Luneburg lens.

is constructed of shells in each of which the refractive index has a particular law. The most important of these is the lens of two regions in which the outer shell has a constant refractive index and the core then has a derived refractive index law. It is the derivation of these laws for particular outer layer parameters that is the concern of the reference given.

The method used is that of the appendix as before, but as basic study of the entire subject of spherical non-homogeneous lenses, the analysis and conclusions of the reference are fundamental reading to anyone intending the design of such an antenna.

Specifically once the refractive lens for the outer layer is specified, the central region law is determined. The different lenses that are obtained from different specified outer layers give rise to different aperture amplitude distributions, when non-isotropic sources are used and thus affect the radiation patterns eventually achieved.

Considering the case where the outer layer is of a constant refractive index, then since this cannot be unity a discontinuity in refractive index now occurs at the surface and a refraction takes place. This lens as shown in Fig. 2.9(a)

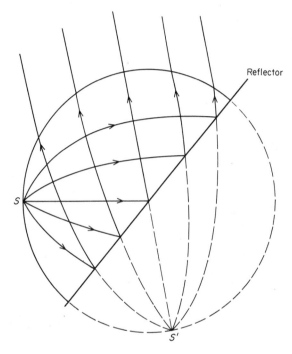

Reflector

FIG. 2.8(b). Virtual source Luneburg lens.

then strictly only applies to the rays contained between the rays tangent to the inner core as shown. As such it is applicable to a microwave source with pattern width less than $\pm 90°$. The design no longer falls within the present theoretical study of perfectly focusing systems, but is an approximation, which however remains perfectly adequate for most practical designs as will be shown later. We wish in this respect to note the conclusion that is reached in this treatment namely that a uniform outer layer results in a refractive index law for the central core that has a *higher* value at the centre than did the original Luneburg law, but that the *overall* variation in refractive index becomes smaller. Some typical results are shown in Fig. 2.9(b). Most practically designed lenses omit to take these conclusions into account as will be discussed subsequently.

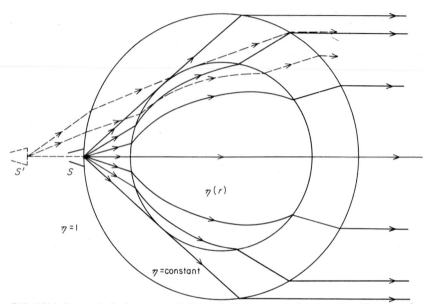

FIG. 2.9(a). *Ray paths for lens surrounded by a shell of constant refractive index. The source has pattern width less than* $\pm 90°$. *S' is an alternative design for a source not on the actual surface of the lens.*

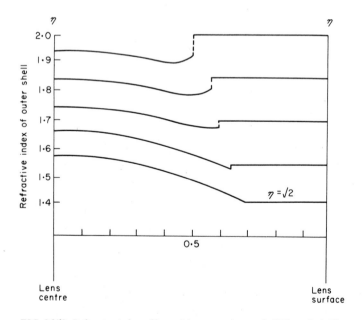

FIG. 2.9(b). *Refractive index of lens with constant outer shell (from Ref. 19).*

The second extension to the Luneburg result is the refractive index law obtained when the source is within the unit sphere iteslf. For a focus at infinity, the solution has been obtained by Gutman[10] and the law derived has unity value at the surface of the unit sphere, which therefore becomes the exit surface. The rays thus have to have tangents, at their intersections with this sphere, which are parallel to the x axis, as in the original Luneburg design. Gutman derives these ray paths by yet another method which,

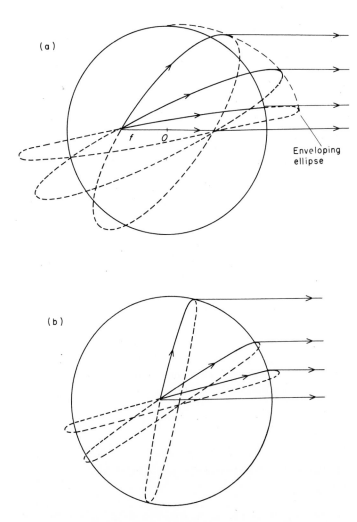

(a)

Enveloping ellipse

(b)

FIG. 2.9(c). Ray paths in a Gutman lens. (i) $f = \frac{1}{2}$, (ii) Small f.

although given by Luneburg (Ref. 2, p. 84) is not in fact used in designing lenses. The principle is to consider a light ray as the path of a particle in a potential field. For a refractive index η this potential field is given by

$$\phi = -\tfrac{1}{2}\eta^2.$$

By this association it can be seen that a great deal of the theory of particle dynamics, can be applied to the theory of optics and the whole of the Hamiltonian method based on Fermat's principle applicable to either. The value of this approach is a further part of the discussion of the final chapter of this book.

For a source of rays at a distance f from the origin of a spherically symmetric medium, where f is essentially less than unity (lens surface is taken to be $r = 1$) the refractive index law obtained by Gutman is

$$\eta = (1 + f^2 - r^2)^{\frac{1}{2}}/f. \tag{2.34}$$

The ray paths in the entire space with such a law are again elliptical, this time with equation

$$x^2 + (\cot^2 \alpha + f^2 \operatorname{cosec}^2 \alpha)y^2 - 2xy \cot \alpha = f^2. \tag{2.35}$$

Both these results can be seen to be generalizations of the equivalent relations (2.32) and (2.33) For a fixed value of f the envelope of these ellipses is the ellipse

$$x^2/(1 + f^2) + y^2 = 1. \tag{2.36}$$

This shows that the rays in the complete space do not penetrate beyond the circle of radius $(1 + f^2)^{\frac{1}{2}}$. As the value of f is descreased the rays become more and more confined to the unit sphere while the refractive index at the centre increases. In the limit of zero f the refractive index at the centre becomes infinite and the ellipses degenerate into radial lines trapped within the sphere by total internal reflection at the surface which has become a singularity of Eqn (2.34). This result has analogies to the Schwarzschild singularity in relativity theory, which gives rise to the proposition regarding the existence of black holes[11] (Fig. 2.9c).

2.5 THE GENERAL THEORY OF FOCUSED RAYS IN A SPHERICALLY SYMMETRICAL MEDIUM

As can be seen from the foregoing, the spherically symmetrical non-homogeneous medium has been the focus of attention with regard to applications to microwave lenses. We can thus consider the properties of a total space in which the refractive index law is spherically symmetric with regard to a fixed origin of coordinates, and we allow in this space the entire range of fictitious refractive indices including values that are less than unity, negative, zero and infinite (but *not* complex). The refractive index is a continuous function of the radial coordinate r and has a continuous first derivative. Real lenses can be constructed in this space by defining a region with "realistic" refractive index. If, in such a medium perfect focusing is possible, then all the rays intersecting at a given point, the source, will intersect again at a second point, the focus and hence all the rays must form closed loops. If we

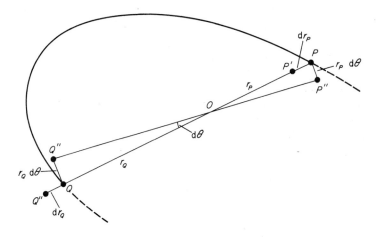

FIG. 2.10. Source and image on a closed ray path.

include in this space (or the plane cross-section we need only consider) the circle at infinity then curves such as hyperbolas and parabolas can be considered also as closed loops. The plane is then the projective plane.

The optical distance from the source to the focus along any one such closed loop must be the same for both paths around the loop. In the sense of the dynamical analogue of particle trajectories, the field is conservative and hence describable by a potential function. This too can be seen from the fact that the curl of Eqn (2.1) is identically zero, implying that the entire optical path around a loop in one direction of travel is zero.

We consider points P and Q on such a loop to be image points. Then, by virtue of the equivalence of the *optical* path from P to Q, for the infinitesimally small triangle at P imaged in the small triangle at Q (Figure 2.10) we have

$$\eta_P PP' = \eta_Q QQ' \quad \text{and} \quad \eta_P PP'' = \eta_Q QQ''$$

that is

$$\eta_P r_P d\theta = \eta_Q r_Q d\theta \qquad (2.38a)$$

and

$$\eta_P dr_P = \mp \eta_Q dr_Q \qquad (2.38b)$$

the ambiguity in sign referring as in Chapter 1 to real or virtual focusing.
Division of these relations gives

$$d[\log r_P] = \mp d[\log r_Q]$$

and hence either

$$r_P r_Q = a^2 \qquad \text{or} \qquad r_P/r_Q = b^2 \qquad (2.39)$$

where a and b are arbitrary constants.

If each source has only a single focus then P and Q can be interchanged hence $r_Q/r_P = b^2$ and thus for this situation $b^2 = 1$.

We now show that the first relation in Eqn (2.39) gives the law for Maxwell's fish eye and the second the Luneburg and Gutman lenses.

Ray paths obeying the law $r_P r_Q = a^2$ where P and Q are any two points on the same radius vector through the origin are inverse curves with respect to a circle radius a. We wish to derive the refractive index law for which the ray paths are circular. Let one such ray be the circle radius R centre M Fig. 2.11 where OM is the length D. Then $2a$ is the length of the chord at right

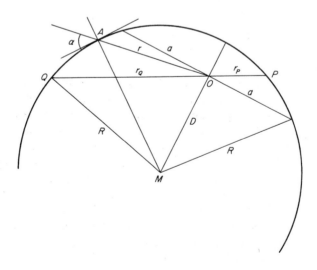

FIG. 2.11. *Circular ray in spherically symmetrical medium.*

angles to OM and thus $r_P r_Q = a^2$. For a general point such as A on the ray we have

$$D^2 = R^2 + r^2 - 2Rr \sin \alpha. \tag{2.40}$$

From Eqns (2.38) we can derive Bouguer's law namely

$$\eta r \sin \alpha = \text{constant along a ray} = C.$$

Then substituting into Eqn (2.40),

$$D^2 = R^2 + r^2 - 2RC/\eta,$$

or

$$\eta = \frac{2RC}{r^2 + R^2 - D^2}. \tag{2.41}$$

This is the law of refraction for Maxwell's fish eye with constants chosen so that $RC = 1, R^2 - D^2 = 1$, that is for a refractive index that is unity at the unit circle.

In the second situation the ray paths satisfy $r_P = r_Q$ where P and Q are on the opposite sides of a radius vector through the centre of spherical symmetry. The simplest curves for this case are thus the central conics. We choose those ellipses which pass through the points ± 1 on a diameter which we choose as the x axis. We further require the tangents to these ellipses to be parallel to the x axis at the points where they intersect the unit circle. The ray paths can be recognised as those of the Luneburg lens. These ellipses, with the conditions prescribed above must have equations

$$x^2 + a^2 y^2 + 2bxy = 1 \tag{2.42}$$

If the ellipse makes the angle α with the x axis at the point $x = -1$ and hence $\pi + \alpha$ at the point $x = 1$, and has the tangent property required the constants in this equation can be evaluated to give

$$x^2 + y^2(1 + 2\cot^2\alpha) - 2xy \cot \alpha = 1.$$

If the tangent at any point P with coordinate r on this ellipse makes an angle ϕ (Fig. 2.12(a)) with the radius vector, then it can be shown, with some fairly complicated algebra, that the equation to the ellipse can be expressed as

$$r^2 = 1 \pm (1 - \sin^2\alpha/\sin^2\phi)^{\frac{1}{2}} \tag{2.43}$$

the sign depending upon the parts of the ellipse internal or external to the unit circle.

If then $\eta r \sin \phi = \text{constant}$ for the ray as Bouguer's theorem states, $\eta r \sin \phi = \sin \alpha$ since $\eta = 1$ when $r = 1$ at which point $\phi \equiv \alpha$.

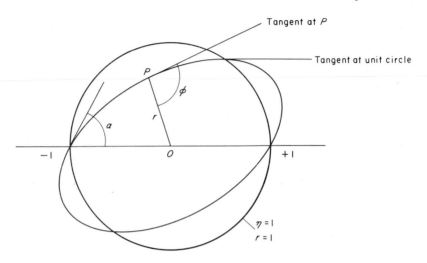

FIG. 2.12(a). *Geometry of elliptical ray Luneburg lens.*

Substituting for sin α/sin φ in Eqn (2.43) gives the refractive index law

$$\eta^2 = 2 - r^2.$$

required for the Luneburg lens.

The Gutman modification can be obtained in a similar manner. If the source of rays be placed at a point F a distance $f < 1$ from the origin, and the same conditions of tangency made for the ellipse, the equation for the ellipse becomes

$$x^2 + y^2 \left(\frac{f^2 + \cos^2 \alpha}{\sin^2 \alpha} \right) - 2xy \cot \alpha = f^2, \qquad (2.44)$$

from which can be derived after considerably more complex algebra the relation (Fig. 2.12(b))

$$2r^2 = 1 + f^2 \pm \{(1 + f^2)^2 - 4f^2 \sin^2\alpha/\sin^2\phi\}^{\frac{1}{2}} \qquad (2.45)$$

Substituting $\eta(r)$ $r \sin \phi = \eta(f) f \sin \alpha = \sin \alpha$ the first equation relating to the ray at the source and the second to the ray at its intersection with the unit circle, results in the law

$$\eta^2 = (1 + f^2 - r^2)/f^2 \text{ of the Gutman lens.}$$

Thus particular solutions of Eqn (2.39) gives the lenses with which we are now familiar.

The general solution is a solution of the *functional* relation obtained by combining equation 38a with the result $r_P r_Q = a^2$ to give

$$\frac{r_P}{a} \eta\left(\frac{r_P}{a}\right) = \frac{a}{r_P} \eta\left(\frac{a}{r_P}\right)$$

(2.46)

We can without loss of generality take the constant to be unity and the general solution of Eqn (2.46) is seen to be

$$r\eta(r) = f\left\{\phi(r) \otimes \phi\left(\frac{1}{r}\right)\right\}$$

(2.47)

where f and ϕ are arbitrary functions of their arguments and $\otimes$ implies an associative law of combination. Taking this to be summation as the simplest instance then

$$r\eta(r) = f\left\{\phi(r) + \phi\left(\frac{1}{r}\right)\right\}$$

(2.48)

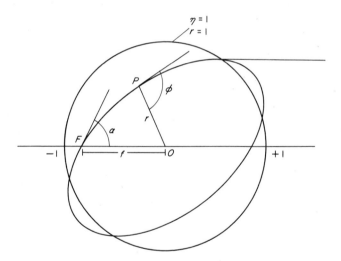

FIG. 2.12(b). Geometry of elliptical ray Gutman lens.

Maxwell's fish-eye is given by $\phi(r) = r$; $f(x) = A/x$. Luneburg (Ref. 2, p. 179) gives a generalization of Maxwell's fish-eye obtained by a direct conformal mapping of the cross section of the problem described above. This has the refractive index law

$$\eta(r) = \frac{2r^{\gamma-1}}{1 + r^{2\gamma}}$$

This, in terms of Eqn (2.48) can be seen to be obtained from the function

$$f(x) = \frac{1}{x}, \qquad \phi(x) = r^\gamma.$$

Further generalization will be noted in Section 2.8.

No similar general solution can be obtained for Eqn (2.38b) other than the trivial $r\,\eta(r) = f(r)$ and hence no generalizations of the Luneburg lens can be obtained from this equation.

As a consequence of this theory the existence of a spherically symmetrical medium in which the ray paths are hyperbolae, the remaining form of central conic, can be postulated. Since these are outward curving as shown in Fig. 2.12(c) the refractive index law will be an increasing function of the radius

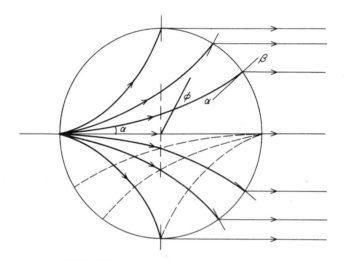

FIG. 2.12(c). *Lens with hyperbolic ray paths.*

r and a real lens, η always larger than unity, will thus require a refraction at the surface $r = 1$. The refractive index there will be taken to have the value $\eta(1) = \kappa$. If this refraction is taken to turn each ray to the horizontal the lens would have a focusing property similar to that of the Luneburg lenses, as is shown.

For central hyperbolae intersecting the axis at points $x = \pm 1$ and with the refractive property prescribed at the circle $r = 1$ the equations will be

$$x^2 + y^2 \left\{ 1 + \frac{2 \cot \alpha (1 - \kappa^2 \sin^2 \alpha)^{\frac{1}{2}}}{\kappa \sin \alpha} \right\} - 2xy \cot \alpha = 1 \qquad (2.49a)$$

With ϕ the angle between the radius vector and the tangent to the ray path as in the previous studies (Fig. 2.12(c)) we can obtain

$$2xyp + y^2 - x^2 = \tan\alpha/\tan\phi,$$

and   .

$$r^2 \cot\alpha = 2x^2 \cot\alpha - 2xyp \cot\alpha + \cot\phi,$$

$$p = (1 - \kappa^2 \sin^2\alpha)^{\frac{1}{2}}/\kappa \sin\alpha, \tag{2.49b}$$

resulting in the (r, ϕ) relation

$$(r^2 - \tfrac{1}{2})^2 = \tfrac{1}{4} + \frac{r^4 \cosec^2\alpha - \cosec^2\phi}{2(p \cot\alpha + 1)}. \tag{2.49c}$$

Substitution of Bouguer's law however does not lead to the complete elimination of α and ϕ simultaneously, that is, putting $\eta r \sin\phi = \kappa \sin\alpha$ results finally in

$$(r^2 - \tfrac{1}{2})^2 = \tfrac{1}{4} + \frac{\kappa^2 r^4 - \eta^2 r^2}{2\{\kappa \cos\alpha(1 - \kappa^2 \sin^2\alpha)^{\frac{1}{2}} + \kappa^2 \sin^2\alpha\}} \tag{2.49d}$$

As a check we find that the value $\kappa = 1$ gives the Luneburg lens with *elliptical* ray paths. As will be seen for the uniform sphere (p. 154) for a linear ray $\kappa = 2$ thus for an inwardly curving ray κ has to be greater than 2. For these values of κ and over a small angular range of α the complicated denominator in Eqn (2.49d) can be given the value 2κ and the resulting *approximate* law of the lens is

$$[\eta(r)]^2 = 2\kappa + (\kappa^2 - 2\kappa)r^2, \tag{2.49e}$$

which still incorporates the Luneburg result. For larger values of κ the approximation is valid over a decreasing range of α, but due to the increasing curvature of the rays this can still create an exit aperture of a considerable fraction of the total lens diameter. With the rays given by equation 2.49a this can be seen to be

$$2y_{max} = 2\kappa \sin\alpha_{max}.$$

The substitution of this refractive index law into the ray path integral of equation II.31 p. 399 shows that to the first order in $\kappa^2 \sin^2\alpha$, that is for the conditions of our approximate solution, the rays do follow the required trajectory.

That an approximation has to be resorted to is only an indication that the solution to this problem is unlikely to exist for an algebraic form of refractive index law. Solving this problem by the use of Abel's integration formula as shown in Appendix II will probably result in a refractive index law involving elliptic integrals. A similar situation arises with the axially defocused Luneburg lens and computational processes have to be invoked.

A similar lens in which the *angles* α and β in Fig. 2.12(c) are related by the factor κ, instead of their *sines* as required by refraction, has been investigated by Toraldo di Francia[13] (Appendix II).

The law of proportionality between these angles is taken to be

$$\kappa = 2(1 - p),$$

and the refractive index law of the lens is found to be given by

$$\eta^2 r^2 = \eta^{1/p}(2 - \eta^{1/p}). \tag{2.50}$$

When $p = \frac{1}{2}$ we obtain Luneburg's result and when $p = 1$ we obtain Maxwell's fish-eye. This is then another generalization of this form of lens.

2.6 THE LEGENDRE TRANSFORM

The Legendre transformation is a relation between wave fronts in two different systems of coordinates, and is used by Luneburg to demonstrate the connections between the characteristic functions of Hamilton for an optical system (Ref. 2, p. 102). It is also used by Luneburg to derive the refractive index law of Maxwell's fish-eye from a different refractive index law. In this case we shall show the procedure in reverse and start from the fish-eye.

If the surfaces $S(x, y, z) = $ constant are the wave fronts associated with a system of rays in x, y, z space with refractive index $\eta(x, y, z)$ and $T(\lambda, \mu, \nu)$ = constant are the wave fronts in λ, μ, ν space with refractive index $N(\lambda, \mu, \nu)$ then these are connected by a Legendre transformation if

$$\frac{\partial S}{\partial x} = \lambda; \quad \frac{\partial S}{\partial y} = \mu; \quad \frac{\partial S}{\partial z} = \nu$$

and

$$\frac{\partial T}{\partial \lambda} = x; \quad \frac{\partial T}{\partial \mu} = y; \quad \frac{\partial T}{\partial \nu} = z. \tag{2.51}$$

Since S and T are wave fronts the eikonal equation Appendix I Eqn (A.I. 27) applies and thus

$$\left(\frac{\partial S}{\partial x}\right)^2 + \left(\frac{\partial S}{\partial y}\right)^2 + \left(\frac{\partial S}{\partial z}\right)^2 = \eta^2,$$

$$\left(\frac{\partial T}{\partial \lambda}\right)^2 + \left(\frac{\partial T}{\partial \mu}\right)^2 + \left(\frac{\partial T}{\partial v}\right)^2 = N^2. \tag{2.52}$$

With S and T linearly independent the relations in Eqn (2.51) can be summed up by the equation

$$S + T = x\lambda + y\mu + xv.$$

If in the first space the refractive index law of Maxwell's fish eye applies then

$$\eta(x, y, z) = \frac{2}{1 + r^2} = \frac{2}{1 + x^2 + y^2 + z^2}; \qquad r^2 = x^2 + y^2 + z^2,$$

that is

$$\left\{\left(\frac{\partial S}{\partial x}\right)^2 + \left(\frac{\partial S}{\partial y}\right)^2 + \left(\frac{\partial S}{\partial x}\right)^2\right\}^{\frac{1}{2}} = \frac{2}{1 + x^2 + y^2 + z^2},$$

which transformed becomes

$$(\lambda^2 + \mu^2 + v^2)^{\frac{1}{2}} = \frac{1}{(1 + N^2)},$$

that is

$$N^2 = \frac{2}{\rho} - 1 \qquad \rho^2 = \lambda^2 + \mu^2 + v^2. \tag{2.53}$$

Again the refractive index is unity at the unit circle. The ray paths are ellipses all with one focus at the origin and all with horizontal tangents at their *two* intersections with the unit circle (Fig. 2.13)† If at this point the ray makes an angle α with the radius vector then the major axis is always of magnitude 2 and the minor axis $2 \sin \alpha$. Hence the distance along the x axis from the unit circle at D to the centre of the ellipse E is equal to the distance from the origin with the intersection of the ellipse with the axis at C. Rays parallel to the axis therefore are rotated once about the origin and returned in their original direction. This form of lens is called an Eaton lens.[14]

We can now see that the two relations are in fact connected by the simple transformation $\eta \leftrightarrow r$ first noticed in the comparison between Eqns (2.24) and (2.25).

It is simple to show that under a Legendre transformation or under the transformation $\eta \leftrightarrow r$ the refractive index law for the Luneburg lens $\eta^2 = 2 - r^2$ is invariant, and that of Gutman's lens becomes

$$\eta^2 = 1 + f^2 - f^2 r^2.$$

† The reader may find it of interest to obtain the form $r = 1 - \cos \alpha \sqrt{(1 - \tan^2 \alpha / \tan^2 \phi)}$ for these ellipses and hence by substituting $\eta r \sin \phi = \sin \alpha$ (condition at $\eta = r = 1$) obtain $\eta^2 = (2/r) - 1$.

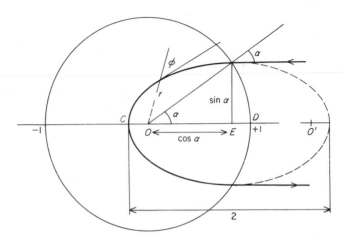

FIG. 2.13. *Elliptical ray in Eaton lens. O and O' are foci. Law of refractive index is* $\eta^2 = (2/r - 1)$.

2.7 THE BENDING OF LIGHT RAYS NEAR THE SUN

The presence of a centre of gravitational attraction distorts the local space time geometry so that the increment of path length of a world line changes from the Lorentz line-element,

$$ds^2 = c^2 \, dt^2 - dr^2 - r^2 \, d\theta^2 - r^2 \sin^2 \theta \, d\phi^2, \qquad (2.54)$$

to that of the Schwarzschild element,

$$ds^2 = \left(1 - \frac{2Km}{c^2 r}\right) c^2 \, dt^2 - \frac{dr^2}{\{1 - (2Km/c^2 r)]\}} - r^2 \, d\theta^2 - r^2 \sin^2 \theta \, d\phi^2,$$

$$(2.55)$$

where K is the universal gravitational constant, m the mass of the centre of gravitational attraction and c is the velocity of light. The effect of this change is to cause a curvature of the geodesics of the geometry. If these are light rays, this curvature is most apparent when they just graze a massive body such as the sun. The observation of this effect is a well documented procedure during eclipses of the sun.

The coordinate velocity of light is given by $ds = 0$ and in the radial direction ($d\theta = d\phi = 0$) this is $dr/dt = c$ for Lorentz space and

$$\frac{dr}{dt} \simeq \left(1 - \frac{2Km}{c^2r}\right)c \equiv c'$$

for Schwarzschild space for $r \neq 0$ and $2Km/(c^2r) \ll 1$.

Thus in the presence of a gravitational field we can specify a radial variation of refractive index given by

$$c/c' = \eta(r) = \left(1 + \frac{2Km}{c^2r}\right). \tag{2.54}$$

To the same degree of approximation this is of the form

$$\eta^2 = 1 + \frac{A}{r}$$

which Luneburg (Ref. 2, p. 171) shows to give ray trajectories which are hyperbolae with focus at the origin. Putting $2Km/c^2 = \phi$ for brevity, then in accordance with Eqn (2.24) the angle turned through by the radius vector deflected from the straight line by an amount δ is

$$\pi + \delta = 2 \int_{\infty}^{r_m} \frac{-h \, dr}{r(\eta^2 r^2 - h^2)^{\frac{1}{2}}} \tag{2.55}$$

where r_m is the radial distance of the nearest approach of the ray to the origin. This we take to be $r_\odot$ the radius of the sun since we consider the ray to be just grazing the surface. The ray constant is h so that at the closest approach to the sun

$$h = \eta(r) \, r \sin \alpha$$

where $r = r_\odot$ and $\alpha = \pi/2$, and

$$h = r_\odot\left(1 + \frac{2Km}{c^2r_\odot}\right) \text{ or } r_\odot = h - \phi.$$

Transforming the integral by the substitution $v = 1/r$ gives

$$\pi + \delta = 2 \int_0^{1/(h-\phi)} \frac{h \, dv}{(1 + 2\phi v + (\phi^2 - h^2)v^2)^{\frac{1}{2}}}$$

$$= \frac{2h}{(h^2 - \phi^2)^{\frac{1}{2}}} \left\{ \sin^{-1}\left(\frac{v(h^2 - \phi^2) - \phi}{h}\right) \right\}_0^{1/(h-\phi)}$$

$$= \frac{2h}{(h^2 - \phi^2)^{\frac{1}{2}}} \left\{ \frac{\pi}{2} + \sin^{-1}\frac{\phi}{h} \right\}$$

F

This, to the permitted order in ϕ, $\simeq \pi + (2\phi/h)$.
Hence the angle of deflection

$$\delta = \frac{2\phi}{h} = \frac{4Km}{c^2 r_\odot}.$$

This is the classical result for the bending of light rays near to the sun.

In a similar way refraction through a variable atmosphere could be assessed for its effects upon the required beam shape of a satellite antenna for example or for propagation beyond the optical horizon.

2.8 OTHER COORDINATE SYSTEMS—SEPARABILITY

It can be seen from the foregoing discussion that there exist only a few laws of refractive index which lead to integrable forms of the ray path equation in terms of elementary functions, and that these laws are themselves algebraic and of a fairly elementary nature. In considering, in particular the elementary parabolic law giving rise to helical rays as in Section 2.3(c), Buchdahl[15] was led to investigate separability conditions that could be applied to the eikonal equation for rays in a generally varying medium

$$|\nabla s|^2 = \eta^2$$

(Appendix I) in different coordinate systems which would lead to algebraically simple results. By this method he formulated the simplest laws of refractive index that would arise in these coordinate systems. What is most remarkable is that this appears to be the first consideration of some of the more unusual coordinate systems that can support this form of lens construction. Any one of these could lead to the design of lenses with desirable qualities by the methods previously formulated, that of determining the ray paths and a suitable delimiting surface at which refraction, if it occurs, can be used to focus the rays in the required manner. The results of Buchdahl's investigations are outlined here, with a necessary change of notation for conformity with the previous sections. For the method the reader is referred to the original work.

(*a*) *Cartesian coordinates*

$$\eta^2 = f(x) + g(y) + h(z),$$

where for symmetry about the z axis

$$f(x) + g(y) \equiv a(x^2 + y^2) \qquad\qquad a \text{ constant,}$$

$$h(z) = (\alpha + \beta e^{-\gamma z}) \qquad\qquad \alpha, \beta, \gamma \text{ constant.}$$

(b) Cylindrical polar coordinates

$$\eta^2 = f(\rho) + \rho^{-2} g(\phi) + h(z).$$

Axial symmetry requires $g(\phi)$ to be constant and for regularity along the axis (not essential however) this constant would be zero. $h(z)$ is also taken to be constant and the result for $f(\rho)$ is either

$$f(\rho) = \alpha \rho^2 + \beta + \frac{\gamma}{\rho^2} \quad \text{or} \quad \alpha + \frac{\beta}{\rho} + \frac{\gamma}{\rho^2} \qquad \alpha, \beta, \gamma \text{ constant.}$$

This contains the parabolic law for helical rays, but not the sech $a\rho$ law of Section 2.3(b) which essentially applies to rays in meridional planes alone and not general rays as in the separable solutions.

(c) Parabolic coordinates

Parabolic coordinates defined by

$$x = \mu v \cos \theta, \qquad y = \mu v \sin \theta, \qquad z = \tfrac{1}{2}(\mu^2 - v^2),$$

has the separable condition for the refractive index

$$\eta^2 = (\mu^2 + v^2)^{-1} [f(\mu) + g(v)] + (\mu v)^{-2} h(\theta).$$

Axial symmetry makes $h(\theta)$ constant and since $\mu^2 v^2 = x^2 + y^2 = \rho^2$ the problem reverts to cylindrical symmetry with

$$\eta^2 = a + (b + cz)(\rho^2 + z^2)^{-\frac{1}{2}} \qquad a, b, c, \text{ constant.}$$

(d) Spherical polar coordinates

$$\eta^2 = f(r) + r^{-1} g(\theta),$$

where

$$g(\theta) = (\alpha + \beta \sin^2 \theta + \gamma \sin^4 \theta)/(\sin^2 \theta \cos^2 \theta)$$

which if regular becomes

$$g(\theta) = \beta_1 + \beta_2 \sec^2 \theta,$$
$$f(r) = r^{-2}(\alpha_1 + \alpha_2 r^\kappa + \alpha_3 r^{2\kappa})/(1 + \alpha_5 r^\kappa)^2,$$

all Greek letters are constant and κ is an arbitrary constant. Maxwell's fish eye is given by

$$\kappa = 2, \qquad \beta_1 = \beta_2 = \alpha_1 = \alpha_3 = 0.$$

It is interesting to observe that this result can be made to agree with the general solution given in Eqn (2.48) by putting $\alpha_1 = \alpha_3$ in the above and

$$f(x) = \left(A + \frac{B}{x^2}\right)^{\frac{1}{2}}, \qquad \phi(r) = r^{\kappa/2},$$

in Eqn (2.48).

(e) Cardioid coordinates

Cardioid coordinates are defined by

$$x = \mu v(\mu^2 + v^2)^{-2} \cos\theta, \qquad y = \mu v(\mu^2 + v^2)^{-2} \sin\theta,$$

$$z = \tfrac{1}{2}(\mu^2 - v^2)(\mu^2 + v^2)^{-2},$$

then, omitting refractive laws giving rise to singularities in the medium

$$\eta^2 = (\mu^2 + v^2)^3 \left[f(\mu) + g(v)\right],$$

and $f(\mu) + g(v) = \alpha + \beta\mu^2 + \gamma v^2$ for simple solutions.

If $r^2 = x^2 + y^2 + z^2$ the result is

$$\eta^2 = r^{-5}(ar^2 + br + cz) \qquad a, b, c, \text{constant}.$$

For each of these coordinate systems the fundamental law

$$\frac{\mathrm{d}}{\mathrm{d}s}\left(\eta \frac{\mathrm{d}r}{\mathrm{d}s}\right) = \nabla\eta \tag{2.56}$$

can be reduced to ray equations and conditions of symmetry, and regularity applied. For given constant coordinate surfaces the ray paths can be determined and the above method of Buchdahl gives those refractive index laws which can be expected to yield analytically integrable results from which ray paths and focusing systems can be determined.

2.9 SPHERICAL SHELL LENSES

The construction of non-uniform spherical lenses normally takes the form of a series of hemispherical shells, each of a material with a constant refractive index, nested together to form a layered sphere. The refractive index variation is then obtained by approximating the smooth refractive index law required by a stepped approximation. It is usually taken that with a sufficiently large number of steps the approximate curve is an adequate approximation for the required curvature of the rays (Ref. 7, p. 135). This procedure disregards two basic principles of antenna design. Firstly it implies that each and every ray from the source within the forward hemisphere is required for full illumination of the aperture. This is so since rays with commencement angles $\pm \alpha = 90°$ are included in the design as can be seen from Fig. 2.8. In any practical situation a finite sized microwave source will have an optimum angular distribution of less than this angle and so rays outside this range are not only ignorable, but if included can reduce the final efficiency of the system. This case would be best approached by a design such as that shown in Fig. 2.9(a).

The second principle is that which has been established by the work of S. P. Morgan (Ref. 9) and others that the presence of an outer spherical layer of constant refractive index changes the refractive index law for the entire interior of the lens inside this layer as was shown in Fig. 2.9(b). This is so for even a thin layer, and the attempts to construct this layer with very light density materials has led to unnecessary complications such as surface coatings of actually very high density materials.

It would appear obvious therefore that a lens with these two *a priori* conditions should be designed as a layered structure with a specified angular width of rays issuing from the source. The analysis required turns out to be a relatively simple geometrical study and results in far fewer constant index layers being required than the arbitrary stepped version. In fact for lenses of medium apertures, approximately 10λ in diameter, 2 or 3 shells only are necessary.

We illustrate the geometry first by deriving Bouguer's theorem in differential form from such a layered medium.[16] If as in Fig. 2.14 a ray is refracted at successive layers at the points $P_1 P_2$ etc. where the radii of the layers are respectively $r_1, r_2 \ldots$ and the incident angles of the rays $\psi_1 \psi_2 \ldots$ we have by virtue of Snell's law at each point P,

$$\eta_1 \sin \psi_1 = \eta_2 \sin \psi_1',$$

or generally

$$\eta_i \sin \psi_i = \eta_{i+1} \sin \psi_i' \qquad (2.57)$$

From the triangle $P_i O P_{i+1}$ we have

$$r_i \sin \psi_i' = r_{i+1} \sin \psi_{i+1}$$

which substituted into Eqn. (2.57) gives

$$\eta_i r_i \sin \psi_i = \eta_{i+1} r_{i+1} \sin \psi_{i+1}$$

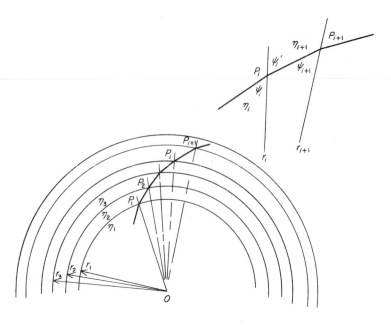

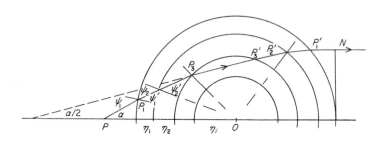

FIG. 2.14. Ray paths in spherically stratified medium.

that is

$$\eta r \sin \psi \text{ is a constant at each radial layer.}$$

This for a continuous smooth distribution becomes Bouguer's law.

Turning to the related problem of the field angle of the rays from the source, we consider the source P (Fig. 2.15a) to be at a distance h from the centre of a sphere of unit radius. A ray from the source making an angle α with the axis is refracted twice in its passage through the sphere at points Q and R. If this single ray emerges from the sphere parallel to the axis then the line QR can be shown quite simply to make an angle $\alpha/2$ with the axis and the required refractive index at the points of refraction has to be

$$\eta = \frac{h \sin \alpha}{\sin(\sin^{-1}(h \sin \alpha) - \alpha/2\}} \tag{2.58}$$

In effect the ray is refracted by equal amounts of $\alpha/2$ at each of the two points.

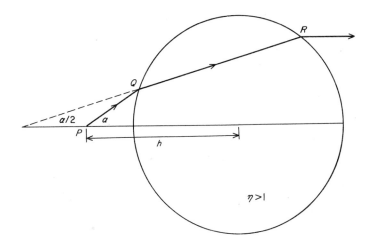

FIG. 2.15(a). *Single exact ray for homogeneous sphere.*

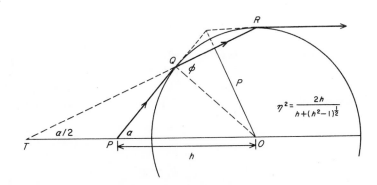

FIG. 2.15(b). *Limit ray for homogeneous sphere.*

If the ray is confined to a uniform shell with this refractive index the ray will be "exact" in the sense of its parallelism with the axis on emergence.

In the limiting case, when the full aperture of the lens is required to be used the geometry is as shown in Fig. 2.15b. Then $PQ = PT = (h^2 - 1)^{\frac{1}{2}}$ and since the ray is tangential both at Q and at R

$$\eta = \frac{1}{\sin \phi} = \frac{1}{p}, \qquad \frac{p}{h + (h^2 - 1)^{\frac{1}{2}}} = \sin \frac{\alpha}{2}$$

and hence

$$\eta = \left[\frac{2h}{h + (h^2 - 1)^{\frac{1}{2}}} \right]^{\frac{1}{2}}. \qquad (2.59)$$

For a source on the surface of the lens therefore an outer shell of refractive index $\eta = \sqrt{2}$ can be used with a source of angular width $\pm 45°$.

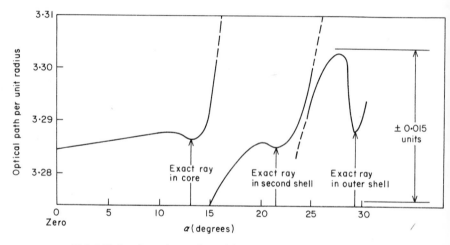

FIG. 2.16. *Resultant phase in lens of three layers; field angle of source* $\pm 30°$.

The same analysis[17] shows that when a ray is refracted at more than one spherical interface as in Fig. 2.14, it will be transmitted horizontally if the sum of the angles turned through over *half* of its passage through the lens is half the angle of emission at the source. That is for a ray with angle α

$$\psi_1 - \psi_1' + \psi_2 - \psi_2' \ldots = \alpha/2$$

or

$$\sum_1^N \psi_i - \sum_1^N \psi_i' = \alpha/2$$

for a lens with N refracting surfaces (including the outer surface). .

It is obvious therefore that all the rays cannot be exact[18] for a discretely layered spherical lens. In microwave practice it is not possible to establish a ray angle criterion up to which departures from "exactness" can be permitted. We have instead to calculate permitted departures from the planeness of the wave front. That is we require to calculate the optical path length

$$PP_1 + \eta_1 P_1 P_2 + \eta_2 P_2 P_3 \ldots \eta_1 P'_2 P'_1 + P'_1 N \qquad \text{(Fig. 2.14)}$$

and maintain its value by the choice of radii and refractive indices, within a prescribed tolerance. This can be done empirically with the concept of the exact ray, since such a ray will automatically have the correct optical path.

The design procedure is thus to establish the angular width at the source and thus the source position and the refractive index of the outer shell for an outer exact ray. The diameter of this shell is then taken to be that for which a tangential ray, now no longer exact, has optical path length within the limit of phase prescribed. A refractive index is then obtained for the next layer for which an exact ray occurs *near* to the interface and a check carried out that rays nearer to the interface are still within the phase tolerance. The next diameter can be obtained by calculation of ray phases until a further layer has to be introduced. The final phase distribution has the form shown in Fig. 2.16 which was obtained for a lens of diameter 8λ constructed in three

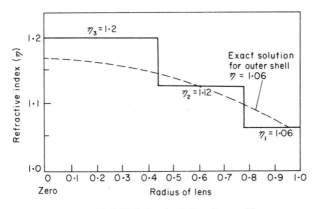

FIG. 2.17. *Refractive index of stepped lens.*

layers[17]. Comparison of the resulting refractive index steps for this lens with the exact solution, given an outer shell of the same refractive index (Fig. 2.17) shows how far a stepped solution departs from a presupposed smooth relation.

The same analysis can be used to determine the depth of focus for such a

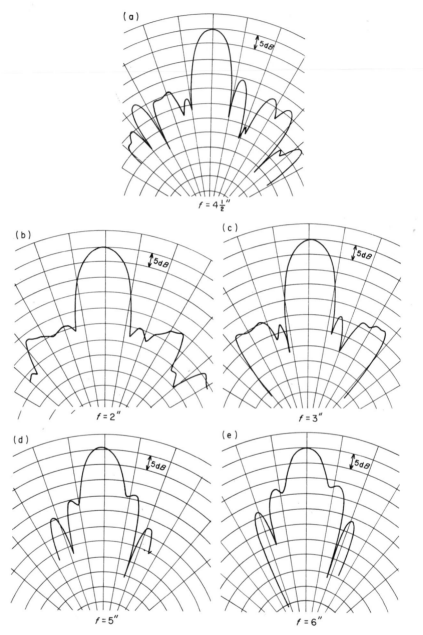

FIG. 2.18. Patterns of lens for varying focal position. (a) Design focus $f = 4.5$, (b) $f = 2$, (c) $f = 3$, (d) $f = 5$, (e) $f = 6$.

lens, which must, in principle be similar for all such non-uniform lenses. It is found that an axial defocusing movement of magnitude ε produces a phase error *differential* between the centre ray and a marginal one of 0.06ε. This allows a considerable source movement before noticeable deterioration of the radiated pattern is observed. Some measured results are shown in Fig. 2.18.

A similar design procedure can be followed for lenses with refractive index law that increases outwardly from the centre. Some complexity occurs in that the field angle of the source rays has to be confined to a much smaller angle to prevent stray anomalous rays from interfering with the pattern. The dependence upon total internal reflection is also a hazardous procedure with regard to phase, but in this design, Fig. 2.19, can be confined to the single most marginal ray. Such a lens, first suggested by Toraldo di Francia and Zoli,[19] has apparently a greater aperture efficiency for even fewer layers

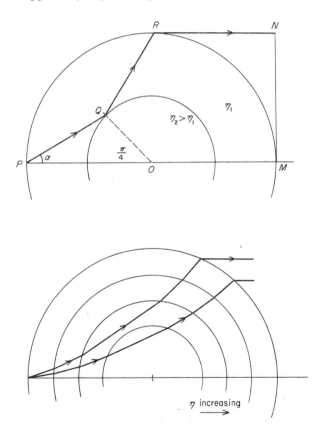

FIG. 2.19(a). *Ray path in lens of two layers and extrapolation toward hyperbolic rays.*

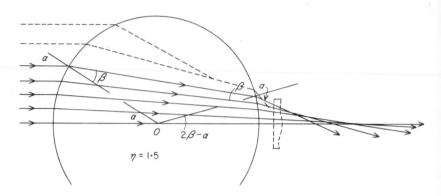

FIG. 2.19(b). *Rays through a solid sphere with possible caustic corrector.*

than the previous design. Both in this reference and in a later study[20] a two layer lens is given with very similar characteristics. That is an outer layer of refractive index 3·4 (Ref. 19) (3.236 Ref. 20) and inner core of refractive index 2·665 (Ref. 19) (2·618 Ref. 20).

If the full aperture is required to be used then by the geometry of Fig. 2.19 such a lens can be designed upon the basic criterion that the optical length of the marginal ray be the same as that of the central ray.

From Fig. 2.19 and the refraction of the ray at R we find $\eta_1 = 1/\sin \alpha$ for a marginal ray making an angle α with the axis at the source.

Then

$$PQ = \eta_1 r/\sqrt{2},$$

and hence

$$r = \sqrt{2}/[1 + (\eta_1^2 - 1)^{\frac{1}{2}}]. \tag{2.60}$$

Thus r and α are determined by the specification of the refractive index η_1 For the ray $PQRN$ to be of the same optical length as the ray POM we have simply (for a lens of unit radius)

$$2\eta_1 PQ + 1 = 2\eta_2 r + 2\eta_1(1 - r), \tag{2.61}$$

which together with Eqn (2.60) gives

$$\eta_2 = \frac{\eta_1^2}{\sqrt{2}} + \eta_1 - \frac{(2\eta_1 - 1)\left[1 + \sqrt{(\eta_1^2 - 1)}\right]}{2\sqrt{2}} \tag{2.62}$$

This is only one of the many criteria that could be used however, and others may optimize the planeness of the complete wave front.

The interest in this design of lens lies mainly in the extrapolation of it to a lens of many thin layers. It will be seen then that the rays will become smooth outwardly bending curves of precisely the kind hypothesized as the hyperbolae given by Eqn (2.62).

Finally, we consider the simplest construction of all, that of a uniform sphere, and it is found by comparatively simple geometrical ray tracing, that over a limited range of paraxial rays, a caustic is created of a similar nature to that for a spherical mirror. This can be corrected in a similar way therefore by a Schmidt plate designed as before. As shown in Fig. 2.19(b) the correcting plate has to be situated so that no ray intersection can occur in the space between it and the sphere. This limits the aperture available. It is also found that this region itself is limited and varies with refractive index. For, in the notation of the figure we have $\sin \beta = (1/\eta) \sin \alpha$ from the law of refraction at the surface and thus the angle $2\beta - \alpha$ shown will be zero when

$$\frac{\alpha}{2} = \cos^{-1}\left(\frac{2}{\eta}\right)$$

limiting the refractive index to be less then 2 for rays refracting correctly in the upper hemisphere.

The construction shown is for a practicable refractive index of $\eta = 1\cdot5$, which covers the range of the common plastic materials, and shows that an acceptance angle $\alpha = \pm 45°$ is feasible.

2.10 NON-HOMOGENEOUS PARALLEL PLATE LENSES AND GEODESIC LENSES

Full discussion of the methods whereby non-uniform media can be created in the two dimensional space can be found in the standard literature of the subject. These are referred to in general as "configuration lenses" (Ref. 1, p. 101 and Ref. 7, p. 106) and include "conflection doublets".[21]

The first method is to make a two dimensional analogue of the lens concerned by confining the wave between parallel conducting plates and varying the propagation properties of the medium to create the necessary curvature of the rays. For the TE_{01} mode of propagation this can be done by varying the plate spacing as by Jones[22] or the plates can be kept parallel and the dielectric medium varied in an appropriate manner as Gutman (Ref. 10). In the former some use can be made of the cut-off phenomenon to produce refractive indices in the range $0 \leqslant \eta \leqslant 1$ but a problem of refraction at the periphery has to be contended with. The latter can be made with plane surfaces and several layers can be stacked to produce a cylindrical lens. All such lenses can be designed directly from standard principles.

The geodesic lenses are fundamentally more important. The waves are now confined between a pair of parallel plates which are curved in such a way that the rays, being geodesics of the surfaces so formed, are curved in the manner required for the focusing property. This is obtained by a transformation between the ray paths as designed in a plane (cross section of a three dimensional lens) and the geodesics on a surface. So far the subject has been entirely confined to systems with rotational symmetry and then concerns the one dimensional description of the refractive index law $\eta = \eta(r)$ with a two dimensional curve (ρ, z) which on rotation about the z axis creates the required surface. It is thus a transformation between spaces of different dimension.

One particular version of this concept has been met already when it was shown (Section 2.4) that the ray paths of Maxwell's fish eye are the stereographic projection of the great circles of the sphere. The geodesic lens corresponding to the fish eye is thus a pair of concentric hemispherical conducting surfaces.

For completeness the method of transformation has to be given. The increment of distance between two near points on a surface of revolution created by the rotation of a curve $\rho = \rho(z)$ about the axis (Fig. 2.20) is given by

$$ds^2 = d\sigma^2 + \rho^2 d\phi^2. \tag{2.63}$$

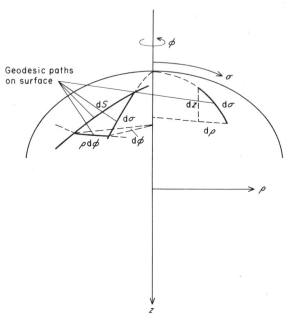

FIG. 2.20. *Geometry of geodesics on a surface of revolution.*

The radial coordinate can equally be expressed as a function of the distance from the apex measured along the natural geodesic lines ϕ = constant. Then since the propagation is assumed to be T.E.M., the refractive index is unity and (2.63) can be written

$$ds^2 = d\sigma^2 + \rho^2(\sigma)\,d\phi^2. \tag{2.64}$$

The optical distance between two points in the plane with variable refractive index $\eta(r)$ is (Ref. 2, p. 176)

$$ds^2 = \eta^2(r, \theta)\,\{dr^2 + r^2\,d\theta^2\} \tag{2.65}$$

and for these two relations to compare, it can be seen by inspection that

$$d\phi = d\theta, \qquad \rho(\sigma) = r\eta(r),$$

and

$$d\sigma = \eta(r)\,dr. \tag{2.66}$$

The first can be integrated directly to give $\theta = \phi$ as the constant of integration can be made zero. This illustrates that the transformation is one between the linear variation $\eta(r)$ with the curved line (ρ, σ).

Solutions have been given in particular cases notably by Rinehart.[23] For the Luneburg law of refractive index

$$\eta(r) = \{2 - r^2\}^{\frac{1}{2}}$$

we obtain

$$d\sigma = (2 - r^2)^{\frac{1}{2}}\,dr \qquad \text{and} \qquad (\rho(\sigma))^2 = 2r^2 - r^4,$$

i.e.

$$r^2 = 1 - (1 - \rho^2)^{\frac{1}{2}}.$$

Elimination of r and dr leads to the differential equation relating σ and ρ which is

$$2d\sigma = \{1 + (1 - \rho^2)^{-\frac{1}{2}}\}\,d\rho, \tag{2.67}$$

or

$$\sigma = \tfrac{1}{2}\{\rho + \sin^{-1}\rho\}, \tag{2.68}$$

the constant again being zero on the assumption that $\rho = 0$ when $\sigma = 0$.

Transforming this solution to $z = z(\rho)$ however leads to a much more complicated result[24]

$$z = \frac{2}{2\sqrt{3}}\kappa - \sqrt{3} - \log\left(\frac{\kappa + 3\sqrt{1 - \rho^2} + 2}{2 + \sqrt{3}}\right),$$

$$\kappa = \{12 - 9\rho^2 + 12(1 - \rho^2)^{\frac{1}{2}}\}^{\frac{1}{2}}, \tag{2.69}$$

This is obtained by the integration (Ref. 24, p. 227) from Fig. 2.20 of

$$dz^2 = d\sigma^2 - d\rho^2,$$

that is

$$z(\rho) = \int_\rho^1 \left\{ \left(\frac{d\sigma}{d\rho}\right)^2 - 1 \right\}^{\frac{1}{2}} d\rho. \qquad (2.70)$$

The same procedure shows quite readily that the fish-eye ray paths give rise to geodesics on a sphere. For with $\eta = 2/(1 + r^2)$ Eqns (2.66) give

$$r = \{1 - (1 - \rho^2)^{\frac{1}{2}}\}/\rho$$

and hence

$$d\sigma = d\sigma/(1 - \rho^2)^{\frac{1}{2}}$$

Thus

$$\sigma = \sin^{-1}\rho \qquad (2.71)$$

the equation to a circle.

The refractive index law obtained from the fish-eye by the transformation $\eta \leftrightarrow r$ for the Eaton lens, is

$$\eta = 2/r - 1$$

for which

$$r = 1 - (1 - \rho^2)^{\frac{1}{2}}$$

hence

$$d\sigma = \{1 + (1 - \rho^2)^{-\frac{1}{2}}\} d\rho$$

or

$$\sigma = \rho + \sin^{-1}\rho \qquad (2.72)$$

The generalized law derived by Luneburg by a conformal mapping (Ref. 2, p. 179) is

$$\eta = 2r^{\gamma-1}/(1 + r^{2\gamma}),$$

leading to

$$r^\gamma = \{1 - (1 - \rho^2)^{\frac{1}{2}}\}/\rho,$$

and

$$d\sigma = \frac{1}{\gamma}\frac{d\rho}{(1 - \rho^2)^{\frac{1}{2}}} \qquad \text{or} \qquad \sigma = \frac{1}{\gamma}\sin^{-1}\rho. \qquad (2.73)$$

The result of applying this procedure to the non-uniform lenses discussed in this chapter are given in the accompanying table. In this we have included the plane surface $\sigma = \rho$ and the right circular cone $\sigma = a\rho$ ($a = \operatorname{cosec}\psi$, ψ = cone half angle).

TABLE 2.1. Lenses and Geodesic Analogues

Luneburg	$\eta(r) = (2 - r^2)^{\frac{1}{2}}$	$\sigma = \frac{1}{2}\rho + \frac{1}{2}\sin^{-1}\rho$
Maxwell	$\eta(r) = 2/(1 + r^2)$	$\sigma = \sin^{-1}\rho$
Eaton	$\eta(r) = (2/r - 1)^{\frac{1}{2}}$	$\sigma = \rho + \sin^{-1}\rho$
Generalized Luneburg	$\eta(r) = 2r^{\gamma-1}/(1 + r^{2\gamma})$	$\sigma = \dfrac{1}{\gamma}\sin^{-1}\rho$
Gutman	$\eta(r) = (1 + f^2 - r^2)^{\frac{1}{2}}/f$	$\sigma = \frac{1}{2}\rho + \left(\dfrac{1 + f^2}{4f}\right)\sin^{-1}\dfrac{2f}{1 + f^2}\rho$
Transformed Gutman	$\eta(r) = 1 + f^2 - f^2 r^2$	$\sigma = \rho + \dfrac{1 + f^2}{2f}\sin^{-1}\dfrac{2f}{1 + f^2}\rho$
Plane	$\eta(r) = 1$	$\sigma = \rho$
Right circular cone (half angle ψ)	$\eta(r) = r^{1/a-1}$	$\sigma = a\rho \quad a = \operatorname{cosec}\psi.$
Approximate lens with hyperbolic rays	$\eta(r) = 2\kappa + (\kappa^2 - 2\kappa)r^2$ $\kappa > 2$	$\sigma = \frac{1}{2}\rho + \dfrac{1}{2(1 - 2/\kappa)^{\frac{1}{2}}}$ $\times \cosh^{-1}\{1 + \rho^2(1 - 2/\kappa)\}^{\frac{1}{2}}$

For the cone we have derived the analogous refractive index law by reversing the procedure. Thus with $\sigma = a\rho, a > 1$

$$d\sigma = ad\rho = a(rd\eta + \eta dr) = \eta dr$$

the solution of which is $\eta = r^{1/a-1}$ as given.

Hence any refractive index law that can be constructed piecewise by segments with the law $r^{1/a-1}$ each segment having the appropriate value of the parameter a can be formed in its geodesic analogue by piecewise straight conical sections with half angles corresponding to the same values of $a = \operatorname{cosec}\psi$. The centre of the geodesic lens would then be a plane section corresponding to the (assumed) flat region of the refractive index law.

The generic form of all these lenses with the exception of the last one given would appear to be

$$\sigma = A\rho + B\sin^{-1}C\rho \tag{2.74}$$

with A, B and C constants. It should thus be possible to obtain a generalized differential equation for the $\eta(r)$ relation containing these constants and including all the lenses shown by using the reverse procedure given above for the cone.

One other method becomes possible whereby lenses may be transformed into other lenses, possibly new lenses, by means of the geodesic analogue. This is by applying transformations to the geodesic surfaces themselves. Conformal mapping of geodesic surfaces is a well established procedure and led to a method for studying the trajectories of particles in a central force field. Since this is already analogous to the ray in a non-uniform medium as shown by the Gutman design, its applicability to this problem is readily appreciated. Not much work has as yet been done on this topic but many of the relevant papers are listed in the bibliography section on geodesic surfaces.

The actual geodesic structure once manufactured can be tested by the exact analogy to the particle in a potential field, for if the surface of revolution be placed on a horizontal surface, its actual shape is then the potential function. An incident particle projected as a light ray toward this surface takes the form of a small sphere with the appropriate energy (Ref. 2, p. 6) and will roll over the surface with a trajectory identical to the geodesic ray path in a lens of the same shape.

This principle has been used in fact to demonstrate such particle trajectories for various potential distributions in physics lecture rooms.[25] The surface is formed of a rubber-like membrane drawn up into the shape of the required potential distribution. Particle trajectories, scattering and reflection are shown by rolling small spheres toward the potential distribution. The properties of refraction, reflection and tunnelling can likewise be demonstrated. On consideration, it appears as good a method as any for actually designing geodesic lenses.

PART II   MULTILAYER STRUCTURES RADOMES AND ELECTROMAGNETIC WINDOWS

The applications of the plane stratified medium as an optical element in microwave practice are mainly, but not exclusively, confined to the design of transparent windows or radomes. These consist essentially of few surfaces when compared with their optical counterpart the optical filter. Other devices to which the same design method can be applied include frequency filters, dielectric layers on conducting surfaces to enhance or reduce the reflectivity, absorbing multilayers and the microwave equivalent of the half-silvered mirror.

In all these applications it is required to determine the field properties of reflection coefficient, transmission coefficient, phase and polarization and therefore the ray optic approach is no longer suitable.

On the other hand the complete solution of the field equations, while giving the exact solution for a pre-determined non-homogeneous or stratified layer, cannot of itself be used as a design method for that layer, to give the system some required property of reflection or transmission. Having regard therefore to the eventual construction of practical layers of this kind, we consider the reflection and refraction properties of a plane multi-layered medium and modifications that can be sought for a transition to a continuously varying medium. The reflection and transmission of plane waves through a curved layer, in which the radius of curvature is a large number of wavelengths is customarily taken to be that of the plane structure tangent to the surface at the point of incidence of a ray. This is much the same approximation as is taken in the physical optics studies of curved reflectors where the induced currents are taken to be those of a local tangent plane. The validity of this approximation in the case of curved radomes has not been tested either theoretically or by experiment.

The analogy of a microwave multilayer radome with an optical filter as band-pass frequency devices is well-known. The optical filter is also an analogue of the wave guide filter which itself is a frequency analogue of the spatial form which compares with the problem of the diffraction grating and linear arrays. Some of these subjects have been treated in the literature to a much greater depth than have others and a considerable body of mathematical method has developed around some of them, which by virtue of the analogies, could be applied to many of the others. The reader may enjoy discovering for himself those areas where a large fund of theory can be readily interchanged between the various problems.

The method to be employed in the following is the standard procedure in optics[26] of multiplying the transmission matrices of the individual layers that go to make up the multilayered structure. It is considered to be well known that complete transmission does not occur unless the multi-layer is fundamentally symmetrical (unless each layer is separately reflectionless). Advantage can be taken of this symmetry to simplify certain of the procedures. The degrees of freedom available then increase by one for each symmetrically placed pair of surfaces. Since the three layer structure is then the simplest to consider, we illustrate the different uses to which the additional degree of freedom can be put, with variations on that system.

To make the transition to a smoothly varying medium the same transmission matrix method can be used. However, it can be shown that the usual procedure whereby the smoothly varying medium is approximated by a series of infinitesimally small steps, is only amenable to this treatment if the resulting "infinitesimal" matrices are correctly multiplied. The approximate multiplication of the matrices gives a result which is only valid for a layer whose *entire* overall optical width is only a small fraction of a wavelength. The proper multiplication of the matrices is highly complex and greatly reduces the benefit that is to be derived from the process.

An improvement of this matrix method is presented which gives an exact result in the limiting process and is applicable to a layer of any width of a plane stratified medium.

The procedure whereby losses of the materials and conductivity can be included into the analysis is standard even if mathematically complex, and some new forms of radome structure can be assessed by these general methods.

2.11 SINGLE LAYER TRANSMISSION MATRIX

2.11.1 Uniform plane layer

The transmission matrix for a uniform layer of finite width of homogeneous dielectric material, is determined by the solution for the continuity of the field components at the two surfaces of the layer (Ref. 26). This method is standard in most text books on electromagnetic theory. In terms of this matrix the fields on the incident side of the layer, for a normally incident plane wave in free space, are given by

$$
\begin{pmatrix} E_{\text{inc}} \\ H_{\text{inc}} \end{pmatrix} = \begin{pmatrix} \cos \beta d & \dfrac{i\omega\mu}{\beta} \sin \beta d \\ \dfrac{i\beta}{\omega\mu} \sin \beta d & \cos \beta d \end{pmatrix} \begin{pmatrix} 1 \\ \dfrac{\gamma'}{i\omega\mu'} \end{pmatrix} \tag{2.75}
$$

where $\beta = \omega\sqrt{\mu\varepsilon}$ is the propagation constant for a plane wave in an infinite medium of the same material as the layer and for increased generality the medium on the transmission side (Fig. 2.21) has been made complex with medium constants ε', μ' and propagation constant γ'.

For oblique incidence we have to distinguish between the two polarization states of the wave, perpendicular when the electric vector is perpendicular to the plane containing the wave and surface normals (the plane of incidence) and parallel when the magnetic vector is perpendicular to this plane and hence the electric vector parallel to it.

The relative permittivity K_c and permeability K_m are defined by

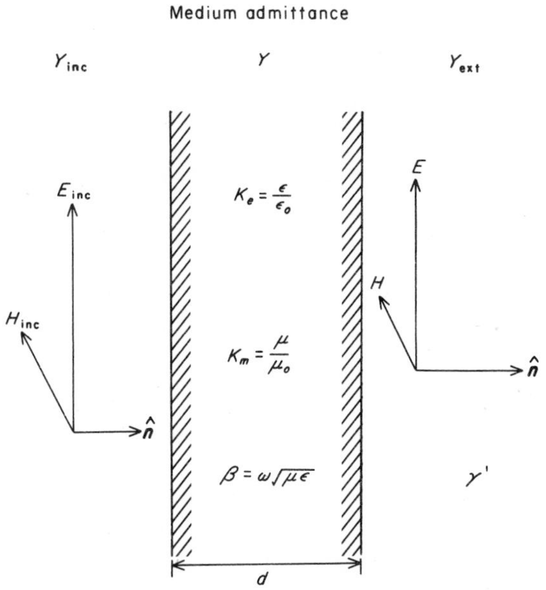

Medium admittance

FIG. 2.21. *Plane wave normally incident upon a uniform layer.*

$K_e = \varepsilon/\varepsilon_0$ $K_m = \mu/\mu_0$ and the media admittances by $Y = (\varepsilon/\mu)^{\frac{1}{2}}$; $Y_0 = (\varepsilon_0/\mu_0)^{\frac{1}{2}}$ where ε_0 and μ_0 are the free space constants.

Then Eqn (2.75) can be written

$$\begin{pmatrix} E_{\text{inc}} \\ H_{\text{inc}} \end{pmatrix} = \begin{pmatrix} \cos A & \dfrac{i}{Y}\sin A \\ iY\sin A & \cos A \end{pmatrix} \begin{pmatrix} 1 \\ Y_{\text{ext}} \end{pmatrix} \qquad (2.76)$$

where A is the phase angle

$$A = \frac{2\pi d}{\lambda}(K_e K_m - \sin^2\theta)^{\frac{1}{2}}$$

For a perpendicularly polarized incident wave Y becomes

$$Y_\perp = \frac{Y_0[(K_e/K_m) - \sin^2\theta]^{\frac{1}{2}}}{\cos\theta}$$

and for a parallel polarized wave Y becomes

$$Y_\parallel = \frac{Y_0(K_e/K_m)\cos\theta}{[(K_e/K_m) - \sin^2\theta]^{\frac{1}{2}}}$$

$$(2.77)$$

Y_{ext} is the admittance of the exterior medium on the transmission side.

Assuming temporarily for complete generality that the admittance of the medium on the incident side is Y_{inc} then the reflection coefficient[27] referred to the incident surface is

$$\rho = \frac{Y_{inc} E_{inc} - H_{inc}}{Y_{inc} E_{inc} + H_{inc}}, \tag{2.78}$$

and the transmission coefficient referred to the *same* surface is

$$\tau = \frac{2Y_{inc}(Y_{ext} + Y^*_{ext})^{\frac{1}{2}}}{(Y_{inc} E_{inc} + H_{inc})(Y_{inc} + Y^*_{inc})^{\frac{1}{2}}} \tag{2.79}$$

where * refers to complex conjugate.

In most of the cases to be considered, the incident and external media will both be free space and the relations (2.73) and (2.74) simplify to

$$\rho = \frac{Y_0 E_{inc} - H_{inc}}{Y_0 E_{inc} + H_{inc}}; \qquad \tau = \frac{2Y_0}{Y_0 E_{inc} + H_{inc}}. \tag{2.80}$$

This result can be simplified further since the magnitude of the reflection and transmission coefficients are obviously only dependent upon the *relative* admittance between the media. Thus defining the relative admittance of the homogeneous layer by

$$Y' = Y/Y_0$$

the *same* result as in equation 2.75 can be obtained from the transmission matrix

$$\begin{pmatrix} E_{inc} \\ H_{inc} \end{pmatrix} = \begin{pmatrix} \cos A & (i/Y') \sin A \\ iY' \sin A & \cos A \end{pmatrix} \begin{pmatrix} 1 \\ 1 \end{pmatrix} \tag{2.81}$$

and the coefficients

$$\rho = \frac{E_{inc} - H_{inc}}{E_{inc} + H_{inc}}; \qquad \tau = \frac{2}{E_{inc} + H_{inc}}.$$

Throughout the foregoing the medium has been considered to be lossless. Such a loss can be introduced by the modification of the various parameters to give K_e complex values that is

$$K_e = K'_e - iK''_e = K_e(1 - i\tan\delta_e)$$

$$= K_m/\varepsilon_0 \left(\varepsilon - \frac{i\sigma}{\omega} \right), \tag{2.82}$$

where σ is the conductivity of the medium and $\tan\delta_e$ the dielectric loss factor.

For a highly conductive medium the approximation for large σ may be made in Eqn (2.82) to give

$$(K_e)^{\frac{1}{2}} = \left(\frac{\mu\sigma}{2\omega\mu_0\varepsilon_0}\right)^{\frac{1}{2}} (1 - i). \tag{2.83}$$

In this case $Y_\parallel$ and $Y_\perp$ become complex, as do the propagation constants. The trigonometric functions in the transmission matrix then have to be replaced by hyperbolic functions. An application of this result is given in Section 2.12e.

2.11.2 The reflection coefficient of a single homogeneous layer

The properties of reflection and transmission through a single homogeneous layer of width d are summarized here for future reference when comparing the limiting effects in otherwise non-uniform layers. From Eqn (2.80) or (2.81)

$$\rho = \frac{i \sin A[(1/Y') - Y']}{2 \cos A + i \sin A[(1/Y') + Y']}, \tag{2.84}$$

$$\tau = \frac{2}{2 \cos A + i \sin A[(1/Y') + Y']}.$$

These values are periodic with frequency, being respectively zero and unity for those values of A for which

$$A = \frac{2\pi d}{\lambda}(K_e - \sin^2 \theta)^{\frac{1}{2}} = n\pi.$$

The lowest order for a reflectionless layer is thus the value of d for $n = 1$ namely

$$d = \lambda / \{2(K_e - \sin^2 \theta)^{\frac{1}{2}}\} \tag{2.85}$$

It is however, the effect of oblique radiation that limits the operational frequency band of a panel of this type. If such a radome is to accommodate all angles of incidence then both polarization states have to be considered. It is then found that the frequency of maximum transmission for the two states become separate from that for normal incidence, as illustrated in Fig. 2.22. It thus becomes impossible to obtain a single frequency with maximum transmission over anything but a narrow range of incidence angles.

This effect can be demonstrated by plotting

$$|\tau|^2 = \frac{4}{4 \cos^2 A + [Y' + (1/Y')]^2 \sin^2 A}$$

in a polar (τ, A) coordinate system (Fig. 2.23).

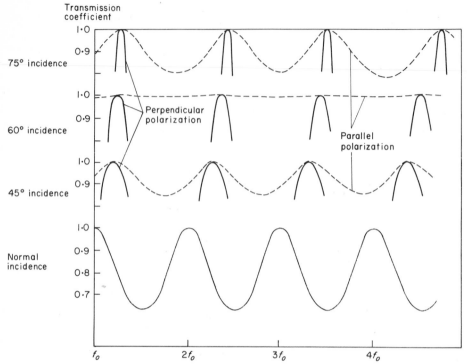

FIG. 2.22. Periodicity of transmission maxima for uniform layer.

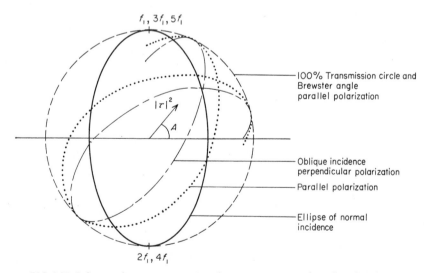

FIG. 2.23. Polar coordinate representation of power transmission through uniform layer.

For normal incidence the graph is an ellipse. With oblique incidence both the minor axis and the radius vector are scaled by a factor $[Y' + (1/Y')]$ and the periodicity of A changes. Thus the ellipse separates into two systems, one of increasing eccentricity and the other with decreasing. The former corresponds to perpendicular polarization and the latter to parallel polarization, and which becomes circular at the Brewster angle of the medium. At the same time, because of the change of periodicity, successive orbits of the ellipse with change of frequency rotates the major axis giving an effect akin to the advance of the perihelion in a planetary orbit. This causes the maxima of the transmission coefficients to disperse from their original first order agreement with the normal incidence ellipse.

2.12 THE THREE LAYER RADOME

The properties of multilayer radomes can be obtained directly by the product of the respective transmission matrices of each layer. Outside the basic half-wave homogeneous layer, the first extension giving a degree of freedom that can be applied to improvement of performance, is the symmetrical structure of three layers. The basic three layer radome, called a sandwich radome, is designed for complete transmission at a specified angle of incidence in one of the polarization states, usually the perpendicular, since the parallel state still gives a useful Brewster angle effect and can usually be left to take care of itself. A basic discussion of the subject can be found in Ref. 28.

For the structure shown in Fig. 2.24, M and K are the relative admittances of the layers whose refractive indices are $\sqrt{K_{e_1}}$ and $\sqrt{K_{e_2}}$ respectively. Then from Eqn (2.81) the fields on the incident side are given by

$$\begin{pmatrix} E_{\text{inc}} \\ H_{\text{inc}} \end{pmatrix} = \begin{pmatrix} \cos \alpha & (i/K)\sin \alpha \\ iK \sin \alpha & \cos \alpha \end{pmatrix} \begin{pmatrix} \cos \beta & (i/M)\sin \beta \\ iM \sin \beta & \cos \beta \end{pmatrix} \begin{pmatrix} \cos \alpha & (i/K)\sin \alpha \\ iK \sin \alpha & \cos \alpha \end{pmatrix} \begin{pmatrix} 1 \\ 1 \end{pmatrix}$$

$$(2.86)$$

where

$$\alpha = \frac{2\pi d_1}{\lambda} (K_{e_1} - \sin^2 \theta)^{\frac{1}{2}} \quad \text{and} \quad \beta = \frac{2\pi d_2}{\lambda} (K_{e_2} - \sin^2 \theta)^{\frac{1}{2}}.$$

For perpendicularly polarized wave, $K_\perp = (K_{e_1} - \sin^2 \theta)^{\frac{1}{2}}/\cos \theta$ and for parallel polarization, $K_\parallel = K_{e_1} \cos \theta/(K_{e_1} - \sin^2 \theta)^{\frac{1}{2}}$ with similar relations for $M_\perp$ and $M_\parallel$.

The relection coefficient is then found to be

$$
|\rho|^2 = \left\{ \sin \beta \cos^2 \alpha \left(\frac{1}{M} - M \right) + \sin 2\alpha \cos \beta \left(\frac{1}{K} - K \right) \right.
$$
$$
\left. - \sin^2 \alpha \sin \beta \left(\frac{M}{K^2} - \frac{K^2}{M} \right) \right\}^2
$$
$$
\times \left[\left\{ 2 \cos \beta \cos 2\alpha - \sin \beta \sin 2\alpha \left(\frac{K}{M} + \frac{M}{K} \right) \right\}^2 \right.
$$
$$
+ \left\{ \sin \beta \cos^2 \alpha \left(\frac{1}{M} + M \right) + \sin 2\alpha \cos \beta \left(\frac{1}{K} + K \right) \right.
$$
$$
\left. \left. - \sin^2 \alpha \sin \beta \left(\frac{M}{K^2} + \frac{K^2}{M} \right) \right\}^2 \right]^{-1}
\tag{2.87}
$$

and the phase of the transmission coefficient referred to the incident face

$$
\arg \tau = \tan^{-1} \left\{ - \sin \beta \cos^2 \alpha \left(\frac{1}{M} + M \right) - \sin 2\alpha \cos \beta \left(\frac{1}{K} + K \right) \right.
$$
$$
+ \sin^2 \alpha \sin \beta \left(\frac{M}{K^2} + \frac{K^2}{M} \right)
$$
$$
\left. \times \left[2 \cos \beta \cos 2\alpha - \sin \beta \sin 2\alpha \left(\frac{K}{M} + \frac{M}{K} \right) \right]^{-1} \right\}
\tag{2.88}
$$

From Eqn (2.87) we find $|\rho| = 0$ when

$$
\beta = n\pi - \tan^{-1} \left\{ \frac{2KM(K^2 - 1) \sin 2\alpha}{(M^2 - K^2)(K^2 + 1) + (M^2 + K^2)(K^2 - 1) \cos 2\alpha} \right\}
\tag{2.89}
$$

which is the result given in Ref. 28, p. 533

Several options now exist for the utilization of the degrees of freedom inherent in this structure, four of which will be considered here. A final example will be given of the applications of a conducting centre layer which modify somewhat the above relations but use the identical process.

We consider the following cases[29]

(a) A three-layer structure which is reflectionless at normal incidence for two frequencies f_0 and nf_0 and with $K_{e_1} > K_{e_2}$ (this type of structure is commonly called an A sandwich).

(b) A structure reflectionless at two frequencies and for which $K_{e_2} = K_{e_1}^2$ (a B sandwich).

(c) A structure which at a given frequency is reflectionless both for normal incidence and for oblique incidence of both polarizations.

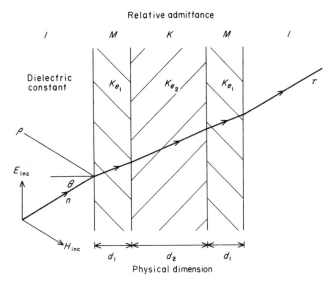

FIG. 2.24. *Three layer symmetrical sandwich.*

(d) A structure which is reflectionless at normal incidence and has zero phase delay at a given frequency.

Finally we will consider a three layer structure in which the centre layer is an extremely thin film of conducting material. These cases, as will be seen subsequently, do not exhaust all the possibilities inherent in even a three layer structure.

(a) The A *Sandwich*

A double frequency design can be simply effected by choosing d_1 and K_{e_1} to be those values for which the layer in isolation would be a half-wave reflectionless sheet at the higher frequency and then, with a value of K_{e_2} sufficiently near to unity say $1 < K_{e_2} < 1.2$, the three layer structure can be made reflectionless at the second frequency by the use of Eqn (2.89). With a low value of K_{e_2} little or no redesign of the outer layers is found to be required. The results for two frequencies f_0 and $5f_0$ in which $K_{e_1} = 4$ and $K_{e_2} = 1.1$ are shown in Fig. 2.25. Typical effects can be noted, that is additional "windows" in the frequency bands at $3f_0$ and $4f_0$, but an unexplained destructive interference at $2f_0$. A Brewster angle effect exists at the angle appropriate to the material of the exterior layers, and a similar separation of the frequencies of the maximum transmission peaks, with angle of incidence occurs as in the homogeneous layer. This latter effect makes it extremely

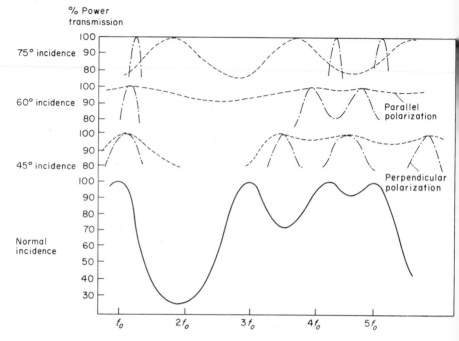

FIG. 2.25. *Three layer. A sandwich computed transmission curves. The sandwich designed for 100% transmission at f_0 and $5f_0$.*

difficult to obtain reflectionless operation over a wide range of incidence angles at any one of the selected frequencies.

(b) The B Sandwich

Choosing the same two frequencies as in the previous example, we make the outer layers equivalent to $\frac{1}{4}$ wavelength matching surfaces to the high refractive index centre layer Ref. 26, p. 64. That is $K_{e_2} = K_{e_1}^2$ and each surface of the centre layer is made to be independently reflectionless at the higher frequency. The centre thickness can then be chosen quite arbitrarily and can thus be chosen so that the overall *equivalent* layer is a half-wave layer at the lower frequency. For parameters $K_{e_2} = 16, K_{e_1} = 4$ the transmission peaks are shown in Fig. 2.26. Notable differences have now arisen between this result and the previous one. Transmission bands now occur at many more multiples of f_0 (up to six were encountered in the range investigated) and at the higher frequency chosen two of these coalesce to give a broad band of operation. The peaks in themselves although narrower than for the A-sandwich, remain fixed in frequency with change of incidence angle for a wide range of the latter, and there is absence of any Brewster angle effect.

(c) Radome with zero insertion phase delay[30]

At any given angle of incidence the insertion phase delay is the phase difference at the transmission surface between an unimpeded plane wave and the wave in the presence of the layer. This then is

$$\psi = \arg \tau - \frac{2\pi}{\lambda}(2d_1 + d_2)\cos\theta, \tag{2.90}$$

where $\arg \tau$ is given in Eqn (2.88).

For this to be made zero it is intuitively obvious that the centre medium must be a "phase-advance" material if the outer layers of dielectric are "phase-retarding". Such a material exists in the form of the two dimensional array of square waveguides such as is used for the microwave waveguide lenses discussed in Chapter 1. The obvious advantage of such a construction is in those applications where the insertion phase of the radome is a major cause of beam distortion or displacement. The use of a waveguide medium changes the description of the relative admittance since for the constrained field the admittance will be the same both for perpendicular and parallel polarization. That is for symmetrical waveguides of square or circular

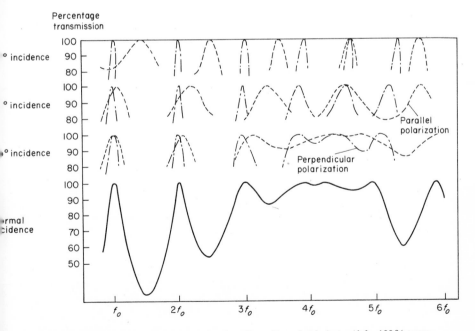

FIG. 2.26. *Theoretical transmission curves for three layer B sandwich designed for* 100% *transmission at* f_0 *and* $5f_0$.

cross-section and approximately for hexagonal cross-section

$$M_{\parallel} = M_{\perp} = \sqrt{[1 - (\lambda/\lambda_c)^2]}$$

and

$$\beta = \frac{2\pi d_2}{\lambda}\sqrt{[1 - (\lambda/\lambda_c)^2]}$$

(2.91)

where λ_c is the cut-off frequency for the dominant mode of the waveguide. Typical values for M in this case would be in the range 0·5 to 0·75 and for K_{e_1} approximately 4. Simultaneous solution of Eqns (2.90) with $\psi = 0$ and Eqn 2.89 is now possible. With the choice of $M = 0·6$ $K_{e_1} = 4$ this is found to give identical values for α and β namely $\alpha = \beta = 28·6°$. This gives $d_1 = 0·1192$ cm $d_2 = 0·3986$ cm, at a design wavelength of $\lambda = 3$ cm. The result (Fig. 2.27) shows that a very broad band radome results and that over the larger part of this band and for a wide range of incidence angles the insertion phase delay can be kept within ± 1 radian (Fig. 2.28).

The applicability of a medium with relative admittance less than unity gives rise to a further interesting phenomenon. We note that with M greater

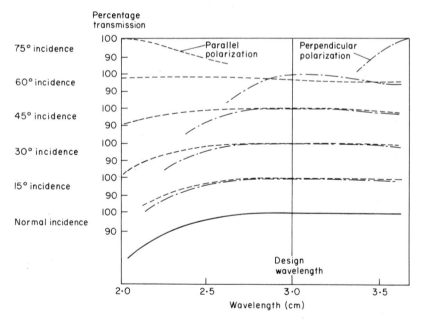

FIG. 2.27. *Theoretical transmission curves for waveguide sandwich.* $Ke_1 = 4, M = 0·6, d_1 = 0·12$cm, $d_2 = 0·4$ cm.

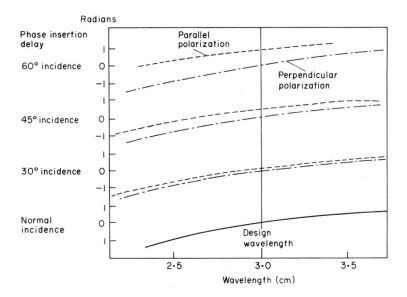

FIG. 2.28. *Phase insertion of waveguide sandwich.*

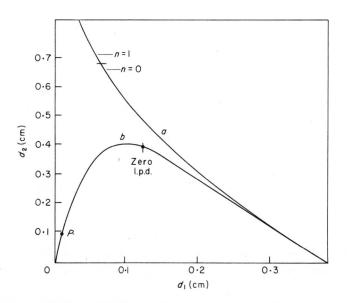

FIG. 2.29. *Dimensions for reflectionless sandwich constructions.* (a) *A sandwich,* $Ke_1 = 4, Ke_2 = 1\cdot2;$ (b) *waveguide sandwich* $Ke_1 = 4, M = 0\cdot6.$

than unity the denominator in Eqn (2.89) has a real root at which point an integral multiple of π has to be added to obtain a solution for β. Physically this means that as the surface layers become infinitely thin the three layer A sandwich structure tends towards a half wave wall of the material of the centre layer. If this refractive index is very close to unity, this half-wave width can be nearly as large as a half wavelength of the radiation itself, and hence does not tend to zero, which would be the most logical result for a vanishing reflectivity overall.

With the waveguide medium M is less than unity, and this condition no longer arises (Fig. 2.29). With increasingly thin surface layers, the central core becomes equally thin and there arises a system of very thin structures which are reflectionless. We then observe that the frequency band of these structures can be extended beyond the cut off frequency for the waveguide medium. This makes both M and β in Eqn (2.91) complex, but since $\sin \beta$, and either M or $1/M$ always appear as a product, the result remains real with some changes of sign. Then for wavelengths far beyond cut off β tends to a limit value, M tends to infinity and the relection coefficient tends to unity. The rate at which it does so, is determined naturally by the thickness of the cut-off medium, and this as has been shown, can be made as small as required.

Computation shows that for a sandwich of very small dimension the reflection loss can be kept at less than 10% over a 10:1 frequency band. This requires, at a frequency of 3 cm, a core thickness of 1 mm and skin thickness of 0·1 mm P in fig. 2.29. A radome of double these dimensions has approximately half the bandwidth. The theoretical transmission is shown in Fig. 2.30.

(d) The two angle radome

Equation (2.84) gives the condition for zero reflectivity in terms of the four parameters α, β, M and K each of which appears in the relation in the form appropriate to a given angle of incidence and prescribed polarization.

The equation is soluble for these parameters under a variety of *a priori* conditions. We choose for this example a radome which besides being reflectionless at normal incidence is reflectionless for *both* polarizations at one other angle of incidence at a single given frequency. This could be generalized to be a reflectionless condition for both polarizations at two prescribed angles of incidence.

Using the subscripts 0, $\perp$ and $\parallel$ to indicate normal incidence, oblique incidence with perpendicular polarization and oblique incidence with parallel polarization the latter at a fixed incidence angle θ, we require to satisfy the three conditions

$$\tan \beta_0 = \frac{-2K_0 M_0 (K_0^2 - 1) \sin 2\alpha_0}{(M_0^2 - K_0^2)(K_0^2 + 1) + (M_0^2 - K_2^2)(K_0^2 - 1) \cos 2\alpha_0}, \qquad (2.92a)$$

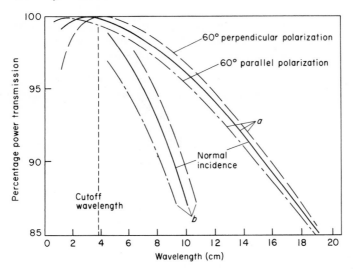

FIG. 2.30. *Broadband thin waveguide sandwich.* (a) $d_1 = 0{\cdot}01$ cm, $d_2 = 0{\cdot}10$ cm; (b) $d_1 = 0{\cdot}02$ cm, $d_2 = 0{\cdot}185$ cm.

$$\tan \beta_\theta = \frac{-2K_\parallel M_\parallel (K_\parallel^2 - 1) \sin 2\alpha_\theta}{(M_\parallel^2 - K_\parallel^2)(K_\parallel^2 + 1) + (M_\parallel^2 + K_\parallel^2)(K_\parallel^2 - 1) \cos 2\alpha_\theta}, \quad (2.92b)$$

and

$$\tan \beta_\theta = \frac{-2K_\perp M_\perp (M_\perp^2 - 1) \sin 2\alpha_\theta}{(M_\perp^2 - K_\perp^2)(K_\perp^2 + 1) + (M_\perp^2 + K_\perp^2)(K_\perp^2 - 1) \cos 2\alpha_\theta}, \quad (2.92c)$$

From Eqn (2.77) we have $Y_\parallel = K_e/Y_\perp$ for *all* angles of incidence. Using this relation for $M_\parallel$ and $K_\parallel$ the equality of Eqns (2.92b and c) gives by division

$$\begin{aligned}
\cos 2\alpha_\theta = {}& (K_\perp^2 - 1)(K_{e_2}^2 K_\perp^2 - K_{e_1}^2 M_\perp^2)(K_{e_1}^2 + K_\perp^2) \\
& - K_{e_1} K_{e_2}(K_{e_1}^2 - K_\perp^2)(M_\perp^2 - K_\perp^2)(K_\perp^2 + 1) \\
& \times \left[(K_\perp^2 - 1)(K_{e_1}^2 - K_\perp^2)(K_{e_1} - K_{e_2})(K_{e_1} M_\perp^2 - K_{e_2} K_\perp^2)\right]^{-1} \quad (2.93)
\end{aligned}$$

Since we also have the relations

$$K_\perp = (K_{e_1} - \sin^2 \theta)^{\frac{1}{2}}/\cos \theta, \qquad K_0 = (K_{e_1})^{\frac{1}{2}},$$

and

$$M_\perp = (K_{e_2} - \sin^2 \theta)^{\frac{1}{2}}/\cos \theta, \qquad M_0 = (K_{e_2})^{\frac{1}{2}},$$

Eqn (2.93) is a single relation between K_{e_1} and K_{e_2} for a given angle α_θ and Eqn (2.92a) is then a second relation between these two parameters, but with

G

α_0 determined by being the zero value of α_θ and β_0 by β_θ. An iterative procedure thus suggests itself which need only be continued up to the point where
the transmission band required has been achieved. The results of a partial

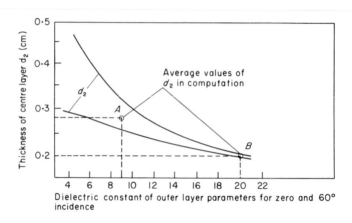

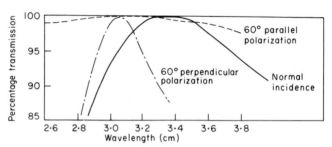

(a) $Ke_1 = 9$, $Ke_2 = 1\cdot6$, $d_1 = 0\cdot137cms$, $d_2 = 0\cdot29$ cm.

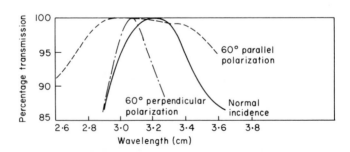

(b) $Ke_1 = 20$, $Ke_2 = 2\cdot8$, $d_1 = \cdot09cms$, $d_2 = 0\cdot2$ cm.

FIG. 2.31. *Reflectionless wall at zero and 60° incidence.*

solution to this problem are shown in Fig. 2.31 for a reflectionless wall at zero and 60° angles of incidence at a wavelength of 3·2 cm. It is found that better results are obtained with high dielectric constant values in the outer layers and the examples have constants appropriate to ceramic materials. The computed results show an optimum design at 3·1 cm based on an average value of the ambiguity d_2.

There exists in radome theory an angle/frequency relationship whereby any example involving a wide range of incidence angle could *instead* apply to a wide range of frequencies. We see that this last example is an angle version of the A sandwich, in the form of a generalized two frequency design.

(e) Radome with central conducting layer
The identical procedure for a three layer radome in which the centre layer is a very thin sheet of highly conducting material requires that the central matrix in Eqn (2.86) be replaced by the matrix

$$\begin{pmatrix} \cosh \beta & (1/M) \sinh \beta \\ M \sinh \beta & \cosh \beta \end{pmatrix} \tag{2.94}$$

where β by virtue of Eqn (2.83) is

$$\beta = \frac{2\pi d_2}{\lambda} \left\{ \left(\left[\frac{\mu\sigma}{2\omega\mu_0\varepsilon_0} \right]^{\frac{1}{2}} - i \left[\frac{\mu\sigma}{2\omega\mu_0\varepsilon_0} \right]^{\frac{1}{2}} \right) - \sin^2 \theta \right\}^{\frac{1}{2}}, \tag{2.95}$$

with corresponding complex forms for M and $M_{\parallel}$. The condition for zero reflection then becomes

$$\tanh \beta = \frac{[(1/K) - K] \sin 2\alpha}{[(K^2/M) - (M/K^2)] \sin^2 \alpha + [(1/M) - M] \cos^2 \alpha}, \tag{2.96}$$

with M complex.
Using

$$\beta = \tanh^{-1} ix = \tfrac{1}{2} \log \left(\frac{1 - x^2}{1 + x^2} + \frac{i2x}{1 + x^2} \right)$$

and equating real and imaginary parts, this can be solved for given realistic values of K_{e_1} and σ to obtain d_1 and d_2 the thickness of the respective layers. The theoretical centre thickness for highly conducting materials such as tin or aluminium are of the order of 10–20 angstrom units, the higher value relating to exterior layers of higher dielectric constant. The theoretical reflectivity for a design at $\lambda = 4·5$ cm is shown in Fig. 2.32. The exterior layers are of ceramic with $K_{e_1} = 9$ and the centre layer tin for which $\sigma = 0·87 \times 10^7$ mho m^{-1}. The computed thicknesses are then $d_1 = 3 \times 10^{-3}$ m and

$d_2 = 24·42 \times 10^{-9}$ m. The change in form from the standard reflectivity curves shown in Fig. 2.32 is remarkable and there is no evidence of a Brewster angle effect. Similar curves with much greater bandwidth, approximately 2:1 can theoretically be obtained with exterior layers of dielectric constant $K_{e_1} = 4$ but this requires the even more impractical centre layer of $9·15 \times 10^{-9}$ m thickness. At these thicknesses other thin film effects in the metal layer have to be taken into consideration before a complete appreciation of the structure can be made.

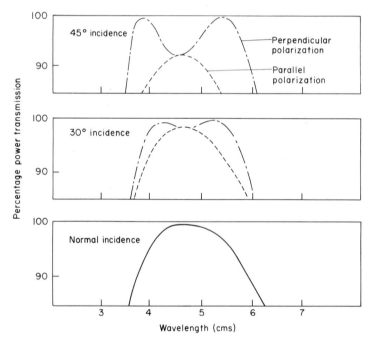

FIG. 2.32. Transmission characteristics of three layer sandwich with central conducting layer $Ke_1 = 9$, $d_1 = 0·3$ cm, $d_2 = 2442 \times 10^{-7}$ cm. Condictivity $\sigma = 0·87 \times 10^7$ mho m^{-1}.

2.13 MULTILAYERED AND STRATIFIED MEDIA

2.13.1 Multilayered media

From the foregoing it can be seen that numerous possibilities will be created by the utilization, in a similar fashion, of the additional degrees of freedom that become available in using five and seven layer structures. Formulae for the reflection and transmission coefficients for symmetrical radomes of this kind are given by Kaplun.[31] The examples include a single layer radome with wire grating reactive grid in the central position. The formulae are presented in terms of the intermediate surface Fresnel coefficients. In Chapter 5 we shall be giving the transmission matrix for reactive wire gratings. The inclusion of such gratings in radome walls can then be carried out by the same matrix procedure as in the previous sections.

Theories relating to multilayered structures which appear in the literature of optics usually have to do with structures which, translated into the microwave region, would be of impractical size. There has been developed in this theory methods applicable to the multiplication of large numbers of transmission matrices of the symmetrical kind that is considered here. This theory is therefore applicable to microwave practice, where the multilayer can be considered to be constructed, as it would be in practice, from a large number of very thin layers. The process then lends itself to the design as well as the analysis of such radome walls.

There is firstly a theorem by Herpin which states that any symmetrical multilayer can always be represented by a single equivalent transmission matrix. The additional statement that any non-symmetrical multilayer can be represented by at most two transmission matrices is also valid but not of our immediate concern. The former has been given an elegant proof by Young.[32]

Transmission matrices are of the form

$$M = \begin{pmatrix} c & is/n \\ isn & c \end{pmatrix},$$

and can be transformed to a symmetric matrix

$$M' = \begin{pmatrix} isn & c \\ c & is/n \end{pmatrix},$$

by the product $M' = AM$ where A is the spin matrix

$$\begin{pmatrix} 0 & 1 \\ 1 & 0 \end{pmatrix}$$

A symmetric multilayer whose matrix is

$$S = M_1 M_2 \dots M_N \dots M_2 M_1$$

becomes

$$S = AM'_1 \ AM'_2 \dots AM'_2 \ AM'_1$$
$$= A\{M'_1 AM'_2 \dots AM'_2 \ AM'_1\} = AS'$$

Now S' is symmetric since, because of the symmetry of M' and A, $S'^T = S'$. Therefore S is the equivalent matrix of the *single* layer that would represent the entire multilayer.

Symmetric periodic multilayers have been studied by many authors[3] through the method of multiplication of a repetitive transmission matrix. In particular Mielenz[27] shows that for a symmetrical periodic structure of $2n$ or $2n + 1$ layers the unitary matrix representing the repeating *pair* of matrices has to be raised to the nth power. This is done as follows. If M and N are the individual transmission matrices of the repeating pair and

$$U = MN = \begin{pmatrix} a_{11} & a_{12} \\ a_{21} & a_{22} \end{pmatrix},$$

we require to evaluate U^n. Other more complex forms of symmetry lead to a similar mathematical requirement.

Since U is unitary it can be put in the form

$$U = a\sigma_0 + b\sigma_1 + c\sigma_2 + d\sigma_3$$

where

$$\sigma_0 = \begin{pmatrix} 1 & 0 \\ 0 & 1 \end{pmatrix}, \quad \sigma_1 = \begin{pmatrix} 0 & 1 \\ 1 & 0 \end{pmatrix}, \quad \sigma_2 = \begin{pmatrix} 0 & -i \\ i & 0 \end{pmatrix} \quad \text{and} \quad \sigma_3 = \begin{pmatrix} 1 & 0 \\ 0 & -1 \end{pmatrix}$$

are the Pauli spin matrices, and $a^2 - b^2 - c^2 - d^2 = 1$.†

An inductive proof then shows that

$$U^n = \begin{pmatrix} a_{11} S_{n-1}(x) - S_{n-2}(x) & a_{12} S_{n-1}(x) \\ a_{21} S_{n-1}(x) & a_{22} S_{n-1}(x) - S_{n-2}(x) \end{pmatrix}, \quad (2.97)$$

where $S_n(x)$ are Chebychev polynomials (the same as appear in Herpin, Ref. 34) given by

$$S_n(x) = \frac{\sin(n+1)\psi}{\sin\psi}, \quad x = 2\cos\psi, \quad x \leqslant 2$$

or   (2.98)

$$S_n(x) = \frac{\sinh(n+1)\phi}{\sinh\phi}, \quad x = 2\cosh\phi, \quad x \geqslant 2$$

and

$$x = a_{11} + a_{22}$$

These results simplify a great deal under conditions appropriate to optical

† Since a, b, c and d *may* be complex U can, with identical analysis and final results, be expanded in quaternion form discussed in Appendix II.

work such as a multilayer of individual $\frac{1}{4}$ wave layers, but which become impracticable at microwave frequencies. However, they have not been fully assessed recently, with the possibility of the application of computing techniques.

The method provides a link between optics and other branches of microwave activity. The same analysis applies to the design of quarter wave coupled filters and loaded line synthesis[34] and non-uniform waveguide transmission. In these fields Chebychev polynomials and spin matrices make their appearance. Chebychev polynomials are also a key design method for linear arrays of sources (Ref. 24, Vol. II) in which problem spin matrices, or their equivalent the quaternions, have not yet made an appearance. Other methods in filter analysis, for example Walsh functions,[35] would presumably have applications in array theory and hence in optical filter theory.

2.14 STRATIFIED MEDIA

There are two methods by which the matrix theory of a multilayered medium can be extended to a medium with a continuously variable refractive index. The first is to consider the continuous variation to be made up of discrete infinitesimal steps and to multiply the corresponding "infinitesimal" matrices. This in principle is no great advance since the multiplication of such matrices is a highly complex procedure. If an approximate method is employed for this *multiplication* process, the degree of approximation is similarly multiplied until in the end result the solution can only be applied to a very small overall (in optical wavelengths) thickness of the medium. As such it is analogous to the Born approximation, as it would be applied to the actual differential equation governing the propagation in such a medium.[36] We include this method for illustrative purpose.

For a layer of thickness δx the transmission matrix is

$$\Delta M = \begin{pmatrix} \cos(2\pi\delta x\alpha/\lambda) & \dfrac{i}{Y(x)}\sin(2\pi\delta x\alpha/\lambda) \\ iY(x)\sin(2\pi\delta x\alpha/\lambda) & \cos(2\pi\delta x\alpha/\lambda) \end{pmatrix}, \qquad (2.99)$$

where α is the phase angle $(K_e(x) - \sin^2\theta)^{\frac{1}{2}}$ and $Y(x)$ the variable admittance of the medium.

Then if the small angle approximation is made

$$\Delta M \simeq \begin{pmatrix} 1 & \dfrac{i}{Y(x)}\dfrac{2\pi\delta x}{\lambda}\alpha \\ i\,Y(x)\,2\pi\delta x\alpha/\lambda & 1 \end{pmatrix}, \tag{2.100}$$

and multiplication of a large number of such matrices results in the complete scattering matrix of the layer

$$M \simeq \begin{pmatrix} 1 & \dfrac{2\pi i}{\lambda}\Sigma\dfrac{\alpha\delta x}{Y(x)} \\ \dfrac{2\pi i}{\lambda}\Sigma\alpha Y(x)\,\delta x & 1 \end{pmatrix}, \tag{2.101}$$

where powers of δx greater than the first have been ignored at each matrix multiplication.

Proceeding to the limit this becomes

$$M \simeq \begin{pmatrix} 1 & \dfrac{2\pi i}{\lambda}\displaystyle\int_0^x \dfrac{\alpha}{Y(x)}\,\mathrm{d}x \\ \dfrac{2\pi i}{\lambda}\displaystyle\int_0^x \alpha Y(x)\,\mathrm{d}x & 1 \end{pmatrix}. \tag{2.102}$$

The error in this result can be illustrated by applying it to a *uniform* layer of thickness D. Then with α and $Y(x)$ constant M in Eqn (2.102) becomes

$$M \simeq \begin{pmatrix} 1 & \dfrac{i}{Y}(2\pi D\alpha/\lambda) \\ iY2\pi D\alpha/\lambda & 1 \end{pmatrix},$$

which has now lost any connection with the periodic result of Eqn (2.81)

$$M = \begin{pmatrix} \cos(2\pi D\alpha/\lambda) & \dfrac{i}{Y}\sin(2\pi D\alpha/\lambda) \\ iY\sin(2\pi D\alpha/\lambda) & \cos(2\pi D\alpha/\lambda) \end{pmatrix}.$$

In fact the two results only compare for values of D for which

$$\cos(2\pi D\alpha/\lambda) \simeq 1 \quad \text{that is} \quad D\alpha \ll \lambda$$

Hence an improved method is required that could apply to a layer of any width. We proceed by diagonalizing the matrix of a single thin uniform layer. The transmission matrix of Eqns (2.81) and (2.77) then separate into surface "admittance" matrices separated by a spacing or phase angle matrix as follows

$$M = \frac{1}{2Y}\begin{pmatrix} 1 & -1 \\ Y & Y \end{pmatrix}\begin{pmatrix} e^{iA} & 0 \\ O & e^{-iA} \end{pmatrix}\begin{pmatrix} Y & 1 \\ -Y & 1 \end{pmatrix}, \qquad (2.103)$$

$$A = 2\pi D\alpha/\lambda$$

Multiplication of a number N of such layers to produce a layer of thickness D is achieved through

$$M_D = M_1 M_2 \dots M_N,$$

or

$$
\begin{aligned}
M_D = {} & \begin{pmatrix} 1 & -1 \\ Y_1 & Y_1 \end{pmatrix}\begin{pmatrix} e^{iA_1} & 0 \\ 0 & e^{-iA_1} \end{pmatrix}\begin{pmatrix} Y_1 & 1 \\ -Y_1 & 1 \end{pmatrix}\left(\frac{1}{2Y_1}\right) \\
& \times \begin{pmatrix} 1 & -1 \\ Y_2 & Y_2 \end{pmatrix}\begin{pmatrix} e^{iA_2} & 0 \\ 0 & e^{-iA_2} \end{pmatrix}\begin{pmatrix} Y_2 & 1 \\ -Y_2 & 1 \end{pmatrix}\left(\frac{1}{2Y_2}\right) \dots \text{etc.} \qquad (2.104)
\end{aligned}
$$

The internal products of the "admittance" matrices are of the form

$$\begin{pmatrix} Y_1 & 1 \\ -Y_1 & 1 \end{pmatrix}\frac{1}{2Y_1}\begin{pmatrix} 1 & -1 \\ Y_2 & Y_2 \end{pmatrix} = \frac{1}{2Y_1}\begin{pmatrix} Y_1 + Y_2 & Y_2 - Y_1 \\ Y_2 - Y_1 & Y_1 + Y_2 \end{pmatrix}. \qquad (2.105)$$

If now Y_2 is incrementally different from Y_1 and A_2 likewise incrementally different from A_1 Eqn (2.105) becomes

$$\frac{1}{2Y_1}\begin{pmatrix} Y_1 + Y_1 + \delta Y_1 & Y_1 + \delta Y_1 - Y_1 \\ Y_1 + \delta Y_1 - Y_1 & Y_1 + Y_1 + \delta Y_1 \end{pmatrix} = \begin{pmatrix} 1 + \dfrac{\delta Y_1}{2Y_1} & \dfrac{\delta Y_1}{2Y_1} \\ \dfrac{\delta Y_1}{2Y_1} & 1 + \dfrac{\delta Y_1}{2Y_1} \end{pmatrix} \quad (2.106)$$

In the limit $\delta Y_i \to 0$ and all the internal products tend to the unit matrix. Equation (2.104) then becomes

$$M_D = \begin{pmatrix} 1 & -1 \\ Y_1 & Y_1 \end{pmatrix}\begin{pmatrix} \exp(iA_1) & 0 \\ 0 & \exp(-iA_1) \end{pmatrix}\begin{pmatrix} \exp[i(A_1 + dA_1)] & 0 \\ 0 & \exp[-i(A_1 + dA_1)] \end{pmatrix} \dots$$

$$\begin{pmatrix} Y_N & 1 \\ -Y_N & 1 \end{pmatrix}\frac{1}{2Y_N} \qquad (2.107)$$

$$= \begin{pmatrix} 1 & -1 \\ Y_1 & Y_1 \end{pmatrix}\begin{pmatrix} \exp\left(i\int_0^D A\,dx\right) & 0 \\ 0 & \exp\left(-i\int_0^D A\,dx\right) \end{pmatrix}\begin{pmatrix} Y_N & 1 \\ -Y_N & 1 \end{pmatrix}\frac{1}{2Y_N}. \qquad (2.108)$$

Reconstituting M_D gives the final result

$$M_D = \begin{pmatrix} \cos\Phi & \dfrac{i}{Y_N}\sin\Phi \\ iY_1\sin\Phi & \dfrac{Y_1}{Y_N}\cos\Phi \end{pmatrix}, \tag{2.109}$$

where Φ is now the phase integral[37]

$$\Phi = \int_0^D \frac{2\pi}{\lambda}\alpha\,dx = \frac{2\pi}{\lambda}\int_0^D \{K_e(x) - \sin^2\theta\}^{\frac{1}{2}}\,dx. \tag{2.110}$$

Multiplication of two such matrices produces a third of the identical form, and it can be seen that a matrix of this kind implies an infinitesimal layer in which the medium properties vary in a *linear* manner from the incident to the transmission sides. Such a layer would have characteristic matrix

$$\Delta M = \begin{pmatrix} \cos\dfrac{2\pi\delta x}{\lambda}\alpha & \dfrac{i}{Y_2}\sin\dfrac{2\pi\delta x}{\lambda}\alpha \\ iY_1\sin\dfrac{2\pi\delta x}{\lambda} & \dfrac{Y_1}{Y_2}\cos\dfrac{2\pi\delta x}{\lambda}\alpha \end{pmatrix} \qquad †(2.111)$$

Multiplication of a string of such matrices and proceeding to the limit produces the same result as in Eqn (2.109). Thus the result is in effect that of making the refractive index profile up by incremental linear sections with no discontinuities at the surfaces between the layers.

These phase integrals were introduced by Eckersley and Budden into their solution for the propagation of radio waves in a variable atmosphere[38] but the introduction appears to be somewhat arbitrary. More recently Jacobsson[39] arrives at the same result and shows that it can be explained as an averaging process of the Fresnel reflection coefficients of the discontinuously stepped approximation. The derivation is from the differential equation governing the propagation and the observation is made that the approximation concerned is the WKJB approximation.[40]

It is clear from the description of the phase integral in Eqn (2.110) that there exist conditions on the thickness of the layer D and the law of the refractive index $K_e(x)$ for the case when the integrand becomes complex. This physically would imply a total internal reflection in a large depth of material, or at very oblique incidence. These problems are more pertinent

---

† To preserve unitarity ΔM in Eqn (2.111) may be multiplied throughout by $(Y_2/Y_1)^{\frac{1}{2}}$ with no loss of generality in the subsequent analysis

to the study of radio wave propagation and hence to all the theory that has been applied to it. The matrix analysis here forms a positive link between these theories and the other branches of microwaves discussed previously. It can be applied particularly to the theory of non-uniform waveguides.

As stated in Ref. 39, "In principle inhomogeneous films, possibly combined with homogeneous films, can be used to realise most of the spectral requirements of modern thin film optics". The additional degrees of freedom obtained by dissociating the reflections at interfaces from the phase length between them, and in addition by having different values of Fresnel coefficients at these interfaces, permits a large potential of radome designs to be considered. Some of the many possible variations are shown in Fig. 2.33.

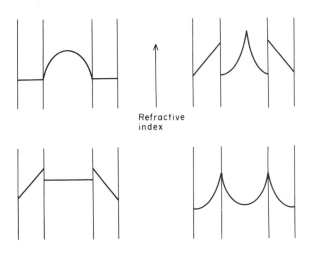

Refractive
index

FIG. 2.33. Non-uniform three layer sandwich designs.

REFERENCES

1. J. Brown. "Microwave Lenses", Methuen, 1953, p. 89.
2. R. K. Luneburg. "The Mathematical Theory of Optics", University of California Press, 1964, p. 180.
3. A. Fletcher, T. Murphy and A. Young. Solutions to two optical problems, *Proc. Roy. Soc.* **223A** (1954), 216.
4. F. P. Kapron. Geometrical optics of parabolic index-gradient cylindrical lenses, *Jour. Opt. Soc. Amer.* **60** (1), (1970), 1433.
5. E. W. Marchand. Ray tracing in cylindrical gradient index media, *Applied Optics*, **11** (5), (1972), 1104.

6. See Ref. 16, Chapter 6.
7. R. E. Collin and F. J. Zucker. "Antenna Theory", Part II, Chapter 18.6. "Lens Antennas", by J. Brown, McGraw Hill, 1969, p. 133.
8. O. N. Stavroudis. "The Optics of Rays, Wavefronts and Caustics", Academic Press, London and New York. Chapter IV, 1972, p. 42.
   J. Brown. Ref. 1, p. 83.
   R. K. Luneburg. Ref. 2, p. 172.
   C. T. Tai. "Dyadic Green's Functions in Electromagnetic Theory", Chapter 12, "Inhomogeneous and Moving Media", Intext Press, 1971.
9. S. P. Morgan. General solution of the Luneburg lens problem. *Jour. Appl. Physics*, **29** (9), (1958), 1358.
10. A. S. Gutman. Modified Luneburg lens, *Jour. Appl. Physics*, **23** (7), (1954), 855.
11. H. A. Atwater, "Introduction to General Relativity", Pergamon Press, 1974, p. 145.
12. W. Lenz, Theory of optical images (in German) "Probleme der Modernen Physik" (P. Debye, Ed.) Hirzel Press Leipzig, 1928, p. 198.
13. G. Toraldo di Francia. A family of perfect configuration lenses of revolution, *Optica Acta*, **1** (4), (1955), 157.
14. J. E. Eaton. On spherically symmetric lenses, *Trans. I.R.E.* **AP4** (1952), 66.
15. H. A. Buchdahl. Rays in gradient index media: separable systems, *Jour. Opt. Soc. Amer.* **63** (1), (1973), 46.
16. P. Prache. Lenses and dielectric reflectors of homogeneous spherical layers, *Annales de Telecomm.* **16** (3–4), 1961, 85.
17. S. Cornbleet. A simple spherical lens with external foci, *Microwave Journal* (*May* 1965), 65.
18. H. F. Mathis, Checking the design of stepped Luneburg lens, *Trans. I.R.E.* **AP8** (1960), 342.
19. G. Toraldo di Francia and M. T. Zoli. Perfect Concentric Systems with an Outer Shell of Constant Refractive Index", Pubblicazione dell' Istituto Nazionale di Attica, Serie II, No. 827, Firenze, 1958.
20. T. L. Ap Rhys. The design of radially symmetric lenses, *Trans. I.E.E.E.* **AP18** (4) (1970), 497.
    G. Toraldo di Francia. Spherical lenses for infrared and microwaves, *Jour. Appl. Physics*, **32** (1961), 2051.
21. G. Toraldo di Francia. Conflection doublets, *Jour. Opt. Soc. Amer.* **45** (8), 1955, 621 and Ref. 13.
    K. S. Kunz. Applications de la geometrie differentielle à l'optique des micro ondes, *Supplement Nuovo Cimento* **9**, Series IX No. 3 (1952), 322.
22. S. S. D. Jones. A wide angle microwave radiator, *Proc. I.E.E.* **97** (1947), Part III, p. 255.
23. R. F. Rinehart. A solution of the problem of rapid scanning for radar antennae, *Jour. Appl. Phys.* **19** (1948), 80, also, A family of designs for rapid scanning radar antennas, *Proc. I.R.E.* **40** (1952), 686.
    S. B. Myers. Parallel plate optics for rapid scanning, *Jour. App. Phys.* **18** (1947), 221.
24. R. C. Hansen (Ed.). "Microwave Scanning Antennas", Vol. 1, Chapter 3. Optical scanners by R. C. Johnson, Academic Press, London and New York, 1964, p. 224 and Refs. therein.
25. J. W. McClain. Simulated nuclear potential well, *Amer. Jour. Phys.* **31** (11) (1963), 888.

C. M. Anderson and H. C. von Bayer. Theory of a ball rolling on a $1/\rho$ surface of revolution, *Amer. Jour. of Phys.* **38** (1), (1970), 140.

26. M. Born and E. Wolf. "The Principles of Optics", Pergamon Press, 1959, p. 57–65. See also Ref. 37, p. 346.

27. K. D. Mielenz. The use of Chebychev Polynomials in thin film computations, *Jour. Res. Nat. Bur.* Stand A November–December (1959), Vol. 63, No. 3, p. 297.

28. S. Silver. "Microwave Antenna Theory and Design", M.I.T. Radiation Laboratory Series, Vol. 12, 1949, p. 522.

29. S. Cornbleet. Multifrequency operation of sandwich radomes, *Microwave Journal*, August (1968), 59.

30. S. Cornbleet. Waveguide sandwich electromagnetic window, *Electronics Letters*, **3** (12), (1967), p. 540.

31. V. A. Kaplun. Nomograms for determining the parameters of plane dielectric layers of various structure with optimum radio characteristics, *Radioteknikha i Elektronika*, **9** (1964), 81.

32. P. A. Young. Extension of Herpin's theorem, *Jour. Opt. Soc. Amer.* **60** (10), (1970), 1422.

33. A. Herpin. Sur une nouvelle methode d'introduction des polynomes de Lucas (Chevychev polynomials), *Comptes Rendus* **225** (1), (1947), 17.
    F. Abeles. Transmission de la lumiere a travers un systeme de lames minces alternees, *Comptes Rendus* **226** (22) ,(1948), 1809.
    L. I. Epstein. The design of optical filters, *Jour. Opt. Soc. Amer.* **42** (11), (1952), 806.

34. L. Young. Multilayer interference filters with narrow stop-bands, *Applied Optics*, **6** (2), (1967), 297 (*and Refs. therein*).
    S. Reed. A note on loaded line synthesis, *Trans. I.R.E. PGMTT* (March, 1961), 201.

35. K. G. Beauchamp. "Walsh Functions and Their Applications", Academic Press, London and New York, January, 1976.

36. D. S. Jones. "Theory of Electromagnetism" Pergamon Press, 1964, Section 6.21, p. 351.

37. J. Heading. "Phase Integral Methods", Methuen Monogtaphs, 1962.

38. T. L. Eckersley. Radio transmission problems treated by phase integral methods, *Proc. Roy. Soc. A* **136** (830), (1932), 499.
    K. G. Budden. "Radio Waves in the Ionosphere", Cambridge University Press, 1961.

39. R. Jacobsson. Inhomogeneous and co-evaporated films for optical applications. *In* "Physics of Thin Films", Academic Press, Vol. 8, 1975, p. 51 and Light-reflection from thin films of continuously varying refractive index. *In* "Progress in optics", ed. E. Wolf, **V** (1966) 249.

40. J. H. Richmond, The WKB approximation for transmission through inhomogeneous plane layers, *Trans. I.R.E.* **AP10** (1962), 472.

BIBLIOGRAPHY

SPHERICAL MEDIA
G. Bekefi and G. W. Farnell. A homogeneous dielectric sphere as a microwave lens, *Canad. Jour. Phys.* **34** (1956), 790.

S. M. Harris. Refraction compensation in a spherically stratified ionosphere, *Trans. I.R.E.* **AP9** (2), (1961), 207.

A. F. Kay. Spherically symmetric lenses, *Trans. I.R.E.* **AP7** (1959), 32.

E. W. Marchand. Ray tracing in gradient index media, *Jour. Opt. Soc. Amer.* **60** (1) (1970), 1.

S. P. Morgan. Generalizations of spherically symmetric lenses, *Trans. I.R.E.* **AP** (1959), 342.

G. D. M. Peeler and H. P. Coleman. Microwave stepped index Luneburg lens, *Trans. I.R.E.* **AP6** (1958), 202.

J. H. Richter. Application of conformal mapping to earth flattening procedures in radio propagation problems, *Radio Sci.* **1** (12), (1966), 1435.

R. Stettler. Über radial symmetrische optische medien, *Optica Acta*, **3** (3), (1955), 101.

C. T. Tai. The electromagnetic theory of the spherical Luneburg lens, *Applied Sci. Res.* **B7** (1958), 113.

P. L. E. Uslenghi and N. G. Alexopoulos, A special class of spherically inhomogeneous dielectrics, *Alta Frequenza*, **38** (Special issue), p. 65, U.R.S.I. Symposium on E.M. waves, STRESA, 1968.

METAL FILMS

D. Marcuse. "Light Transmission Optics", Van Nostrand, 1972.

Z. H. Meiskin. Discontinuous and Cermet films, "Physics of Thin Films", Vol. 8, 1975, Academic Press, London and New York, p. 99.

J. Shamir. Optical parameters of partially transmitting thin films, *Applied Optics*, **15** (1), (1976), 120.

H. Tischer. The electrical properties of a thin evaporated layer of silver at 3,000 mc/s, *Zeit Ag. Phys.* **5** (11), (1953), 413.

A. Väsiček. "Optics of Thin films", North Holland, 1960, p. 5.

RADIALLY INHOMOGENEOUS MEDIA

A. K. Ghatak, D. P. S. Malik and M. S. Sodha. Path of rays in conical Selfoc fibres, *Jour. Opt. Soc. Amer.* **62** (4), (1972), 594.

J. P. Gordon, Optics of general guiding media, *Bell Syst. Tech. Jour.* **45** (1966), 321.

S. Kawakami and J. Nishizawa, An optical waveguide with the optimum distribution of refractive index with reference to waveform distortion *Trans. I.E.E.E. PG MTT*, **16** (10), (1968), 814.

E. T. Kornhauser and A. D. Yaghjian. Modal solution of a point source in a strongly focusing medium, *Radio Sci.* **2** (1967), 299.

K. B. Paxton and W. Streifer. Propagation in radially inhomogeneous media (abstract only), *Jour. Opt. Soc. Amer.* **60** (5) (1970), 738.

E. G. Rawson and D. R. Herriott. Analysis of graded index glass rods used as image relays (abstract only), *Jour. Opt. Soc. Amer.*   (11), (1969), 1520.

M. S. Sodha, A. K. Ghatak and I. C. Goyal. Series solution for E.M. wave propagation in radially and axially non-uniform media: Geometrical optics approximation, *Jour. Opt. Soc. Amer.* **61** (11), (1971), 1492; also **62** (12), (1972), 963.

W. H. Southwell. Sine wave optical paths in gradient index media, *Jour. Opt. Soc. Amer.* **61** (12), (1971), 1715.

W. Streifer and C. N. Kurtz. Scalar analysis of radially inhomogeneous guiding media, *Jour. Opt. Soc. Amer.* **56** (6), (1967), 779.

GENERAL MEDIA

N. G. Alexopoulos. On the refractive properties of media with poles or zeros in the index of refraction ,*Trans. I.E.E.E.* **AP22** (2), (1974), 242.

D. W. Berreman. Optics in stratified and anisotropic media: 4 × 4 matrix formulation, *Jour. Opt. Soc. Amer.* **62** (4), (1972), 502.

G. Eichmann. Quasi-geometric optics of media with inhomogeneous index of refraction, *Jour. Opt. Soc. Amer.* **61** (2), (1971), 161.

J. J. Gibbons and R. L. Schrag. Method of solving the wave equation in a region of rapidly varying complex refractive index, *Jour. Appl. Phys.* **23** (1972), 1139.

S. Gorn. Series expansions of rays in isotropic non-homogeneous media, *Jour. Quart. App. Math.* **11** (1953), 355.

W. N. Hansen, Electric fields produced by the propagation of plane coherent electromagnetic radiation in a stratified medium, *Jour. Opt. Soc. Amer.* **58** (3) ,(1968), 380.

E. W. Marchand. Gradient index lasers, *Progress in Optics* (Ed. E. Wolf), Vol. XI.

D. Marcuse. "Theory of Dielectric Optical Waveguides", Academic Press, London and New York, 1974.

M. Matsuhara, Analysis of electromagnetic wave modes in lens like media *Jour. Opt. Soc. Amer.* **63** (2), (1973), 135, also **60** (1) 1970 p. 1.

L. Montagnino. Ray tracing in inhomogeneous media, *Jour. Opt. Soc. Amer.* **58** (12) (1968), 1667.

S. Nakao, S. Fujimoto, R. Nagata and K. Iwata. Model of the refrective index distribution in the Rabbit crystalline lens (an elliptical parabolic distribution), *Jour. Opt. Soc. Amer.* **58** (8), (1968), 1131.

R. Sedney. Geometrical optics of angular stratified media, *Quart. Jour. Appl. Maths,* **14** (1956), 225.

A. Walther. Lenses, wave optics and eikonal functions, *Jour. Opt. Soc. Amer.* **59** (10) (1969), 1325.

THEORY OF STRATIFIED MEDIA

V. M. Agafonov. Polynomial filters from symmetrical units of the same system, *Radio Eng. and Electronic Physics,* **14** (6), (1969), 962.

H. G. Booker, J. A. Fejer and K. F. Lee. A theorem concerning reflection from a plane stratified medium, *Radio Sci.* **3** (3), (1968), 207.

L. M. Brekhovskikh. "Waves in Layered Media", Academic Press, New York and London, 1960.

H. Bremmer, The propagation of E.M. waves through a stratified medium and its WKB approximation for oblique incidence. *In* "Handbuch der Physik", Vol. 16, Springer-Verlag, Berlin, 1958.

K. F. Casey, Application of Hill's functions to problems of propagation in stratified media, *Trans. I.E.E.E.* **AP20** (3), (1972), 368.

C. Favier de Coulomb. Théorie génerale des milieux stratifiés, *Novelle Rev. d'Optique,* **5** (3), (1974), 186.

G. Franceschetti. Scattering from plane layered media, *Trans. I.E.E.E.* **AP12** (6), (1964), 754.

G. Hines. Relection of waves from varying media, *Quart. Jour. App. Maths.* **11** (1953), 9.

H. Levine, Reflection and transmission by layered periodic structures, *Quart. Jour. App. Maths.* **19** (1), (1966), 108.

D. S. Saxon. Modified WKB approximation for the propagation and scattering of E.M. waves. *In* Symposium on E.M. Theory, University of Toronto, June 1959.
F. W. Sluijter. Generalizations of the Brenmer series based on physical concepts, *Jour. of Math. Analysis and Applications*, **27** (2), (1969), 282.
C. T. Tai. "Dyadic Green's Functions in E.M. Theory", Intext Press, 1971, Chapter 12.
J. R. Wait, "Electromagnetic Waves in Stratified Media", Pergamon Press, 1970.

RADOMES AND MULTILAYER SYSTEMS
F. Abeles. Optical properties of metallic films, *In* "Physics of Thin Films", Vol. 6, Academic Press, London and New York, 1973.
P. W. Baumeister, Methods of altering the characteristics of a multilayer stack, *Jour. Opt. Soc. Amer.* **62** (10), (1972), 1149.
J. A. and P. H. Berning, Thin film calculations, *Jour. Opt. Soc. Amer.* **50** (1960), 813.
D. Conti. Third International Colloquium on Electromagnetic Windows", Paris, September 1975.
K. C. Park. The extreme values of reflectivity and the conditions for zero reflection from thin dielectric films on metal, *Applied Optics,* **3** (7), (1964), 877.
A. E. Philippe, Reflection and transmission of radio waves at a dielectric slab with variable permittivity, *Trans. I.E.E.E.* **AP21** (2), (1973), 234.
J. S. Seeley, Multilayer filters, *Jour. Opt. Soc. Amer.* **54** (1964), 342, *also* **52** (1962), 431.
T. Tamir and H. L. Bertoni, Lateral displacement of optical beams at multi-layered and periodic structures, *Jour. Opt. Soc. Amer.* **61** (10), (1971), 1397.
A. Thelen. Multilayer filters with wide transmittance bands, *Jour. Opt. Soc. Amer.* **63**, (1), (1973), 65; Design of multilayer interference filters, *in* "Physics of Thin Films", Vol. 5, Academic Press, London and New York, 1972.
A. Tonkin and R. Graham. Dielectric radome designs for two frequency operation. First international symposium on electromagnetic windows, Paris, September 1967.
L. Young, Multilayer reflection coatings on a metal mirror. *Applied Optics,* **2** (4), (1963), 445, also, Prediction of absorption loss in multilayer interference filters, *Jour. Opt. Soc. Amer.* **52** (7), (1962), 753.

GEODESIC LENSES
S. Adachi, R. C. Rudduck and C. H. Walter. A general analysis of non-planar two dimensional Luneburg lenses, *Trans. I.E.E.E.* **9** (1961), 353.
G. D. M. Peeler and D. H. Archer. A two dimensional microwave Luneburg lens, *Trans. I.R.E.* **AP1** (1953), 12.
R. G. Rudduck and C. H. Walter. A general analysis of geodesic Luneburg lenses, *Trans. I.E.E.E.,* **AP10** (1962), 444.
C. H. Walter. Surface wave Luneburg lens, *Trans. I.R.E.* **AP8** (1960), 508.

GEODESICS AND RAYS
All papers in this section are from the Transactions of the American Mathematical Society.

H. F. Stecker. On the determination of surfaces capable of conformal representation upon a plane in such a manner that geodesic lines are represented by algebraic curves, **2** (1901), p. 152.

H. F. Stecker. Concerning the existence of surfaces capable of conformal representation upon the plane in such a manner that geodesic lines are represented by a prescribed system of curves, **3** (1902), 12.

E. Kasner. Isothermal systems of geodesics, **5** (1904), 56.

E. Kasner. The generalized Beltrami problem concerning geodesic representation, **4** (1903), 149.

E. Kasner. Natural families of trajectories: Conservative fields of force, **10** (1909), 201.

E. Kasner. The problem of partial geodesic representation, **7** (1906), 200.

E. Kasner. The trajectories of dynamics, **7** (1906), 401.

L. P. Eisenhart. Surfaces with isothermal representation of their lines of curvature and their transformations, **9** (1908), 149 and **11** (1910), 474.

G. W. Hartwell. Plane fields of force whose trajectories are invariant under a projective group, **10** (1909), 220.

E. Kasner. Surfaces whose geodesics may be represented by parabolas, **6** (1905), 141.

3

Scalar Diffraction Theory of the Circular Aperture

The ray optical methods of the preceding chapters that have been used to determine the basic focusing properties of the reflector and lens antennas give a qualitative description of the resultant field intensity over the exit aperture, from the presumed or known field distribution of the source. They can no longer apply to the final problem of the determination of the radiated field in space from the antenna, the radiation pattern, without considering the major effect of the finiteness of the aperture, that is, diffraction. The effect is far more noticeable at microwave frequencies by virtue of the smallness (in wavelength number) of the apertures concerned. It is compensated in some measure by requiring the consideration of fewer aberration effects than does the optical system. It also differs from the optical problem in that amplitude distributions can be varied in a very general way as has been shown with resultant variation in the radiation patterns.

The exact solution for this problem requires a knowledge of the total vector field distribution over the infinite surface containing the radiating apertures, and the application of Green's theorem to the half-space, otherwise source free, into which the antenna radiates. Green's theorem, applying as it does to a closed surface, requires the integration over the entire plane containing the sources, that is the radiating aperture and any fields occurring locally because of its presence, and a closing surface which is generally taken to be an infinite hemisphere. By invoking a "radiation condition at infinity" the integration over this hemisphere is shown to tend to zero as its radius becomes infinite.

Even where the sources of the radiating field have been precisely determined, the ensuing integration has to be approximated in some way, and therefore the effects of two approximations have to be taken into account, that of the definition of the source field, and that of its resulting radiation pattern. The procedure has become increasingly refined in recent years. The major problem in both cases is the precise effect of the edges and boundaries that must exist to define the finite radiating optical device, or its aperture.

An attempt to cover the various theories and their solutions is made in the accompanying table. Apart from the actual references given all the names occurring in this chart are to be found in the bibliography to this chapter in the section relating to diffraction theory. In the following survey of these theories the references (A), (B), (C), etc. are to the equations given in the chart.

In the case where the radiating system is based solely upon reflecting surfaces of presumed infinite conductivity, the most fruitful method appears to be that of the physical optics procedure (A), that is the determination of the actual currents arising on the surfaces concerned from the known incident field and their use as the source for the radiated field. The assumption is made that the current distribution arising at any point on the surface from a known value of the incident magnetic field, is the current that would be created by an infinite plane wave, of the same amplitude and polarization, incident upon an infinite conducting plane. The theory is thus most accurate for large reflectors with necessarily large radii of curvature and the contribution from the locality of the edges can still be expected to cause errors. As in what follows, this concern for the edge effect is the major problem in the whole of diffraction theory. In the physical optics method it can be shown to be small and the agreement with the best of observed measurements is the closest of the many theories available.

Unfortunately, it results in a procedure that is basically computational and is applied only to reflector systems. Hence most studies have been applied exclusively to the two reflector Cassegrain antennas[1] subsequent to their design by geometrical optics methods. Since this system is capable of an endless variation of the geometrical parameters and illumination functions, a great deal of work has to be done in order to provide sufficient data for general assessments to be made. It is difficult as yet to see how the method

DIFFRACTION THEORY

G is Green's function for free space $= \dfrac{1}{4\pi}\dfrac{e^{ik|r-r'|}}{|r-r'|}$; R is vector from source point r' to field point r, $R = r - r'$

S completely encloses volume V, A is aperture in S, S–A is conducting surface of infinite plane screen, $\hat{n}$ is *inward* pointing normal.

Symmetrised Maxwell Equations
(Papas, p. 7)

$$\nabla \times E = -j_m - \frac{\partial B}{\partial t} \qquad j_m = -n \times E \qquad \nabla \times H = j + \frac{\partial D}{\partial t} \qquad j = \hat{n} \times H$$

$$\nabla \cdot B = \rho_m \qquad \rho_m = \mu\hat{n}.H \qquad \nabla \cdot D = \rho \qquad \rho = \varepsilon\hat{n}.E$$

	Retarded Potentials			*Retarded Potentials*	
	Volume	Surface		Volume	Surface

$$F = \varepsilon\int_V j_m G\,dv = -\varepsilon\int_S \hat{n} \times E\,G\,ds \qquad\qquad A = \mu\int_V jG\,dv = \mu\int_S \hat{n} \times H\,G\,ds$$

$$X = -\frac{1}{\mu}\int_V \rho_m G\,dv = -\int_S \hat{n}.H\,G\,ds \qquad \Phi = -\frac{1}{\varepsilon}\int_V \rho G\,dv = -\int_S \hat{n}.E\,G\,ds$$

Transformation. $E \to H \qquad H \to -E \qquad \mu \leftrightarrow \varepsilon \qquad j \to j_m \qquad j_m \to -j \qquad \rho \to \rho_m \qquad \rho_m \to -\rho$

then $A \to F;\ F \to -A;\ \Phi \to X;\ X \to -\Phi$ and for all solutions $E_p \to H_p$ etc.

Wave Equation $\nabla^2 E - \mu\varepsilon\dfrac{\partial^2 E}{\partial t^2} = \mu\dfrac{\partial j}{\partial t} + \nabla \times j_m + \dfrac{1}{\varepsilon}\nabla\rho = f(x',y',z',t')$ **Scalar wave equation** $\left(\nabla^2 - \mu\varepsilon\dfrac{\partial^2}{\partial t^2}\right)u = f(x',y',z',t')$

Solution for time harmonic fields $e^{-i\omega t}$ Solution (Appendix II, Stratton p. 470, Jackson, p. 283)

$$\S E_p = -\frac{1}{\varepsilon}\operatorname{curl} F + i\omega A - \operatorname{grad}\Phi$$

$$u_p = \frac{-1}{4\pi}\int_V \frac{[f_{rel}]}{|r-r'|}\,dv \qquad + \qquad \int_S \{G(\hat{n}.\nabla)u - u(\hat{n}.\nabla G)\}\,ds$$

Volume Source Solution   **Surface Source Solution**

$$u_p = -\int_S \left\{ G\frac{\partial u}{\partial n} - u\frac{\partial G}{\partial n}\right\} ds$$

By components

$$E_p = -\int_S \left\{ G\frac{\partial E}{\partial n} - E\frac{\partial G}{\partial n}\right\} ds$$

Silver, p. 108

(Silver, p. 83 complex conjugate)

$$E_p = \int_V \left\{ i\omega\mu jG - j_m \times \nabla G + \frac{\rho}{\varepsilon}\nabla G\right\} dv$$

Stratton p. 466
Jackson p. 285

$$\int_S \{i\omega(\hat{n} \times B)G + (\hat{n} \times E) \times \nabla G + (\hat{n}.E)\nabla G\}\,ds \quad \text{(1)}$$

Far Field Approximation (Silver, p. 88)

$$E_p = \frac{i}{4\pi\omega\varepsilon}\int_V \{k^2 j - \kappa^2(\hat{j}.R)R + \kappa\omega\varepsilon\, j_m \times R\}\,\frac{e^{ikrR - r'.R}}{R}\,dv$$

$$E_p = \frac{i}{\omega\varepsilon}\nabla \times \nabla \times \int_S (\hat{n} \times H)G\,ds + \nabla \times \int_S (\hat{n} \times E)G\,ds$$
Franz
Sommerfeld, p. 325
C. T. Tai, 1972

$$= \frac{ic^2}{\omega}\nabla \times \nabla \times A - \frac{1}{\varepsilon}\nabla \times F \quad \text{from § with div } A = i\omega\Phi$$

Physical optics
Approximation
(Rusch, 1963)

$$j = 2\hat{n} \times H_{inc} \qquad j_m = \rho_m = 0 \qquad \hat{n} \times E = \hat{n}.H = 0$$

Aperture integrals
$S \to A$

Silver, p. 165
Stratton, p. 469

$$E_p = \int\int_A \left\{ G\frac{\partial E}{\partial n} - E\frac{\partial G}{\partial n} \right\} da + \oint_{edge}(E\times\hat{t})G\,dl - \frac{i}{\omega\varepsilon}\oint_{edge}(\hat{t}\cdot H)\nabla G\,dl = \int_A\{i\omega(\hat{n}\times B)G + (\hat{n}\cdot E)\nabla G\}da - \frac{i}{\omega\varepsilon}\oint_{edge}(\hat{t}\cdot H)\nabla G\,dl + \int_A(\hat{n}\times E)\times\nabla G\,da$$

(B) additional edge current Stratton and Chu (C) (D)

$\hat{n}\times E = 0$ on S − A

$\hat{n}\times H = 0$ on S − A

Vector Rayleigh–Sommerfeld

r' is vector from origin to element da $\hat{k} = R/|R|$

$$E_p = \int\int_A\left\{G\frac{\partial E}{\partial n} - E\frac{\partial G}{\partial n}\right\}da$$ (E)

Far Field → $\dfrac{i}{2\pi}\dfrac{e^{ikR}}{R}\hat{k}\times\int_A(\hat{n}\times E(r))e^{-ik\cdot r}da$

Jackson, p. 292

Jackson, p. 282
Silver, p. 109 and p. 166

Kirchhoff approximation

$$u_p \simeq \int\int_A\left\{G\frac{\partial u}{\partial n} - u\frac{\partial G}{\partial n}\right\}da$$

Far Field → $\dfrac{1}{4\pi}\int_A\dfrac{e^{ikR}}{R}\hat{n}\cdot\left[\nabla u + ik\left(1 + \dfrac{i}{kR}\right)\dfrac{Ru}{R}\right]da$

Silver, p. 118

Rayleigh–Sommerfeld

$$u_p = \int_S G\frac{\partial u}{\partial n}ds - \int_S u\frac{\partial G}{\partial n}ds$$

(H) u = 0 on S–A

Heurtley Bouwkamp $\dfrac{\partial u}{\partial n} = 0$ on S–A

Far Field → $u_p = \dfrac{i}{2\lambda}\int_A u\left(1 - \dfrac{\cos\hat{n}\cdot R}{R}\right)da$

Silver, p. 167
Born & Wolf, p. 435

$$u_p = \frac{-ik}{4\pi}\frac{e^{ikR}}{R}(1 + \cos\theta)\int_A u\,e^{ik\cdot R}da$$

$$u_p = (1 + \cos\theta)\int_A F(\xi,\eta)e^{ik\sin\theta(\xi\cos\phi + \eta\sin\phi)}d\xi\,d\eta$$ Silver, p. 173

(G) Circular aperture radius a

$$u_p = \int_0^{2\pi}\int_0^1 f(r,\phi')e^{ika\sin\theta\cos(\phi-\phi')}r\,dr\,d\phi'$$

Jackson, p. 293
Silver, p. 192
≡ Equation 3.1

Diffraction of waves by apertures

Rubinowicz–Miyamoto–Wolf $u_{Kirchhoff} = u_{Geom.\,optics} + Boundary\,Wave$
Marchand $= u_G + u_B$ (K)

Sommerfeld, p. 263
Born and Wolf, p. 562

Miyamoto–Wolf–Sancer $u_K = u_G + u_{B_1} + u_{B_2}$

Geometrical Theory of Diffraction
Keller Sherman E_p = Geom. optics + Edge diffracted wave

Luneburg–Debye expansion
$$\frac{e^{ikr}}{r} = \frac{ik}{2\pi}\int_\Omega \exp\{ik[\alpha(x-x') + \beta(y-y') + \gamma(z-z')]\}d\Omega$$ (L)

Plane Wave Spectrum
Shafer Booker and Clemmow $\Omega = \sin\theta\,d\theta\,d\psi$ $\psi; 0\to 2\pi$ $\theta; 0\to i\infty$

Bouwkamp, p. 41

would apply to optical lenses but a case could be made out for its application to constrained waveguide lenses. In the case of the more complex geometries involved in reflector designs, as shown in Chapter 1, an entirely new computational programme would be required for each design and this fact may inhibit active development in these areas.

Much the same can be said of another recent method adopted specifically to tackle the problem of the effects of the edges in antenna diffraction. This is the geometrical theory of diffraction originated by Keller.[2] This adds to the geometrical optics field an "edge diffracted field" calculated at each point of the periphery as if from the diffraction of a plane wave incident there, with the same amplitude and polarization upon an infinite half-plane. The standard result of Sommerfeld is used for this half-plane effect. Modifications have been found to be necessary to this theory in the most vulnerable regions of the field, the light-shadow boundary or in the region of caustic and foci.[3] Again it remains essentially a computational procedure and no feed-back to the antenna design stage is possible without the collection of a great deal of data.

Earlier work on the subject is to a large extent unsurpassed. In general the integral solution to the problem has to be reduced from that covering the entire half plane as used in the original application of Green's theorem, to the region covered by the antenna aperture alone and over which the total vector field is presumed known. A complete discussion is to be found in Bouwkamp[4] (along with 500 relevant references). Silver[5] shows that the reduction of the infinite integral to a finite integral (B) results in a residual integral which by an application of Stokes' theorem can be converted to a line integral around the periphery of the aperture (C). In addition there has to be a contribution from a line integral around the edge of the aperture from a line source of charge necessary to give continuity of the field between the illuminated aperture and the "dark" side of the plane. This same contour integral arises directly if Kirchhoff approximations (to be discussed) are made to the surface retarded scalar potentials Φ and X (D).

The combined contributions of the line integrals are shown by Silver to be small enough to be ignored. The approximation that results is in effect that the integrand of the aperture integral reverts to the identical form that it had for the infinite plane (E). This is the vector Kirchhoff approximation. Interpretation of the result shows that it implies that the field components over the region of the infinite plane outside the aperture are considered to be zero. In the case of diffraction of waves by an aperture in an infinite plane, the field in the aperture in this approximation is the field that would have existed there in the absence of the plane. This problem is allied to the problem of antenna diffraction and has received far greater attention. It requires adaptation however before it can be applied to the antenna problem with a non-uniform amplitude distribution. For uniformly illuminated apertures both theories can be compared and this is sometimes used as a test in either case.

Other forms of Kirchhoff approximations occur whenever the infinite integral in whatever its form, is applied in that form to the area of the radiating aperture alone, ignoring in effect the residual integrals that result from such a transition. As was shown above, this can apply as far back in the theory as the surface retarded potentials themselves. It is tantamount to a Huygen's construction since it refers to radiating sources of spherical waves, the free-space Green's functions, at points within the illuminated aperture alone.

The scalar Kirchhoff approximation then arises by virtue of the fact that with the omission of the line edge integrals, the vector radiated field can be obtained from the addition of the contributions from each of the components of the aperture field separately. These then obey the scalar form of the approximation (F) which conforms with the earlier optical theory of scalar aperture illumination functions. The requirements of the antennas which are the main consideration of this study are for a circular aperture to which form the scalar diffraction integral is reduced in equation (G). Concerning the allied problem of the diffraction of an incident wave by an aperture in a plane screen, and in particular of a plane wave by a circular aperture other approximate solutions have been obtained. These have a relevance in suggesting a similar treatment for the radiating aperture study.

Transformations of the original infinite integrals both the scalar and the vector forms, can be made which separates them into two parts. Separate boundary conditions can be applied to each part which, when combined over the area of the aperture, gives the Kirchhoff approximation.

In the scalar case, for the diffraction of a plane wave, the two resultant integrals are the Rayleigh–Sommerfeld integrals (H).

In the case of the vector integrals for the radiating antenna a transformation by Franz gives a similar separability (I). Applying boundary conditions to each part does not give the vector form of the Kirchhoff approximation since an edge integral is included. The result could be termed the vector Rayleigh–Sommerfeld integrals (J). The boundary condition chosen closely agrees with the assumptions made in the physical optics approximation which, as stated, is among the best with regard to experimental observation. The resulting integral can thus be expected to give a similar order of agreement.

In considering the diffraction of waves by an aperture in a screen, a physical meaning to the difference that exists between the geometrical optics and the Kirchhoff approximate solution has been sought by Wolf and his coworkers, in the form of a "boundary wave" arising at the edge (K). The first and second orders of this boundary wave have been derived for plane wave illumination.

This effect differs both from the edge current integrals and from the edge diffraction effect of the geometrical theory of diffraction, although it has some aspects in common with the latter.

In all these methods the Green's function for free-space arising from the original solution figures prominently. Methods to associate non-uniform aperture fields with the plane or spherical wave diffraction theory involve expansion of this function as an infinite spectrum of the type of wave concerned. The expansion for plane waves is given (L) but the final results from their application have been shown to be less in agreement with measurements than are the other approximations considered.

Other as yet untried methods to "improve" upon the Kirchhoff approximation can be conjectured. It is known from the experimental results of Andrews and Andrejewski (see bibliography) that the smooth illumination function that would exist in the aperture if the Kirchhoff assumption were indeed correct, is perturbed by a standing wave. This is obviously created by the interaction between the edge currents by whatever means they are assumed to be caused. The magnitude of this standing wave in the aperture is a function of the aperture diameter and the strength of the illumination function at the edge. In principle the standing wave will increase as the aperture decreases and the illumination becomes more uniform until for the very small apertures it becomes the actual mode distribution of a wave guide with the same cross section. In reverse, for apertures of ten wavelengths or over it forms but a slight ripple on the assumed smooth distribution. The diffraction pattern of this standing wave, *in addition* to that of the normal Kirchhoff diffraction pattern, should form the second order correction to the radiation patterns, in the form of the well known high period lobe patterns in the lower field strength regions.

For purely analytical procedures we find that only the scalar Kirchhoff approximation gives a closed form solution which enables an *a priori* assessment of the antenna radiation pattern to be used as a design instruction for the antenna itself. In this manner it applies to diffraction theory in the same way as geometrical optics applies to optical design. Where comparisons with more exact solutions have been made the Kirchhoff solution has always been found to be in close agreement and with experimental results. This is true as we shall show, even in disputed regions such as the neighbourhood of foci and caustics and for apertures of comparatively small dimension. Having regard to all the other contributions to errors of measurement that exist, such as spillover and scattering from support structures, most of which are greater in magnitude than can be attributed to the Kirchhoff approximation alone, it is evident that for a qualitative assessment of antenna performance under quite general conditions of illumination function, the result will be satisfactory.

Further justification for the use of this approximation for design and analysis arises in the sequel. It is found that the closed form series solution that results from its use can be applied to the converse problem, that of pattern synthesis, the design of aperture functions to give prescribed radiation patterns. This will be dealt with subsequently.

3.1 THE SCALAR INTEGRAL TRANSFORMS

Considering mainly the circular aperture the Huygens–Green integral reduces as shown in the table (G), to[5]

$$g(u, \phi) = \int_0^{2\pi} \int_0^1 f(r, \phi')\, e^{iur\cos(\phi - \phi')} r\, dr\, d\phi' \qquad (3.1)$$

in which (Fig. 3.1) r, ϕ' are normalized polar coordinates in the plane of the circular aperture with origin at the centre and θ, ϕ are polar coordinates of a field point at a sufficiently large distance with respect to an axis perpendicular to the plane of the aperture and passing through its centre. The parameter $u = 2\pi a \sin \theta/\lambda$, where a is the radius of the aperture and λ the wavelength of the operating radiation, is a generalized coordinate whose nominal maximum value for real values of θ is $2\pi a/\lambda$. At some stage however u will be considered as taking all possible values on the real axis.

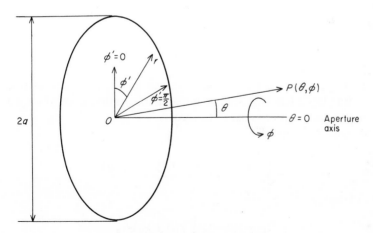

FIG. 3.1. Circular aperture coordinate geometry.

The expression given in Eqn (3.1) is in fact simplified by the exclusion of multiplying constants and more importantly by the factor $(1 + \cos\theta)$. The former are ignored since we are going to be concerned more with pattern shape than with the absolute values of the field in the radiation zone. It is customary to exclude the "obliquity factor" in those cases where the radiated pattern over only small values of θ is of interest. The main beam and side lobes of the focused antennas of a size for which (3.1) is an adequate approximation, usually occur for value of $u < 5$.

In subsequent sections we shall deal analytically with some aspects involving much larger values of u and hence of θ. In these cases, the theory can be modified to the extent that the pattern being considered is not the specific $g(u)$ of Eqn (3.1) but

$$g(u) \rightarrow \frac{g(u)}{(1 + \cos\theta)}.$$

This will be indicated in those situations where it is relevant.

The function $f(r, \phi')$ represents the complex amplitude distribution of field in the aperture. This could be taken to be the separate components of a vector field quantity and added in the result to give a vector solution for the field at the far point. The derivation of Eqn (3.1) makes certain conditions on $f(r, \phi')$ and in particular on the continuity of the imaginary part and on its differential coefficient. These add up to the limitation that sudden changes in amplitude or in phase are not allowable. In what follows however we take the function to be piecewice continuous with no limitation on the variation in amplitudes or phase apart from those which would be impossible in a practical situation.

We can then apply symmetry conditions to the illumination function $f(r, \phi')$ which will reduce the double integration to a single integral over the radial component only.

(a) Circular symmetry

For a perfectly centred circularly symmetric aperture distribution, $f(r, \phi')$ is independent of ϕ' and the integral independent of ϕ, which without loss of generality can be taken to be zero.

The using the relation[6]

$$\frac{1}{2\pi} \int_0^{2\pi} \exp{(iur\cos\phi')}\,d\phi' = J_0(ur), \tag{3.2}$$

Eqn (3.1) becomes

$$g(u, o) = 2\pi \int_0^1 f(r)\, J_0(ur) r\, dr \tag{3.3}$$

We can omit the factor 2π in the following, where its omission leads to no ambiguity.

Equation (3.3) defines the finite Hankel transform of the function $f(r)$ of zero order.[7]

(b) Diametrical antisymmetry[8]

Dividing the aperture into two halves by the diameter $\phi' = 0$ and $\phi' = \pi$, we consider each half to be totally in antiphase. The complex field can then be expressed by

$$f(r, \phi') = f(r) \qquad 0 < \phi' < \pi$$
$$= -f(r) \qquad \pi < \phi' < 2\pi$$

In a practical sense this distribution can only be achieved by the insertion of conducting surfaces into the aperture or by a similar discontinuity created artificially by phase delay media covering half of the aperture.

Equation (3.1) then gives

$$g(u, \phi) = \int_0^\pi \int_0^1 f(r) \exp\left[iur\cos(\phi - \phi')\right] r \, dr \, d\phi'$$

$$- \int_0^\pi \int_0^1 f(r) \exp\left[-iur\cos(\phi - \phi')\right] r \, dr \, d\phi'$$

$$= 2i \int_0^\pi \int_0^1 f(r) \sin\left[ur\cos(\phi - \phi')\right] r \, dr \, d\phi'. \qquad (3.4)$$

In most cases the radiation pattern of most interest is that in the plane where the maximum effect occurs, that is, in the plane perpendicular to the diameter bisecting the aperture. Putting $\phi = \frac{1}{2}\pi$ therefore in Eqn (3.4) results in the separable integral

$$g\left(u, \pm \frac{\pi}{2}\right) = \pm 2i \int_0^1 f(r) \, dr \int_0^\pi \sin(ur\sin\phi') \, d\phi'. \qquad (3.5)$$

The second of these integrals defines the Lommel–Weber function Ω_0 that is from Ref. 6, p. 308

$$\int_0^\pi \sin(ur\sin\phi') \, d\phi' = \Omega_0(ur). \qquad (3.6)$$

It is identical with the zero order Struve function $H_0(ur)$ but higher orders of the two functions differ by a polynomial in the variable (Ref. 6, p. 337).

If we allow positive and negative values of u to correspond with values of

ϕ separated by π then (3.5) gives the result

$$g(u, \pm \tfrac{1}{2}\pi) \equiv g(\pm u, \tfrac{1}{2}\pi) = \pm 2\pi \int_0^1 f(r)\,\Omega_0(ur)\,r\,dr. \qquad (3.7)$$

This defines the zero order Lommel transform of $f(r)$.

In a situation where the two halves are not completely in antiphase the radiation pattern in the plane $\phi = \tfrac{1}{2}\pi$ is expressible as a combination of the results of (*a*) and (*b*). Thus if the two halves are respectively $+\alpha$ and $-\alpha$ radians in phase (Fig. 3.2a)

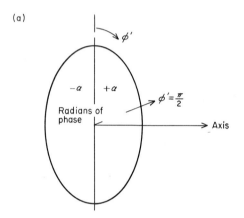

FIG. 3.2(a). *Circular aperture phase distributions. Diametral division of phase*

$$g(u, \phi) = e^{-i\alpha} \int_0^\pi \int_0^1 f(r)\exp\left[iur\cos(\phi - \phi')\right]r\,dr\,d\phi'$$

$$+ e^{i\alpha} \int_0^\pi \int_0^1 f(r)\exp\left[-iur\cos(\phi - \phi')\right]r\,dr\,d\phi', \qquad (3.8)$$

which in the plane $\phi = \tfrac{1}{2}\pi$ reduces to

$$g\left(\pm u, \frac{\pi}{2}\right) = 2\pi\cos\alpha \int_0^1 f(r)\,J_0(ur)\,r\,dr \pm 2\pi\sin\alpha \int_0^1 f(r)\,\Omega_0(ur)\,r\,dr. \qquad (3.9)$$

(*c*) *Cyclic phase variation*

The phase variation termed "cyclic" has p cycles of phase variation in any fixed circular path about the origin where p is an integer. The illumination function can be expressed as

$$f(r, \phi') = f(r)\exp\left(-ip\phi'\right) \qquad (3.10)$$

and

$$g(u, \phi) = \int_0^1 f(r) r \, dr \int_0^{2\pi} \exp\{i[ur \cos(\phi - \phi') - p\phi']\} \, d\phi'. \quad (3.11)$$

The angular integral defines the higher order Bessel function (Ref. 6, p. 20) leaving

$$g(u, \phi) = 2\pi \exp\left(ip\left(\tfrac{1}{2}\pi - \phi\right)\right)\int_0^1 f(r) J_p(ur) r \, dr. \quad (3.12)$$

The factor outside the integral demonstrates that the radiation pattern exhibits the same cyclic variation in phase. The integral in Eqn (3.12) defines the higher order Hankel transforms. This example is included to demonstrate a situation in which such higher order transforms can occur.

We can similarly define the higher order Lommel transforms of a radial function $f(r)$ analogously

$$\mathcal{L}_p[f(r)] \equiv \int_0^1 f(r) \Omega_p(ur) r \, dr. \quad (3.13)$$

(d) Sinusoidal phase variation

A sinusoidal phase variation is a phase surface obtained by rotating a function of the radial coordinate about the origin while varying its magnitude by a factor proportional to $\sin \phi'$. In practical antenna systems a phase front of this type arises when the source is displaced from the focus transversely to the axis of symmetry as derived in Chapter 1, Section 1.7.1. In most instances then the phase variation is approximately cubic in the radial coordinate as shown in Fig. 3.2(b) and gives rise to the aberration of primary coma.

The complex aperture function for this case is

$$f(r, \phi') = f(r) \exp[-i\psi(r) \sin \phi'] \quad (3.14)$$

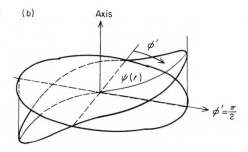

FIG. 3.2(b). *Circular aperture phase distributions. Asymmetric phase surface.*

in which $\psi(r)$ describes the shape of the phase surface in its cross section in the plane $\phi' = \frac{1}{2}\pi$. This would then be the curve given in Chapter 1 p. 61 as the focal curve of a parabolic cylinder with a displaced source. The scalar diffraction integral is then

$$g(u, \phi) = \int_0^{2\pi} \int_0^1 f(r) \exp\{i[ur \cos(\phi - \phi') - \psi(r) \sin \phi']\} \, r \, dr d\phi'. \qquad (3.15)$$

The plane showing the major effects of this perturbation is the plane $\phi = \frac{1}{2}\pi$ hence

$$g(u, \pm \tfrac{1}{2}\pi) \equiv g(\pm u, \tfrac{1}{2}\pi) = 2\pi \int_0^1 f(r) J_0(ur - \psi(r)) \, r \, dr. \qquad (3.16)$$

Using the addition formula for Bessel functions, ref. 6 p. 30

$$g(\pm u, \tfrac{1}{2}\pi) = 4\pi \sum_{s=0}^{\infty}{}' (-1)^s \int_0^1 J_s(\mp \psi(r)) J_s(ur) f(r) \, r \, dr, \qquad (3.17)$$

where the prime in the summation indicates that the *first* term ($s = 0$) is taken at only *half* its value.

For the particular case $\psi(r) = \alpha = $ constant the first Bessel function can be removed from the integration leaving

$$g(\pm u, \tfrac{1}{2}\pi) = 4\pi \sum_{s=0}^{\infty} (-1)^s J_s(\mp \alpha) \int_0^1 f(r) J_s(ur) \, r \, dr. \qquad (3.18)$$

For most cases of practical interest the first Bessel function in Eqn (3.17) can be expanded as a series of powers of r, which can then be included with $f(r)$ in a second series of higher order Hankel transforms.

(e) Radial phase variation

All the phase conditions described have had variation with the angular position in the aperture. A major phase perturbation that has to be considered is that which is constant in the angular coordinate but varies with distance from the centre of the aperture. This condition arises in optical systems when the source is displaced axially from the focus and gives rise to the second major aberration that we need to consider in microwave systems that of spherical aberration (Fig. 3.2(c)). The illumination function is now given by

$$f(r, \phi') = f(r) \exp[-i\psi(r)], \qquad (3.19)$$

where most usually $\psi(r) \equiv br^2$ approximately as shown in Chapter 1, Section 1.7.1.

The methods to be obtained for the integrals of the previous section are found to apply to this phase variation, but are only of a simple enough form for the elementary condition $f(r) \equiv 1$. Other treatments, extend the method to general amplitude distributions but require separate consideration, which will be given in a later section.

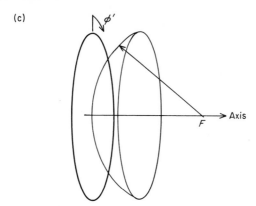

FIG. 3.2(c). *Circular aperture phase distributions. Axisymmetric phase surface.*

(*f*) *Amplitude variation with angular position*

The higher order finite Hankel transforms also occur in the scalar integration when the illumination function has an amplitude variation which is periodic in the angular coordinate although the phase distribution remains constant. This condition can be shown to apply to a particular component of the vector field that is created when a paraboloid is illuminated from a linearly polarized element, a dipole or a waveguide horn, at its focus.[9] This component gives rise to the well known cross polarized lobe patterns of this particular kind of reflector. (Ref. 5, p. 419).

This amplitude distribution of the cross-polarized component can be described by a function of the kind

$$f(r, \phi') = f(r) \sin 2\phi'. \tag{3.20}$$

The maximum perturbations, in this case the maximum values of the cross polarized lobes, occur in the quadrantal planes $\phi = \pm\frac{1}{4}\pi$ and $\frac{3}{4}\pi$.

Substitution into the scalar integral gives

$$g(u, \tfrac{1}{4}\pi) = \int_0^1 \int_0^{2\pi} f(r) \exp\left[i\,ur\cos(\phi' - \tfrac{1}{4}\pi)\right] \sin 2\phi'\,d\phi'\,r\,dr$$

$$\propto \int_0^1 f(r)\,J_2(ur)\,r\,dr. \tag{3.21}$$

From these examples we see that most of the conditions that can be expected to arise in a practical antenna system, affecting the radiation pattern through diffraction or phase and amplitude perturbation can be assessed in the Kirchhoff approximation by the evaluation of integrals of the higher order finite Hankel and Lommel transforms defined by

$$g_n(u) = \int_0^1 f(r) \, J_n(ur) \, r \, dr,$$

$$l_n(u) = \int_0^1 f(r) \, \Omega_n(ur) \, r \, dr, \qquad (3.22)$$

for general real functions of $f(r)$.

3.2 ZERO ORDER TRANSFORMS

The most common requirement is for the evaluation of the unperturbed aperture radiation patterns given by

$$g_0(u) = \int_0^1 f(r) \, J_0(ur) \, r \, dr. \qquad (3.23)$$

The functions $f(r)$ to be considered are in most practical cases smooth monotonically decreasing functions with an edge value at $r = 1$ of between 0·1 and 0·3 (assuming $f(0) = 1$). They can readily be approximated with very close agreement by a few terms of the series

$$f(r) = \sum_{p=0} a_p (1 - r^2)^p. \qquad (3.24)$$

A further degree of flexibility can be obtained by considering binomial expansions of such functions as

$$f(r) = (1 - kr^2)^p, \, (k < 1, \, p \text{ integer})$$

$$= (1 + kr^2)^p, \, (k > 1, \, p > 0)$$

$$= (1 + kr^2)^{-\frac{1}{2}}$$

The resultant series are similar in kind to that in Eqn (3.24). Hence for the result when substituted into Eqn (3.23) we require the relations

$$\int_0^1 (1 - r^2)^p \, J_0(ur) \, r \, dr = 2^p \, p! \, \frac{J_{p+1}(u)}{u^{p+1}} \equiv \frac{\Lambda_{p+1}(u)}{2(p+1)} \, (p = 0, 1, 2) \quad (3.25)$$

Equation (3.25) is the defining relation for the Lambda functions.[10] The nature of the Bessel functions for the larger values of the argument coupled with the power of the argument in the denominator means that only a few terms of any resultant series of the form

$$g_0(u) = \sum_{p=0} a_p \frac{\Lambda_{p+1}(u)}{2(p+1)} \tag{3.26}$$

needs to be considered to reach the degree of approximation to which we are already committed. The results are given in the short table of transforms in Section 3.4.1.

The same quadratic functions have a simple integral transform in the case of the zero order Lommel transforms. This arises from the relation (Ref. 7, p. 160, no. 12)

$$\int_0^1 (1 - r^2)^p \, \Omega_0(ur) \, r \, dr = 2^p \frac{p! \, H_{p+1}(u)}{u^{p+1}}, \tag{3.27}$$

where $H_p(x)$ is the Struve function.

The quadratic functions $(1 - r^2)^p$ are not simply integrable in the same way for the higher order transforms and their use cannot be extended beyond the unperturbed aperture with circular symmetry or antisymmetry as given above. They do not form a orthogonal set in the interval and so cannot be used for the series epxansion of a general function $f(r)$ or for an attack on the synthesis problem. Orthogonalization by the Schmidt method runs into excessive algebraic complexity after a very few terms. We examine means therefore of obtaining a series expansion in terms of more suitable polynomials.

3.3 THE CIRCLE POLYNOMIALS

The Lambda functions defined in Eqn (3.25) contains the argument u in the denominator raised to the same power as the order of the accompanying Bessel function. In nearly all microwave antennas having pencil beam radiation patterns the main radiation falls within the range $-5 < u < 5$ approximately and the first diffraction lobe maxima occur within the adjacent region $5 < |u| < 10$. The effects of the denominator in even the second term of the series of Eqn (3.26) will thus totally mask any pattern shaping effect of the Bessel function in the numerator. This fact was pointed out by McCormick[11] who suggested that a more appropriate series in which the effect of the denominator would be greatly reduced would by the Neumann series (Ref.

H

6, Chapter XVI)

$$g(u) = \sum_{s=0}^{\infty} \alpha_{2s+1} \frac{J_{2s+1}(u)}{u}, \tag{3.28}$$

with a similar series of Struve functions for the Lommel transforms.
  We can consider the more general case

$$g(u) = \int_0^1 f(r) J_0(ur) r \, dr = \sum_{s=0}^{\infty} \alpha(a)_{2s+1} \frac{J_{2s+1}(au)}{au}, \tag{3.29}$$

where a is an arbitrary constant which can later be made equal to unity.
  Taking the Laplace transform of Eqn (3.29) with respect to u then

$$\int_0^1 \frac{r f(r) \, dr}{(t^2 + r^2)^{\frac{1}{2}}} = \sum_{s=0}^{\infty} \alpha(a)_{2s+1} \frac{[(a^2 + t^2)^{\frac{1}{2}} - t]^{2s+1}}{(2s+1) a^{2s+2}}. \tag{3.30}$$

On putting

$$(a^2 + t^2)^{\frac{1}{2}} - t = v^{\frac{1}{2}} \qquad \text{or} \qquad t = \frac{a^2 - v}{v^{\frac{1}{2}}}$$

and

$$(a^2 + t^2)^{\frac{1}{2}} + t = \frac{a^2}{v^{\frac{1}{2}}}$$

then

$$\int_0^1 \frac{2r f(r) v^{\frac{1}{2}} \, dr}{(a^4 - 2a^2 v + v^2 + 4vr^2)^{\frac{1}{2}}} = \sum_{s=0}^{\infty} \alpha(a)_{2s+1} \frac{v^{s+\frac{1}{2}}}{(2s+1) a^{2s+2}}. \tag{3.31}$$

The integrand can be expanded as a series of Legendre polynomials, that is

$$\left[\left(\frac{v}{a^2} \right)^2 - 2v \left(1 - \frac{2r^2}{a^2} \right) + 1 \right]^{-\frac{1}{2}} = \sum_{s=0}^{\infty} P_s \left(1 - \frac{2r^2}{a^2} \right) \left(\frac{v}{a^2} \right)^s,$$

then comparing coefficients of powers of v/a^2 in the two series gives

$$\alpha(a)_{2s+1} = 2(2s+1) \int_0^1 P_s \left(1 - \frac{2r^2}{a^2} \right) f(r) r \, dr. \tag{3.32}$$

We find therefore that for a radiation pattern to be described as a Neumann series of the kind in Eqn (3.28) the coefficients of that series can be obtained directly from the aperture illumination function $f(r)$ by the relation in Eqn (3.32).
  Putting $a = 1$ then

$$\alpha_{2s+1} = 2(2s+1) \int_0^1 P_s(1 - 2r^2) f(r) r \, dr. \tag{3.33}$$

The functions $P_s(1 - 2r^2)$, where P is a Legendre polynomial are the circle polynomials first used by Zernike.[12] They are defined by

$$P_s(1 - 2r^2) = (-1)^s R^0_{2s}(r).$$

A table of polynomials and their properties are included in Appendix II.
With this terminology Eqn (3.33) becomes

$$\alpha_{2s+1} = 2(2s + 1) \int_0^1 R^0_{2s}(r) f(r) r \, dr. \tag{3.34}$$

The higher order circle polynomials can best be described as Jacobi polynomials[13] of which the Legendre polynomials are a subset. That is

$$R^m_n(r) = P^{(m,\,0)}_{\frac{1}{2}(n-m)} (1 - 2r^2)(-1)^{\frac{1}{2}(n-m)} r^m \tag{3.35}$$

$(n - m)$ a positive even integer or zero (Appendix II p. 381). As a consequence of this connection with the orthogonal Legendre and Jacobi polynomials, we have the orthogonality property

$$\int_0^1 R^m_n(r) R^m_p(r) r \, dr = \frac{1}{2(n + 1)} \delta_{n,\,p}, \tag{3.36}$$

$\delta_{n,\,p}$ being the Kronecker symbol, and also (Appendix II)

$$\int_0^1 R^m_n(r) J_m(ur) r \, dr = (-1)^{\frac{1}{2}(n-m)} \frac{J_{n+1}(u)}{u} \tag{3.37}$$

$(n - m)$ an even positive integer or zero.

This last result indicates the form of the appropriate expansion for the aperture distribution $f(r)$ suitable for all orders of Hankel (and Lommel) transforms.

That is we put

$$f(r) = \sum_{t=0}^{\infty} b_{2t+1} R^m_{2t}(r) \tag{3.38}$$

Confining ourselves temporarily to zero order transforms we now have

$$f(r) = \sum_{t=0}^{\infty} b_{2t+1} R^0_{2t}(r) \text{ from Eqn (3.38)} \tag{3.39}$$

$$g(u) = \sum_{s=0}^{\infty} \alpha_{2s+1} \frac{J_{2s+1}(u)}{u} \text{ from Eqn (3.28)}$$

where from an application of Eqn (3.36)

$$b_{2t+1} = 2(2t + 1) \int_0^1 R^0_{2t}(r) f(r) r \, dr.$$

and from Eqn (3.34)

$$\alpha_{2s+1} = (-1)^s \, 2(2s + 1) \int_0^1 R_{2s}^0(r) \, f(r) \, r \, dr,$$

and hence

$$a_{2s+1} = (-1)^s \, b_{2s+1}. \tag{3.40}$$

This recasts Eqn (3.39) into the complete set

$$f(r) = \sum_{s=0}^{\infty} (-1)^s \, \alpha_{2s+1} \, R_{2s}^0(r),$$

$$g(u) = \sum_{s=0}^{\infty} \alpha_{2s+1} \, \frac{J_{2s+1}(u)}{u},$$

and

$$\alpha_{2s+1} = (-1)^s \, 2(2s + 1) \int_0^1 R_{2s}^0(r) \, f(r) \, r \, dr. \tag{3.41}$$

We have therefore a remarkable duality whereby both the aperture illumination function and the resultant radiation pattern can be described by two different series with (apart from an alternation in sign) identical coefficients.

The results can be readily extended to the higher order transforms in which case we have

$$f(r) = \sum_{n=m}^{\infty} a_{n,m} \, R_m^n(r) \quad (n - m) \text{ an even positive integer or zero,}$$

$$g(u) = \sum_{n=m}^{\infty} a_{n,m} (-1)^{\frac{1}{2}(n-m)} \frac{J_{n+1}(u)}{u},$$

$$a_{n,m} = 2(n + 1) \int_0^1 f(r) \, R_n^m(r) \, r \, dr. \tag{3.42}$$

Further interesting properties of the zero order transform now arise. Under the transformation

$$r \to (1 - r^2)^{\frac{1}{2}},$$

we have

$$1 - 2r^2 \to 2r^2 - 1.$$

The circle polynomials being basically Legendre polynomials are alternately odd and even functions of their arguments, that is

$$P_s(1 - 2r^2) = (-1)^s P_s(2r^2 - 1).$$

The following results of the transformation $r \to (1 - r^2)^{\frac{1}{2}}$ then become obvious with

$$\int_0^1 f(r) J_0(ur) r \, dr = \sum_{s=0}^{\infty} \alpha_{2s+1} \frac{J_{2s+1}(u)}{u},$$

then

$$\int_0^1 f[(1 - r^2)^{\frac{1}{2}}] J_0(ur) r \, dr = \sum_{s=0}^{\infty} (-1)^s \alpha_{2s+1} \frac{J_{2s+1}(u)}{u},$$

$$\int_0^1 f(r) J_0[u(1 - r^2)^{\frac{1}{2}}] r \, dr = \sum_{s=0}^{\infty} (-1)^s \alpha_{2s+1} \frac{J_{2s+1}(u)}{u},$$

and

$$\int_0^1 f[(1 - r^2)^{\frac{1}{2}}] J_0[u(1 - r^2)^{\frac{1}{2}}] r \, dr = \sum_{s=0}^{\infty} \alpha_{2s+1} \frac{J_{2s+1}(u)}{u}, \quad (3.43)$$

all with the identical coefficients defined in Eqn (3.41). This particular property does not extend to the higher order transforms.

The circle polynomials have the further property that

$$R_{2s}^0(0) = (-1)^s \quad \text{and} \quad R_{2s}^0(1) = 1.$$

Substitution of these into Eqn (3.41) gives

$$f(0) = \sum_{s=0}^{\infty} \alpha_{2s+1} \quad \text{and} \quad f(1) = \sum_{s=0}^{\infty} (-1)^s \alpha_{2s+1}. \quad (3.44)$$

These are the values of the illumination function at the centre and the periphery of the aperture respectively.

If $f(r)$ is only defined in a circular region of radius $a \leqslant 1$ and is zero outside, then by the transformation $r' = r/a$

$$g(u) = \sum_{s=0}^{\infty} \alpha(a)_{2s+1} \frac{J_{2s+1}(au)}{au}, \quad (3.45)$$

where

$$\alpha(a)_{2s+1} = 2(2s + 1) \int_0^a f(r) R_{2s}^0 \left(\frac{r}{a}\right) r \, dr$$

$$= 2(2s + 1) \int_0^1 f(r) R_{2s}^0 \left(\frac{r}{a}\right) r \, dr,$$

as in Eqn (3.32) and since outside the range $x = \pm 1$

$$R_{2s}^0(x) = P_s(1 - 2x^2) \text{ is zero.}$$

Equation (3.45) is nothing more than the scaling of the radiation pattern by a factor equal to the reduction in size of the aperture and represents the obvious technique for producing patterns of greater width when required.

Analogous results to the above can be obtained in the case of the Lommel transforms. The key relation which parallels that of Eqn (3.37) is[14]

$$\int_0^1 R_n^m(r)\, \Omega_m(ur)\, r\, dr = (-1)^{\frac{1}{2}(n-m)} \left[\frac{\Omega_{n+1}(u)}{u} + \frac{2\cos^2\left(\frac{1}{2}n\pi\right)}{(n+1)\pi u} \right]. \qquad (3.46)$$

The Neumann series then becomes a series with the Bessel functions replaced by the Lommel–Weber functions of Eqn (3.46) but with the identical coefficients α_{2s+1} derived in the same manner as for Eqn (3.41).

For the zero order transform this gives the *odd* function

$$g(u) = \sum_{s=0}^{\infty} \alpha_{2s+1} \left[\frac{\Omega_{2s+1}(u)}{u} + \frac{2}{(2s+1)\pi u} \right]. \qquad (3.47)$$

The original integral Eqn (3.1) is basically a two dimensional Fourier transform, as would have been more apparent had we dealt with a rectangular coordinate system. In that case the integration would have separated into two single dimension Fourier transforms (Ref. 5). In Fourier transform theory a cosine transform can be transformed into a sine transform (and vice versa) through the application of a Hilbert transform. The equivalent even and odd transforms for the circular coordinate system are the Hankel and Lommel transforms. Thus the patterns of the two can be transformed into each other by the same Hilbert transformation. This was first observed by Moss.[14]

Hence if by any method the radiation pattern of a circular aperture antenna can be described by the Neumann series (zero order transform, circular symmetric illumination),

$$g(u) = \sum_{s=0}^{\infty} \alpha_{2s+1} \frac{J_{2s+1}(u)}{u}$$

then the Hilbert transform of this relation gives the radiation in the plane perpendicular to the dividing diameter of the same aperture divided into two antiphase halves. For we have

$$u\, g(u) = \sum_{s=0}^{\infty} \alpha_{2s+1}\, J_{2s+1}(u),$$

the Hilbert transform of which, Ref. 7, p. 243, is

$$yh(y) + \frac{1}{\pi} \int_{-\infty}^{\infty} g(y)\,dy = -\sum_{s=0}^{\infty} \alpha_{2s+1} \Omega_{2s+1}(u)$$

Substituting $g(y)$ into the integral

$$\sum_{s=0}^{\infty} \int_{-\infty}^{\infty} \alpha_{2s+1} \frac{J_{2s+1}(y)}{y}\,dy = 2\sum_{s=0}^{\infty} \frac{\alpha_{2s+1}}{2s+1}$$

giving

$$h(y) = -\sum_{s=0}^{\infty} \alpha_{2s+1} \left[\frac{\Omega_{2s+1}(y)}{y} + \frac{2}{\pi y(2s+1)} \right]$$

which is identical to Eqn (3.47) (the sign is ambiguous as is the definition of the phase in the aperture). The same transformation can be shown to transform the result of Eqn (3.25) into that of Eqn (3.27).

3.4 TRANSFORMS OF APERTURE DISTRIBUTIONS

3.4.1 Uniform phase—zero order transforms

To serve as an illustration of the method, we find the radiation patterns of uniform phase circularly symmetrical distributions with radial amplitude variation of the form which applies to the commonest types of illumination law found in practice.

Binomial expansions
(a) $f(r) = (1 - kr^2)^p$, p integer > 0, $k < 1$,

$$g(u) = \sum_{m=0}^{p} \frac{p!\,(1-k)^{p-m}}{m!\,(p-m)!} k^m \frac{\Lambda_{m+1}(u)}{2(m+1)} \tag{3.48}$$

(b) $f(r) = (1 + kr^2)^{-n}$, $n > 0$ $k > 1$

$$g(u) = \frac{1}{(1+k)^n}\left[\frac{J_1(u)}{u} + \frac{2nk}{1+k}\frac{J_2(u)}{u^2} + 4n(n+1)\left(\frac{k}{1+k}\right)^2 \frac{J_3(u)}{u^3} + . \right] \tag{3.49}$$

Neumann series

$$g(u) = \sum_{s=0}^{\infty} \alpha_{2s+1} J_{2s+1}(u)/u$$

(c) $f(r) = 1$

Putting $\qquad f(r) = 1 = R_0^0(r) \quad$ then $\qquad \alpha_1 = 1; \quad s = 0,$

and all $\qquad\qquad\qquad\qquad \alpha_{2s+1} = 0, s > 0,$

therefore $\qquad\qquad\qquad\qquad g(u) = \dfrac{J_1(u)}{u}$

(d) $f(r) = r^n$

From Ref. 7, p. 278, No. 22 (corrected)

$$\alpha_{2s+1} = \frac{(-1)^s 2(2s+1)[(\tfrac{1}{2}n)!]^2}{[(\tfrac{1}{2}n + s + 1]![(\tfrac{1}{2}n) - s)]!} \qquad (3.50)$$

where $(\tfrac{1}{2}n)!$ can be replaced by the Gamma function equivalent $\Gamma[(\tfrac{1}{2}n) + 1]$ where n is odd. The result

$$g(u) = \sum_{s=0}^{\infty} \alpha_{2s+1} \frac{J_{2s+1}(u)}{u},$$

is also given in Ref. 7, p. p. 22, the form (corrected)

$$g(u) = \frac{1}{n+2} \,_1F_2\left\{\frac{n}{2} + 1; 1, \frac{n}{2} + 2; \frac{-u^2}{4}\right\}, \qquad n > 0$$

where $_1F_2$ is the hypergeometric function.

This key function can now be used for any radial distribution which has a series expansion in powers of r or a good polynomial approximation. For example,

(e) $f(r) = 1 - r$
The "gabled" distribution

$$g(u) = \frac{1}{u}\left\{\frac{1}{3} J_1(u) + \frac{2}{5} J_2(u) + \frac{2}{21} J_5(u) + \frac{2}{45} J_7(u) \ldots\right\}. \qquad (3.51)$$

(f) $f(r) = \tfrac{1}{2}(\cos^2 \pi r)$ (Ref. 8)

$$g(u) \doteq 0.46267 \frac{J_1(u)}{u} + 0.49904 \frac{J_3(u)}{u} + 0.3733 \frac{J_5(u)}{u} + 0.00095 \frac{J_7(u)}{u} \qquad (3.52)$$

(g) $f(r) = \tfrac{1}{2}(\cos^2 \pi r)$

$$g(u) \doteqdot 0.297 \frac{J_1(u)}{u} + 0.473 \frac{J_3(u)}{u} + 0.205 \frac{J_5(u)}{u} + 0.022 \frac{J_7(u)}{u}$$

$$+ 0.002 \frac{J_9(u)}{u} \qquad (3.53)$$

(h) $f(r) = \exp(-2br^2)$

For this function we find

$$\alpha_1 = \frac{1}{2b}[1 - e^{-2b}]$$

$$\alpha_3 = \frac{-3}{2b}\left[\left(1 - \frac{1}{b}\right) + \left(1 + \frac{1}{b}\right)e^{-2b}\right],$$

and the recurrence relation

$$\frac{\alpha_{2s+1}}{2s+1} = \frac{\alpha_{2s-1}}{b} + \frac{\alpha_{2s-3}}{2s-3}.$$

For $b = \frac{1}{2}$ we have $\alpha_1 = 1 - e^{-1}$

$$\alpha_3 = 3(1 - 3e^{-1})$$

$$\alpha_5 = 5(7 - 19e^{-1})$$

$$\alpha_7 = 7(71 - 193e^{-1})$$

$$\alpha_9 = 9(1001 - 2721e^{-1})$$

$$\alpha_{11} = 11(18089 - 49171e^{-1})$$

Since the coefficients decrease monotonically to zero each successive bracket contains a continuously improving rational approximation to e. Similar approximations to powers of e may be derived by inserting other values of b into the recurrence formula. The recurrence formula itself is related to the simple continued fraction of Gauss for the exponential function.

(i) $f(r) = \dfrac{1}{[1 - (r^2/a^2)]^{\frac{1}{2}}} \qquad r < a$

$\quad = 0 \qquad\qquad\qquad\qquad a < r < 1$

An "impossible" aperture distribution due to the divergence to infinity at $r = a$

$$\alpha_{2s+1} = \int_0^1 \frac{R_{2s}^0(r)\,r}{[1 - (r^2/a^2)]^{\frac{1}{2}}}\,dr = \frac{2}{a}\sin[(2s+1)\sin^{-1}a], \qquad a < 1$$

(Ref. 7, p. 278, No. 23)

then $(-1)^s \alpha_{2s+1} = T_{2s+1}(a)$ the Chebychev polynomial of the first kind. Then

$$f(r) = \frac{1}{[1-(r^2/a^2)]^{\frac{1}{2}}} = \sum_{s=0}^{\infty} \frac{2}{a} T_{2s+1}(a) R_{2s}^0(r),$$

and hence

$$g(u) = \sum_{s=0}^{\infty} 2(-1)^s T_{2s+1}(a) \frac{J_{2s+1}(u)}{u}.$$

Since we know the result for $a = 1$ (Ref. 7, p. 7)

$$f(r) = \frac{1}{(1-r^2)^{\frac{1}{2}}}; \qquad g(u) = \frac{\sin u}{u},$$

we obtain the "scaled" result

$$f(r) = \frac{1}{[1-(r^2/a^2)]^{\frac{1}{2}}}; \qquad g(u) = \frac{\sin au}{au}$$

Hence

$$\frac{\sin au}{au} = 2 \sum_{s=0}^{\infty} (-1)^s T_{2s+1}(a) \frac{J_{2s+1}(u)}{u}. \tag{3.55}$$

(j) $f(r) = \dfrac{1}{r}$

We apply the transformation rule given in Eqn (3.43) to the last result to obtain

$$g(u) = 2 \sum_{s=0}^{\infty} \frac{J_{2s+1}(u)}{u}. \tag{3.56}$$

(Since $T_{2s+1}(1) = 1$ and the factor $(-1)^s$ in Eqn (3.54) is cancelled). In this case $g(u)$ is also known (Ref. 7, p. 7) to be

$$g(u) = J_0(u) + \frac{\pi}{2} \{J_1(u) H_0(u) - J_0(u) H_1(u)]$$

(k) $f(r) = (1-r^2)^p$.

The quadratic functions from which were derived the Lambda function in Eqn (3.25).

$$\alpha_{2s+1} = \frac{(-1)^s (2s+1)(p!)^2}{(p+s+1)!(p-s)!}, \text{ (Ref. 7, p. 230, No. 32)} \tag{3.57}$$

The resultant series can be shown to be equivalent to a Lambda function by the application of the reduction formulae for Bessel functions (Ref. 6, p. 143). Non integral values of p can now be considered by replacing the factorials in Eqn (3.57) by their gamma function equivalents.

To this list may be added the considerable number of finite transforms of functions, many of which apply to special cases of amplitude distributions to be found in Erdelyi. (Ref. 7, Sections 8.2, 8.3, 8.5 and 8.7).

3.4.2 Aperture distribution—transformation of series description

The derivation of the radial in-phase amplitude distribution for the circular aperture from a radiation pattern given in the form of a Neumann series, entails the evaluation of the series given by Eqn (3.41) from the known coefficients α_{2s+1}. This procedure can be simplified through an application of the recurrence relation for these coefficients based upon the recurrence relations which exist for the circle polynomials themselves, (Appendix II). For the required distribution $f(r)$ we then have

$$\alpha_1 = 2 \int_0^1 f(r)\, r\, dr,$$

$$\alpha_3 = 12 \int_0^1 f(r)\, r^3\, dr - 3\alpha_1,$$

$$\alpha_5 = 60 \int_0^1 f(r)\, r^5\, dr - 5\alpha_3 - 10\alpha_1,$$

$$\alpha_7 = 280 \int_0^1 f(r)\, r^7\, dr - 7\alpha_5 - 21\alpha_3 - 35\alpha_1.$$

$$(3.58)$$

If now $f(r)$ is assumed to be in the form

$$f(r) = \sum_{p=0} b_p (1 - r^2)^p,$$

substitution into Eqn (3.58) gives

$$\alpha_1 = \sum_p b_p/(p + 1); \qquad \alpha_3 = -\sum_p b_p\, 3p/(p + 1)(p + 2) \qquad (3.59)$$

$$\alpha_5 = \sum_p b_p\, 5p(p - 1)/(p + 1)(p + 2)(p + 3);$$

$$\alpha_7 = -\sum_p b_p\, 7p(p - 1)(p - 2)/(p + 1)(p + 3)(p + 4),$$

where the continuation of the terms is now obvious.

The number of terms in each of the series descriptions must be made the same, making the summations in Eqn (3.59) finite and hence the solution by a simple matrix inversion possible.

By means of a superposition of two radiation patterns the subsidiary lobes of which were substantially out of phase, and each being expressible as a Neumann series the radiation pattern shown in Fig. 3.3(b) can be obtained. The combination is found to be the series[15]

$$g(u) = 0.34\, J_1(u)/u - 0.44\, J_3(u)/u + 0.198\, J_5(u)/u - 0.022\, J_7(u)/u$$

which has residual side lobes less than -47 dB everywhere in the range $u > 7$ and has not resulted in a greatly increased 3 dB beamwidth. Using the above procedure it is found that the amplitude distribution required to produce this highly desirable pattern is the function

$$f(r) = 0.076 - 0.0441(1 - r^2) + 0.528(1 - r^2)^2 + 0.44(1 - r^2)^3$$

and which is shown in Fig. 3.3c.

3.5 PHASE ERRORS

There are two aspects to the inclusion of phase errors into an otherwise aberration free optical system which are those applicable to microwave systems and those of purely optical systems. In the latter the main concern is with the purity of the image formed over a small area at the nominal focus of the system. For this purpose the object illumination is usually that of an incoming plane wave which is approximately uniform in amplitude over the final aperture of the system. The effects of numerous high order phase errors for uniform illumination are studied in great detail in the literature of optics as for example in Linfoot[16] and it is to be noted that this study forms one of the most important applications of the circle polynomials.

In a microwave antenna we are less concerned with the focal distribution than with the radiated pattern and in either case with only the lowest orders of aberration. This could be put simply down to the fact of the much poorer definition that the focal spot has at microwave frequencies than at light frequencies. In the microwave case the non-uniformity of amplitude is the pre-requisite study of the system which is the reason for "switching" the circle polynomial treatment from the phase distribution as in optical theory to the amplitude distributions of the theory presented above. In practice only the two primary aberrations, spherical aberration and coma, cause any concern in a microwave system and in general only make an appearance when the feed

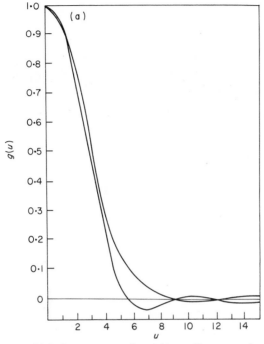

(a) *Radiation patterns from arbitrary Neumann series.*

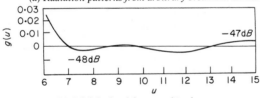

(b) *Sidelobe level from combined patterns.*

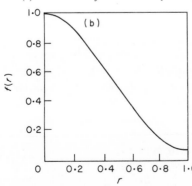

(c) *Aperture distribution for pattern with low side lobes.*

FIG. 3.3. *Radiation Pattern with ultra-low sidelobes.*

is displaced from the focus by a considerable fraction of the wavelength. To this extent it is possible to include in the circle polynomial method for amplitude distributions an additional circle polynomial method for these fundamental phase perturbations. The result, as before, is required in a closed form series expression capable of evaluation from tabulated functions by elementary computational methods.

3.5.1 Quadratic phase distribution

The symmetrical phase distributions can be introduced into the analysis by putting the complex amplitude function in the aperture in the form

$$f(r, \phi') = f(r)\, e^{-i\psi(r)}$$

in which $\psi(r)$ is the phase along any radial coordinate as obtained directly from the optical path length along rays from the source to the aperture of the system (Fig. 3.2(c)).

In the optics of a system of rays tending towards a single point focus, the phase over the spherical wave-front is given by $\psi(r) = br^2$, and in the literature of optics the aberration is considered to be perturbations from the purely spherical, that is primary spherical aberration is $O(r^4)$, secondary $O(r^6)$ and so on. In a microwave antenna focused at infinity the unperturbed wave front is plane, and obtained from an exactly focused source. If this is then perturbed by an axial movement of the source, the phase error introduced is proportional to (by a factor $2\pi/\lambda$) the departure from a straight line of the focal line given in Chapter 1, Section 7.1(e) (page 61). To the first order this line is circular and thus $\psi(r)$ again can be represented by

$$\psi(r) = br^2$$

where b now contains the proportionality factor $2\pi/\lambda$ and the radius of curvature of the wave front. As a consequence of this dual description we are able to use the entire theory of unperturbed optics to investigate the effect of spherical aberration in a microwave system. However, the optical literature rarely includes the non-uniformity of amplitude given by the function $f(r)$ above, and which is essential in a microwave system. This, therefore, requires extensions to the usual optical theory of aberrations.

(a) *Uniform amplitude distribution*
    Putting $f(r) = 1$ we require the evaluation of

$$g(u) = \int_0^1 e^{-ibr^2}\, J_0(ur)\, r\, dr \tag{3.60}$$

In Linfoot (Ref. 16, p. 54), in which it is called the aberration free case, use is made of Bauer's formula (Ref. 6, p. 128)

$$\exp(iz \cos \theta) = \left(\frac{\pi}{2z}\right)^{\frac{1}{2}} \sum_{n=0}^{\infty} (2n + 1)\, i^n\, J_{n+\frac{1}{2}}(z)\, P_n(\cos \theta) \qquad (3.61)$$

then

$$\exp(-ibr^2) = \exp(-\tfrac{1}{2}ib)\exp\left[-\tfrac{1}{2}ib(2r^2 - 1)\right]$$

$$= \exp(-\tfrac{1}{2}ib)\left(\frac{\pi}{b}\right)^{\frac{1}{2}} \sum_{n=0}^{\infty} (2n + 1)(-i)^n\, J_{n+\frac{1}{2}}\left(\frac{b}{2}\right) R_{2n}^0(r) \qquad (3.62)$$

This is then a complex radial amplitude already put in the form of the series of circle polynomials required by the theory of Section 3.3 and the resulting Eqns (3.41). Hence we have directly the coefficients α_{2s+1} and

$$g(u) = \exp(-\tfrac{1}{2}ib)\left(\frac{\pi}{b}\right)^{\frac{1}{2}} \sum_{n=0}^{\infty} (2n + 1)\, i^n\, J_{n+\frac{1}{2}}\left(\frac{b}{2}\right) J_{2n+1}(u)/u \qquad (3.63)$$

the appropriate Neumann series.

It is also to be noted that by operating on the integrand of Eqn (3.60) by the transformation given in Eqn (3.43) that is $r^2 \to (1 - r^2)$ transforms both

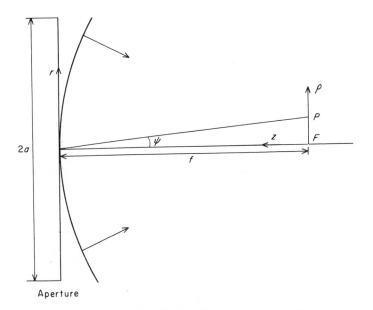

Aperture

FIG. 3.4. Spherical phase front.

Eqn (3.60) and the result in Eqn (3.63) into their respective complex conjugates. The absolute radiated power distribution which is $P(u) = \{g(u)g^*(u)\}^{\frac{1}{2}}$ is then the same for both cases. That is the radiative power pattern is distorted to the same degree by positive or negative values of the parameter b, that is (approximately) by motions of the feed source away from the focus in either a positive or negative axial direction.

Being a situation involving uniform amplitude distribution the major application of this result is in the focal distribution of a circular aperture antenna, which is receiving a plane wave normally incident upon it. This requires a re-interpretation of the parameters u and b. The parameter b is the sphericity of the wave front contracting to the focus (Fig. 3.4) and is given by[16]

$$b = \frac{2\pi\, a^2 z}{\lambda f^2},$$

where a is the radius of the aperture, f the (geometrical) focal length and z the axial coordinate with respect to an origin at the focus. In this coordinate system u becomes

$$u = \frac{2\pi}{\lambda} \cdot \frac{a\rho}{f},$$

where, because of the circular symmetry, (ρ, z) are the polar coordinates required at this origin. This definition of u is not remote from the previous one since if ψ is the angle subtended by an off-axis point in the plane of the focus, at the centre of the aperture, the above definition of u becomes approximately

$$u = \frac{2\pi a}{\lambda} \sin \psi.$$

The complete description of the field in the region of the focus is now obtained from Eqn (3.63). In planes of constant z the pattern is a Neumann series and thus has the appearance of the usual radiation pattern, that of a central peak and decreasing amplitude side lobes. This has the usual appearance of a bright spot surrounded by rings of decreasing illumination. In the particular plane $z = 0$ this pattern is the usual Airy ring pattern proportional to $J_1(u)/u$.

In the axial direction we can put $u = 0$ and the result is proportional to $\sin(b/2)/(b/2)$.

Thus the entire ring pattern obeys a $\sin X / X$ variation along the axis resulting in a system of elongated bright spots surrounded by similarly elongated but decreasingly bright spheroidal like surfaces (Fig. 3.5). The centre of the brightest region is the nominal focus, and is the obvious position for a receiving element. Further contributions to the received energy could be obtained by matching the receiving element to the right pattern. Identical results are obtained from the complete field analysis of receiving antennas.[18]

It can be seen from the above that, whereas the additional contribution to be expected from the first subsidiary ring of the pattern is − 17 dB on the centre illumination, since it is a $J_1(X)/X$ function, the additional energy that could be expected from a subsidiary feed placed at the secondary maximum along the axis would be only − 13 dB since it derives from a sin X/X function. This amounts to only a slight increase in the total of the received energy of approximately $\frac{1}{4}$ dB but the axial distribution can be used to play another important part in the requirements of large antennas, that of defining with great precision the operational axis of the antenna.[19]

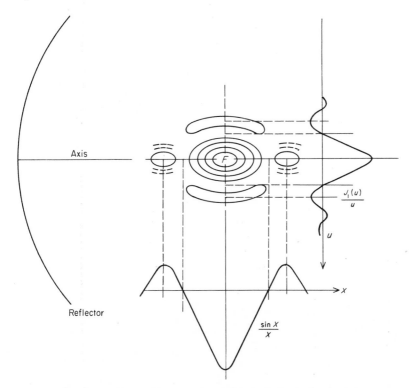

FIG. 3.5. Field in focal region of reflector with incident axial plane wave.

Usually in the case of paraboloidal dishes the pointing direction of the axis of the antenna is defined by the mechanical construction and can be in error in so far as the shape of the reflector surface is not known with great accuracy over its entire area. This is very pronounced in those cases where the antenna changes shape as does a large paraboloid in its different positions for example when used in radio astronomy. The axial distribution of field is however a function of the entire surface at all times and a sampling of this distribution under the condition of reception of a normally incident plane wave, will give an exact location of this axis. To do this, a second subsidiary receiving element is placed behind the main receiving element at a position given by the second maximum of the sin X/X distribution. This position, and X in this instance, is a function of the geometry of the paraboloid and can be sufficiently large to cause only a minor blockage of the second element by the first. Then a comparison of the energy received by the second feed with a sample of equal magnitude of that received by the first will only give precise cancellation when the two feeds are exactly in the direction of the electrical axis of the antenna which in turn has to be exactly normal to the incident plane wave. The apparatus for a small paraboloid is shown in Fig. 3.6(a). The axial field measured is shown in Fig. 3.6(b) and the null effect in Fig. 3.6(c). This null, as indicated on a meter, was sensitive even to the vibration of the paraboloid caused by a light tap on its rim, and thus to extremely small departures of the surface from its true shape.

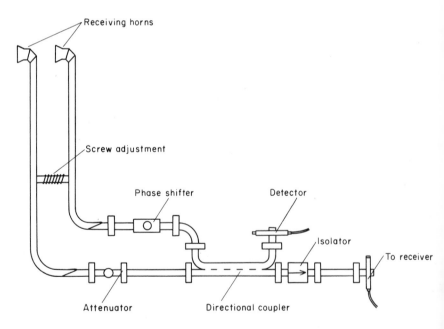

FIG. 3.6(a). *Feed arrangement for receiving secondary maximum for determination of reflector axis.*

The alignment procedure which suggests itself is to focus the antenna in the direction of a distant source, or along the axis of any source sufficiently distant to give an assumed incident plane wave, and to adjust the direction until the null position of the two receiving elements is exactly in line with the distant source. The line defined by the two receivers will then be the exact axis of the optical system and the first of these will be at the exact focus. The procedure being extremely sensitive as illustrated above and requires the use of fine adjustment mechanisms.

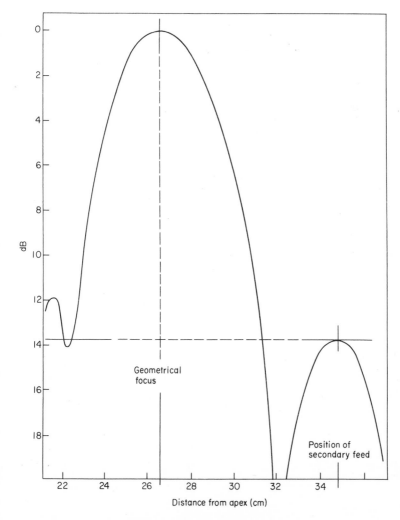

FIG. 3.6(b). Measured field along paraboloid axis.

Subsequently, a change in phase can be made which will allow the addition of the energy in the second receiving element to that of the first instead of the subtraction to give the additional $\frac{1}{4}$ dB of gain.

(b) *Near and far field patterns*
   The two situations in which the quadratic phase error has to be considered when applied to a *non-uniform* amplitude distribution are those of the far field radiation pattern of an axially defocused antenna and also the near field pattern of a correctly focused antenna.
   In the latter case, when the measurement of the field is made in the near

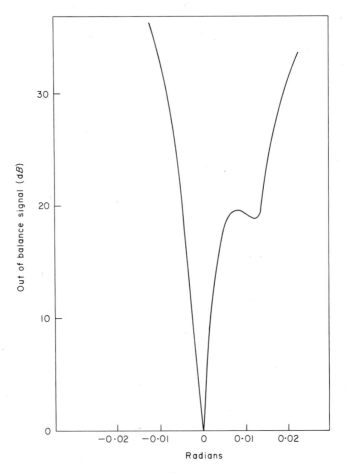

FIG. 3.6(c). Error signal of alignment of two feed system in paraboloid receiver.

field, that is up to approximately $a^2/2\lambda$, the error in phase that is created has the same parabolic form as in the former situation. At such close-in points however modifications to the obliquity factor as applied to the amplitude distribution have also to be considered.[20]

The general series solution for the integral

$$g(u) = \int_0^1 f(r) e^{-ibr^2} J_0(ur) \, r \, dr \qquad (3.64)$$

can be obtained by the classical method involving the use of the Lommel functions of two variables for uniform illumination and modified circle polynomials for the non-uniform extension. This has been shown to be possible by Hu,[21] but still requires the simplification of expressing $f(r)$ as the series of terms $(1 - r^2)^p$. If we have thus to confine ourselves to particular forms of $f(r)$ a method due to de Size[22] is of great value.

We consider amplitude distributions of the form

$$f(r) = \left(\cos \frac{\pi r}{2}\right)^p \qquad \text{for } p = 0, 1, \text{and } 2.$$

For $p = 0$ $f(r) = 1$ and the corresponding radiation pattern is $g(u)$ given in Eqn (3.63) we will term this pattern $g(u, b)$.

For $p = 1$ we can put, with an error of less than 2% in the range $0 < r < 1$

$$\cos\left(\frac{\pi r}{2}\right) \simeq 1 - \sqrt{2} \sin\left(\frac{\pi r^2}{4}\right)$$

Then

$$g(u)_{p=1} = \int_0^1 \left\{1 - \sqrt{2} \sin\left(\frac{\pi r^2}{4}\right)\right\} e^{-ibr^2} J_0(ur) \, r \, dr$$

$$= \int_0^1 e^{-ibr^2} J_0(ur) r \, dr + \frac{i}{\sqrt{2}} \int_0^1 \exp\left[-i(b - \tfrac{1}{4}\pi)r^2\right] J_0(ur) r \, dr$$

$$- \frac{i}{\sqrt{2}} \int_0^1 \exp\left[-i(b + \tfrac{1}{4}\pi)r^2\right] J_0(ur) r \, dr \qquad (3.65)$$

and hence

$$g(u) = g(u, b) + \frac{i}{\sqrt{2}} g(u, b - \tfrac{1}{4}\pi) - \frac{i}{\sqrt{2}} g(u, b + \tfrac{1}{4}\pi)$$

Repeating the procedure for

$$f(r) = \cos^2\left(\frac{\pi r}{2}\right)$$

gives

$$g(u)_{p=2} = g(u, b) + \frac{i}{\sqrt{2}} g(u, b - \tfrac{1}{4}\pi) - \frac{i}{\sqrt{2}} g(u, b + \tfrac{1}{4}\pi)$$

$$-\tfrac{1}{4} g(u, b - \tfrac{1}{2}\pi) + \tfrac{1}{4} g(u, b + \tfrac{1}{2}\pi) \qquad (3.66)$$

This method applies of course to any derivation of $g(u, b)$ that can be made and can be extended if necessary to higher values of p.

(c) Asymmetric phase error

The general description of the phase surface that is created when the source is moved away from its focal position in the plane transverse to the axis, can be given in simplified form as

$$f(r, \phi') = f(r) \exp(-ia_n r^n \sin \phi')$$

where for highly complex shapes more than one value of n may be required and a superposition of terms used. In this expression a_n is a phase angle in radian measure and the phase surface has the appearance given in Fig. 3.2(b).

The lowest order of these terms that is generally needed in a microwave antenna analysis is that given by the function

$$f(r, \phi') = f(r) \exp(-iar^3 \sin \phi'),$$

which approximates the focal line derived in Section 7.1(f) of Chapter I (p. 61).

Substitution into the scalar integral and evaluation in the plane containing the offset source gives

$$g\left(u, \frac{\pi}{2}\right) = \int_0^{2\pi} \int_0^1 f(r) \exp(iur \sin \phi') \exp(-iar^3 \sin \phi') r \, dr \, d\phi'. \qquad (3.67)$$

The procedure adopted in optical theory[23] is to expand the second exponential term as a series of sines and cosines of multiples of ϕ', thus giving rise to a series of higher order Hankel transforms of the function $f(r)$. For uniform amplitude distributions $f(r) = 1$ this series simplifies greatly if the phase function ar^3 is replaced by a circle polynomial $R_3^1(r) = 3r^3 - 2r$ (Ref. 16, p. 57).

The method to be given here is to combine the two exponential terms and expand the resultant Bessel function by the addition theorem. Each term of the resultant series can then be evaluated by the method of circle polynomials individually. This can be done for the more general phase function $a_n r^n \sin \phi'$. Then we have[24]

$$g\left(u, \frac{\pi}{2}\right) = 4\pi \sum_{m=0}^{\infty}{}' (-1)^m \int_0^1 f(r) J_m(a_n r^n) J_m(ur) r \, dr \qquad (3.68)$$

$$= 4\pi \sum_{m=0}^{\infty}{}' \sum_{s=0}^{\infty} \frac{(-1)^s}{s!(m+s)!} \left(\frac{\pm a_n}{2}\right)^{m+2s} \int_0^1 f(r) r^{n(m+2s)} J_m(ur) r\, dr \qquad (3.69)$$

As always Σ' refers to the sum when the first term ($m = 0$) is multiplied by the factor 0·5.

Choosing a short series of terms like $(1 - r^2)^p$ to describe $f(r)$ makes the evaluation of the integrals elementary. Then putting $f(r) = j + kr^2 + lr^4$ we have the final Neumann series

$$g\left(u, \frac{\pi}{2}\right) = A_1 J_1(u)/u \pm A_2 J_2(u)/u + A_3 J_3(u)/u \pm \ldots \qquad (3.70)$$

where the coefficients A_i are given in Table I.

Table 1. Coefficients for asymmetric patterns

$$A_1 = \left(\frac{j}{2} + \frac{k}{4} + \frac{l}{6}\right) - \frac{a_n^2}{8}\left(\frac{j}{(n+1)} + \frac{k}{(n+2)} + \frac{l}{(n+3)}\right) + \frac{a_n^4}{128}\left(\frac{j}{(2n+1)} + \frac{k}{(2n+2)}\right.$$
$$\left. + \frac{l}{2n+3}\right) - \frac{a_n^6}{4608}\left(\frac{j}{3n+1} + \frac{k}{3n+2} + \frac{l}{3n+3}\right)$$

$$A_2 = 2a_n\left(\frac{j}{n+3} + \frac{k}{n+5} + \frac{l}{n+7}\right) - \frac{a_n^3}{4}\left(\frac{j}{2n+3} + \frac{k}{2n+5} + \frac{l}{2n+7}\right)$$
$$+ \frac{a_n^5}{96}\left(\frac{j}{5n+3} + \frac{k}{5n+5} + \frac{l}{5n+7}\right)$$

$$A_3 = -\left(\frac{k}{4} + \frac{l}{4}\right) - \frac{3a_n^2}{4}\left(\frac{3j}{2n+4} + \frac{3k}{2n+6} + \frac{3l}{2n+8} - \frac{j}{2n+2} - \frac{k}{2n+4} - \frac{l}{2n+6}\right)$$
$$- \frac{a_n^4}{64}\left(\frac{4j}{4n+4} + \frac{4k}{4n+6} + \frac{4l}{4n+8} + \frac{j}{2n+2} + \frac{k}{2n+4} + \frac{l}{2n+6} - \frac{j}{4n+2} - \right.$$
$$\left. - \frac{k}{4n+4} - \frac{l}{4n+6}\right)$$

$$A_4 = -4a_n\left(\frac{3j}{n+5} + \frac{3k}{n+7} + \frac{3l}{n+9} - \frac{2j}{n+3} - \frac{2k}{n+5} - \frac{2l}{n+7}\right)$$
$$+ \frac{a_n^3}{3}\left(\frac{5j}{3n+5} + \frac{5k}{3n+7} + \frac{5l}{3n+9} - \frac{j}{3n+3} - \frac{k}{3n+5} - \frac{l}{3n+7}\right)$$
$$- \frac{a_n^5}{96}\left(\frac{7j}{5n+5} + \frac{7k}{5n+7} + \frac{7l}{5n+9} - \frac{4j}{5n+3} - \frac{4k}{5n+5} - \frac{4l}{5n+7}\right)$$

For those situations where the amount of the perturbation is small and the function more accurately a cubic curve, $n = 3$ in the above and the effect

on the pattern is mainly that of a shift of the maximum by a small amount
from the axial direction $u = 0$. The angular amount of this shift does not
agree precisely with the angular offset of the source which creates the cubic
wave-front. This difference or "squint" is of prime interest in those antennas
where exact pointing accuracy is a requisite and which uses more than one
system of sources dispersed about the true focus of the antenna. The amount
of squint can be derived directly from Eqn (3.70) by differentiation and the
evaluation of the coefficients in the above table (see additional bibliography).

(*d*) *Higher order symmetric errors*

An expression has been derived for the evaluation of a uniformly illuminated circular aperture pattern in which the phase error is a symmetrical
function of the radial coordinate alone, that is

$$f(r, \phi') = \exp(-i\beta_n r^n).$$

Substitution is therefore possible into the symmetric integral

$$g(u) = \int_0^1 \exp(-i\beta_n r^n) J_0(ur) r \, dr.$$

Evaluation is obtained in closed form[25] by the expansion of both functions
in the integrand and integration term by term of the double summation resulting. One of the ensuing summations then takes the form of a confluent
hypergeometric function (Ref.6, p. 100). The result is given by

$$g(u) = \pi b^2 \sum_{m=0}^{\infty} \frac{(-1)^m}{(m!)^2} (m+1) \left(\frac{u}{2}\right)^{2m} {}_1F_1(a_m, c_m, -i\beta_n), \tag{3.71}$$

where $_1F_1$ is Kummer's confluent hypergeometric function

$${}_1F_1(a_m, c_m, X) = 1 + \frac{a_m}{c_m} X + \frac{a_m(a_m+1)}{c_m(c_m+1)} \frac{X^2}{2!} + \cdots \tag{3.72}$$

In the analysis given

$$a_m = \frac{2(m+1)}{n} \quad \text{and} \quad c_m = 1 + a_m$$

leading to a considerable simplification of this result.

Results are illustrated in the reference for powers of r up to $n = 5$ and
thus include the important quadratic and quartic terms of spherical abberra-
tion.

3.6 TRANSFORMATION TO ELLIPTICAL APERTURES

A common modification of the purely circular paraboloid antenna is a
section of the paraboloid having an elliptical aperture. All of the foregoing
analysis can be made directly applicable to elliptical distributions by the use
of a projection transformation which converts the elliptical aperture into a
circular one. As illustrated in Fig. 3.7, the elliptical aperture may be taken
to lie in the x, y plane with z axis normal at its centre and with major axis,
2a, along the x axis. The minor axis will then be $2a \cos \alpha$ where α is the angle
between the plane containing the circular projection of the ellipse and the
x, y plane. In the plane containing the circle we use the coordinates (r, ϕ')

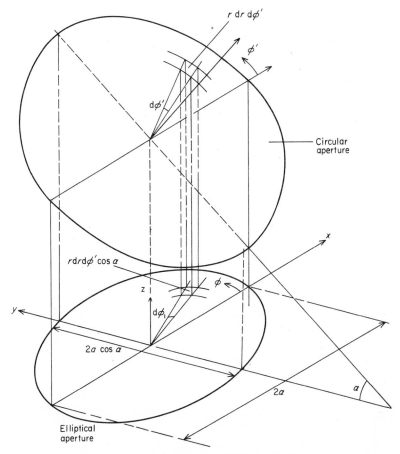

FIG. 3.7. Projection into elliptical aperture.

as in the previous analysis and the field point at infinity has angular coordinates (θ, ϕ) with respect to the z axis. A wave with wave vector $\mathbf{k} = (k_x, k_y, k_z)$ in this coordinate system has

$$k_x = k \sin \theta \cos \phi,$$

$$k_y = k \sin \theta \sin \phi,$$

$$k_z = k \cos \theta, \qquad k = 2\pi/\lambda,$$

and in the plane of the ellipse

$$x = ar \cos \phi',$$

$$y = ar \cos \alpha \sin \phi' \qquad 0 < r < 1.$$

The increment of area in the aperture plane is therefore

$$dA = a^2 \cos \alpha \, r \, dr \, d\phi',$$

and the scalar integral for the aperture is

$$g(\theta, \phi) = \iint_A f(x, y) \exp \left[-i(k_x x + k_y y) \right] dA.$$

The aperture distribution in the elliptical aperture $f(x, y)$ transforms to an aperture distribution $F(r, \phi')$ in the circular aperture, and thus

$$g(\theta, \phi) = a^2 \cos \alpha \int_0^1 \int_0^{2\pi} F(r, \phi')$$

$$\times \exp \left\{ -ikar \sin \theta (\cos \phi \cos \phi' + \sin \phi \sin \phi' \cos \alpha) \right\} r \, dr \, d\phi' \qquad (3.73)$$

Putting

$$u = ka \sin \theta (\cos^2 \phi + \cos^2 \alpha \sin^2 \rho)^{\frac{1}{2}}$$

$$= ka \sin \theta (1 - \sin^2 \alpha \sin^2 \phi)^{\frac{1}{2}},$$

and

$$\phi_1 = \tan^{-1} (\cos \alpha \tan \phi)$$

we find

$$g(\theta, \phi) = a^2 \cos \alpha \int_0^1 \int_0^{2\pi} F(r, \phi') \exp \left[-iur \cos (\phi_1 - \phi') \right] r \, dr \, d\phi' \qquad (3.74)$$

In this form all the previous theory can be adapted to the elliptical case.

3.7 THE MICROWAVE AXICON

As an illustrative example of the foregoing procedures, we investigate the effect of a conical phase distribution upon the radiated pattern of an already focused antenna. In practice this could be obtained by the insertion of a cone

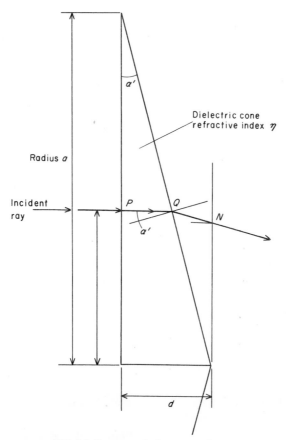

FIG. 3.8. Geometry of microwave axicon.

of wide angle (Fig. 3.8) into the near field of a paraboloid or lens. It has analogies with, but is not identical to, the optical axicons of McLeod.[26] In the optical case the axicon is fed directly from a point-source and consists of a glass cone which forms a continuous line of images from a point source. This effect had been used for alignment with great precision over considerable distances before the advent of the laser. The question arises whether an effect akin to this occurs in the smaller apertures at microwave frequencies.[27]

Considering the dielectric cone of Fig. 3.8 as having a phase effect alone, that is discounting reflections from the surfaces, then for a typical ray *PQN* the phase is given by

$$\Phi = 2\pi(\eta PQ + QN)/\lambda + \text{constant}, \tag{3.75}$$

where η is the refractive index of the material of the cone.
With Snell's law of refraction at Q,

$$\eta PQ + QN = \eta t - \eta r \tan \alpha + \frac{r \tan \alpha}{\cos \alpha \sqrt{(1 - \eta^2 \sin^2 \alpha)} + \eta \sin^2 \alpha}$$

$$= \eta t + b'r \tag{3.76}$$

where b' is defined by Eqn (3.76) and is a constant for any given cone. The phase thus has a linear radial variation. The order of magnitude of b' can be assessed by considering cones of wide vertex angle (i.e. small α), and therefore

$$b' \simeq (1 - \eta) \tan \alpha.$$

Substitution into Eqn (3.1) and with $k = 2\pi/\lambda$ gives

$$g(u) = \exp(-jk\eta t) \int_0^1 a(\rho) \exp(-jkb'a\rho)J_0(u\rho)\rho \, d\rho.$$

Ignoring the constant factor $\exp(-jk\eta t)$,

$$g(u) = \int_0^1 a(\rho) \exp(-jb\rho)J_0(u\rho)\rho \, d\rho, \tag{3.77}$$

where

$$b \simeq \frac{2\pi}{\lambda} a(1 - \eta) \tan \alpha,$$

and is therefore negative for dielectric materials with $\eta > 1$. The analysis, however, is even with respect to b, and the sign can be either positive or negative. The exact value of b can be obtained by inserting into Eqn (3.77) the more complicated expression given by Eqn (3.76). This is immaterial as b is a constant and only its order of magnitude is required. For cones of the usual plastic material, e.g. polythene or polystyrene, $\eta = 1\cdot6$. With $\tan \alpha$ of the order of $0\cdot25$, b has a magnitude of a/λ (i.e. the ratio of aperture radius to wavelength).

The integral in Eqn (3.77) can thus be evaluated by the technique of expansion in terms of the circle polynomials

$$g(u) = \int_0^1 a(\rho) \, e^{-jb\rho} J_0(u\rho)\rho d\rho = \sum_{n \text{ even}} \alpha_n (-1)^{n/2} \frac{J_{n+1}(u)}{u}, \qquad (3.78)$$

where

$$\alpha_n = 2(n + 1) \int_0^1 e^{-jb\rho} a(\rho) R_n^0(\rho)\rho d\rho. \qquad (3.79)$$

In this equation $a(\rho)$ is the real amplitude distribution.

We shall only be considering amplitude distributions $a(\rho)$, which are simple quadratic functions of ρ such as $(1 - \rho^2)^p$ for $p = 0$ and 1. Consequently, Eqn (3.79) can be expanded and integrated term by term. This requires the summation of terms involving powers of ρ of the form

$$I_n = \int_0^1 e^{-jb\rho} \rho^n d\rho, \qquad (3.80)$$

and this gives the reduction formula

$$\left. \begin{aligned} I_n &= -\frac{e^{-jb}}{jb} + \frac{n}{jb} I_{n-1}, \\ I_0 &= \frac{e^{-jb}}{jb} + \frac{jb}{1}, \end{aligned} \right\} \qquad (3.81)$$

hence

$$I_n = -e^{-jb} \left\{ \frac{1}{jb} + \frac{n}{(jb)^2} + \frac{n(n-1)}{(jb)^3} + \dots + \frac{n!}{(jb)^n} + \frac{n!}{(jb)^{n+1}} \right\} + \frac{n!}{(jb)^{n+1}}. \qquad (3.82)$$

3.7.1 Theoretical far-field patterns

Uniform illumination

We have, from Eqns (3.78) and (3.79)

$$g(u) = \alpha_0 \frac{J_1(u)}{u} - \alpha_2 \frac{J_3(u)}{u} + \alpha_4 \frac{J_5(u)}{u}, \qquad (3.83)$$

for the first three terms. This confines the range of validity of u to those values for which $J_7(u)$ can be ignored with respect to $J_5(u)$ and for values of b for which the coefficients are either decreasing or remaining constant. Small values of b are therefore not suitable, since they would require an extension of the series of coefficients and correspondingly higher orders of J_n, possibly without convergence. Values of b greater than 5, however, are suitable for the first few coefficients, and hence for the series given by Eqn (3.83). The range of u that can be then justified is approximately $u < 6$. This covers the region of the main lobe ($u < 4$) and the first side lobe, ($u \simeq 5.5$) adequately. With $a(\rho) = 1$ in Eqn (3.79), we obtain from Eqn (3.82)

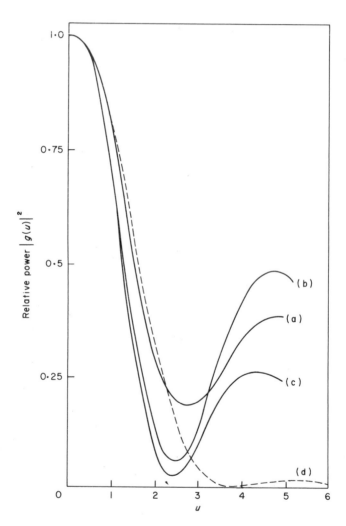

FIG. 3.9. *Microwave axicon. Uniform illumination. Theoretical patterns*; (a) $b_m = 1.5\pi$, (b) $b_m = 3.5\pi$, (c) $b_m = 5.5\pi$, (d) $g(u) \propto J_1(u)/u$.

$$\alpha_0 = 2e^{-jb}\left(\frac{j}{b} + \frac{1}{b^2}\right) - \frac{2}{b^2}$$

$$\alpha_2 = 6e^{-jb}\left(\frac{j}{b} + \frac{5}{b^2} - \frac{12j}{b^3} - \frac{12}{n^4}\right) + 6\left(\frac{12}{b^4} + \frac{1}{b^2}\right)$$

$$\alpha_4 = 10e^{-jb}\left(\frac{j}{b} + \frac{13}{b^2} - \frac{84j}{b^3} - \frac{324 + 720j}{b^4}{b^5} + \frac{720}{b^6}\right)$$

$$+ 10\left(-\frac{720}{b^6} - \frac{36}{b^4} - \frac{1}{b^2}\right)$$

To simplify the computation, and without loss of generality, b can be taken to have the particular values $b_m = (2m + \frac{3}{2})\pi$ for which $\exp(-jb_m) = -j$.
Thus

$$|g(u)|^2 = \left\{\left(\frac{2}{b_m} - \frac{2}{b_m^2}\right)\frac{J_1(u)}{u} - 6\left(\frac{1}{b_m} + \frac{1}{b_m^2} - \frac{12}{b_m^3} + \frac{12}{b_m^4}\right)\frac{J_3(u)}{u}\right.$$

$$+ 10\left(\frac{1}{b_m} - \frac{1}{b_m^2} - \frac{84}{b_m^3} - \frac{36}{b_m^4} + \frac{720}{b_m^5} - \frac{720}{b_m^6}\right)\frac{J_5(u)}{u}\Bigg\}^2$$

$$+ \left\{-\frac{2}{b_m^2}\frac{J_1(u)}{u} - 6\left(-\frac{5}{b_m^2} + \frac{12}{b_m^4}\right)\frac{J_3(u)}{u}\right.$$

$$+ 10\left(-\frac{12}{b_m^2} + \frac{324}{b_m^4} - \frac{720}{b_m^6}\right)\frac{J_5(u)}{u}\Bigg\}^2$$

Patterns for $b_m = \frac{3}{2}\pi$, $\frac{7}{2}\pi$ and $\frac{11}{2}\pi$ are shown in Fig. 3.9.

It can be seen from consideration of the terms in b_m^{-1} that $|g(u)|^2$ will have a minimum value approximately when $J_1(u)/u = 3J_3(u)/u$; i.e. at $u \simeq 2.5$. It is the effect of this forced minimum that causes the beamwidth to be reduced by the ratio of $2.5 : 3.8$, where 3.8 is the first zero value of $J_1(u)/u$ above, although this does not appear to become fully effective until b is greater than 3.5π. This effect and the same ratio will be shown to be obtained with other amplitude distributions.

3.7.2 Tapered illumination

With the illumination law $a(\rho) = 1 - \rho^2$ and the same considerations as above, we obtain

$$\alpha_0 = 2(I_1 - I_3),$$

$$\alpha_2 = 6(2I_5 + 3I_3 - I_1),$$

$$\alpha_4 = 10(-6I_7 + 12I_5 - 7I_3 + I_1),$$

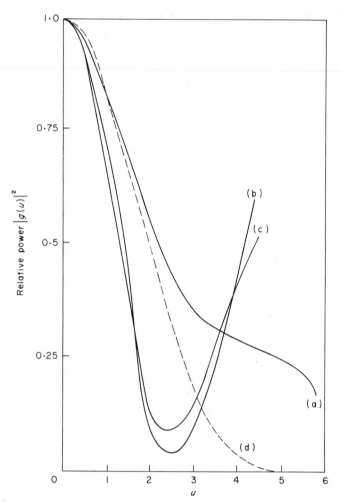

FIG. 3.10. Microwave axicon with tapered illumination $[(f(r) = 1 - r^2]$; (a) $b_m = 1\cdot\partial\pi$, (b) $b_m = 3\cdot 5\pi$, (c) $b_m = 5\cdot 5\pi$, (d) $g(u) \propto J_2(u)/u^2$.

which for $b_m = (2m + \frac{3}{2})\pi$, becomes

$$\alpha_0 = 2\left(-\frac{1}{b_m^2} - \frac{6}{b_m^3} - \frac{6}{b_m^4}\right) + 2j\left(-\frac{2}{b_m^2} + \frac{6}{b_m^4}\right),$$

$$\alpha_2 = 6\left(\frac{1}{b_m^2} - \frac{22}{b_m^3} + \frac{18}{b_m^4} + \frac{240}{b_m^5} + \frac{280}{b_m^6}\right)$$

$$+ 6j\left(\frac{2}{b_m^2} + \frac{102}{b_m^4} - \frac{240}{b_m^6}\right),$$

$$\alpha_4 = 10\left(-\frac{1}{b_m^2} - \frac{54}{b_m^3} - \frac{42}{b_m^4} + \frac{3600}{b_m^5} - \frac{1440}{b_m^6} - \frac{30240}{b_m^7} - \frac{30240}{b_m^8}\right)$$

$$= 10j\left(-\frac{2}{b_m^2} + \frac{582}{b_m^4} - \frac{13680}{b_m^6} + \frac{30240}{b_m^8}\right).$$

With these coefficients, the patterns given by Eqn (3.83) are illustrated in Fig. 3.10.

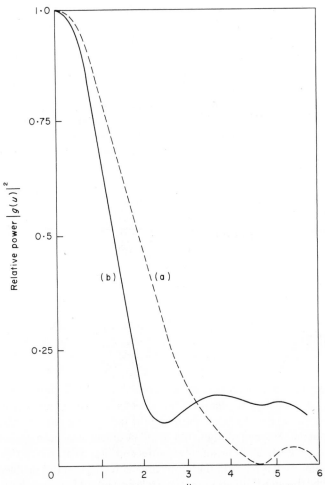

FIG. 3.11. *Microwave axicon experimental results.* (a) *without axicon,* (b) *with axicon.*

I

Again the leading terms in α_0 and α_2 are in the ratio $3:1$, giving the narrowing forced minimum.

Other illumination laws can be contrived to cause this effect. It appears to be general for functions $a(\rho) = (1 - \rho^2)^p$, and it occurs for $a(\rho) = 1 + \rho$, but not, strangely, for $a(\rho) = 1 - \rho$.

3.7.3 Experimental results and conclusions

Two experiments to confirm these effects have been carried out: namely, one in the near field and one in the far field. The radiation patterns evaluated above are essentially far-field patterns. The simple experiment to be performed is therefore to measure the radiation pattern of a focused antenna, and to repeat the measurement with a cone of dielectric supported in front of the antenna and in its near field. This experiment was carried out in x band with a cone of 45 cm. diameter and semivertex angle of approximately $76°$, giving b a value of approximately 9. Although the result shown in Fig. 3.11 confirms the theory precisely in the reduction of the beamwidth factor and forced minimum at $u = 2·5$, the sidelobes are smaller than expected, and a decrease of gain of nearly 5 dB occurred. However, some consideration has to be given to the fact that the actual radiating element was smaller in diameter than the axicon lens and some of the reduction could have been due to internal reflections giving an effectively larger radiating aperture. Nevertheless, the final radiation pattern has a beamwidth of half power, which is much smaller than that which could have been achieved with· an aperture of the largest dimension of the experimental apparatus and with uniform inphase illumination.

3.8 MODES IN CONFOCAL CAVITIES WITH PARABOLIC MIRRORS

It is shown in this section that the iterative integral equation of Fox and Li[28] for an open cavity resonator with confocal parabolic mirrors can be solved by a series approximation involving the use of the circle polynomials. Although not an antenna design, the same problem has been tackled by many workers in the fields of beam waveguides and laser cavities using series of other polynomials. It is relevant therefore, after showing that this problem can be approached through the medium of the circle polynomials, to see whether the other methods have an application to the scalar theory of antenna diffraction.

In order to use the Kirchhoff approximation in the *Fresnel* region we assume:[29]

1. that the dimensions of the resonator are large compared to the wavelength

2. that the field in the resonator is substantially transverse electromagnetic. Hence the scalar field can be represented by

$$U_p = \frac{ik}{4\pi} \int_A U_a \frac{e^{-ikR}}{R} (1 + \cos\theta)\, ds, \tag{3.84}$$

where θ is the angle R makes with the unit normal to the aperture, and where U_a is the amplitude and phase distribution on the surface of one mirror giving rise to a distribution U_p on the other. Taking the case of a symmetrical cavity where the mirrors have the same aperture and curvature, equation (3.84) will take the form

$$v(r_1\phi_1) = Y \int_A K(r_1, \phi_1, r_2, \phi_2) v(r_2, \phi_2)\, ds, \tag{3.85}$$

where

$$K = \frac{ik}{4\pi R} (1 + \cos\theta)\, e^{-ikR}, \qquad R = P_1 P_2,$$

and $v(r, \phi)$ is the self consistent field mode; that is, the radiation on one transit of the cavity will reproduce the amplitude and phase distribution over the mirror surface to within a constant Y, which contains the diffraction loss and phase change.

We require for resonance that the phase change for a single transit be π. The diffraction loss for a single transit $= 1 - 1/|Y|^2$.

Using the notation in Fig. 3.12. b is the separation between the mirrors, $2a$ is the diameter of the circular mirrors, and λ is the wavelength of the radiation in free space. As shown in Ref. 28 the condition governing the dimensions of the resonant cavity is

$$\frac{a^2}{b\lambda} \ll \left(\frac{b}{a}\right)^2,$$

then Eqn (3.85) becomes

$$v(r_1, \phi_1) = Y \int_0^a \int_0^{2\pi} K(r_1, \phi_1, r_2, \phi_2) v(r_2, \phi_2) r_2\, d\phi_2\, dr_2, \tag{3.86}$$

where for the confocal configuration having the foci of the mirrors coincident at the centre of the cavity

$$K(r_1, \phi_1, r_2, \phi_2) = \frac{i}{\lambda b} \exp\left[\frac{-ikr_1 r_2 \cos(\phi_1 - \phi_2)}{b}\right]$$

For circular symmetry

$$v_n(r, \phi) = e^{-in\phi} Z_n(r)$$

Substituting into (3.86)

$$Z_n(r_1) = Y_n \int_0^a \frac{i^{n+1}}{b} k J_n\left(\frac{kr_1 r_2}{b}\right) Z_n(r_2) r_2 \, dr_2,$$

and we obtain

$$Z_n(r_1) = Y_n \int_0^a \frac{k}{b} J_n\left(\frac{kr_1 r_2}{b}\right) Z_n(r_2) r_2 \, dr_2, \tag{3.87}$$

where i^{n+1} has now been included in the factor Y_n and $(k/b)J_n(kr_1 r_2/b)$ is the complete kernel of the equation.

Equation (3.87) is a homogeneous Fredholm integral equation of the second

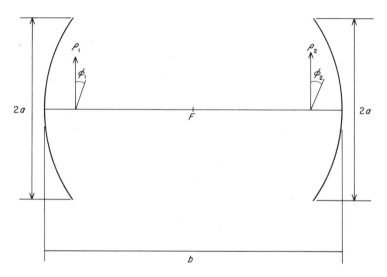

FIG. 3.12. *Confocal mirror resonant cavity. F is common focus of the circular aperture parabolic mirrors at the centre of the cavity.*

kind. As its kernel is continuous and symmetrical in r its eigenfunctions Z_{nm} corresponding to distinct eigenvalues Y_{nm} are orthogonal over the interval $(0, a)$, that is

$$\int_0^a Z_{nm}(r) Z_{np}(r) r \, dr = 0, \qquad m \neq p$$

and

$$\int_0^{2\pi} \int_0^a v_{nm}(r\phi)v_{pe}(r\phi)r\,dr\,d\phi = 0, \qquad \text{if } n \neq p \text{ or } m \neq e.$$

We define $v_{nm}(r\phi)$ to be a normal mode of the cavity. That is $v_{nm}(r\phi) \equiv T_{nm}$ in keeping with the notation used in Ref. 28.

Also from Eqn (3.87) it follows, since the kernel is real and symmetrical, that Y_{nm} must be purely real.

The modal solution of the integral equation is obtained as follows. Let $r_2 = ax$, $r_1 = ay$, and defining

$$N = \frac{ka^2}{b} = \frac{2\pi a^2}{b\lambda} = 2\pi \times \text{(the number of Fresnel zones subtended by one}$$

mirror at the centre of the other)

equation (3.87) becomes

$$Z_n(y) = Y_n \int_0^1 N J_n(Nxy) Z_n(x) x \, dx. \tag{3.88}$$

It is shown in Appendix II, relation (20), that the Bessel function of the first kind can be expanded in the form

$$J_n(Nxy) = \sum_{s=0}^{\infty} 2(n + 2s + 1)(-1)^s \frac{J_{n+2s+1}(Nx)}{Nx} R_{n+2s}^n(y) \tag{3.89}$$

for $0 \leqslant x \leqslant 1, 0 \leqslant y \leqslant 1$.

Substituting Eqn (3.89) into Eqn (3.88) we have

$$Z_n(y) = Y_n N \sum_{s=0}^{\infty} 2(n + 2s + 1)(-1)^s R_{n+2s}^n(y) \int_0^1 \frac{J_{n+2s+1}(Nx)}{Nx} Z_n(x) x \, dx, \tag{3.90}$$

or

$$Z_n(y) = \sum_{s=0}^{\infty} (n + 2s + 1)^{\frac{1}{2}} A_s^n R_{n+2s}^n(y), \tag{3.91}$$

where the constant

$$A_s^n = 2(n + 2s + 1)^{\frac{1}{2}}(-1)^s Y_n \int_0^1 J_{n+2s+1}(Nx) Z_n(x) \, dx. \tag{3.92}$$

the separation of $(n + 2s + 1)$ and the inclusion of $(-1)^s$ in A_s^n is required for eventual symmetry.

From Eqns (3.91) and (3.92)

$$2(n + 2t + 1)^{\frac{1}{2}}(-1)^t Y_n \int_0^1 J_{n+2t+1}(Ny)Z_n(y)\,\mathrm{d}y$$

$$= \sum_{s=0}^{\infty} 2(n + 2t + 1)^{\frac{1}{2}}(n + 2s + 1)^{\frac{1}{2}} Y_n(-1)^t A_s^n \int_0^1 R_{n+2s}^n(y)J_{n+2t+1}(Ny)\,\mathrm{d}y,$$

$$(3.93)$$

giving

$$A_t^n = \sum_{s=0}^{\infty} Y_n A_s^n B_{st}^n. \tag{3.94}$$

From the symmetrical relation $B_{st}^n = B_{ts}^n$ (Appendix II, No. 21) where

$$B_{st}^n = 2(n + 2s + 1)^{\frac{1}{2}}(n + 2t + 1)^{\frac{1}{2}}(-1)^t \int_0^1 R_{n+2s}^n(y)J_{n+2t+1}(Ny)\,\mathrm{d}y, \quad (3.95)$$

Eqn (3.94) may be transposed to give

$$0 = \sum_{s=0}^{\infty} A_s^n \Big| B_{st}^n - \frac{1}{Y_n}\delta_{s,t}\Big|, \tag{3.96}$$

$\delta_{s,t}$ is the Kronecker delta function.

Equation (3.96) represents a system of linear homogeneous equations possessing non trivial solutions if the determinant

$$\left| B_{st}^n - \frac{1}{Y_n}\delta_{s,t} \right| = 0.$$

In the form shown this would be an infinite determinant and thus is insoluble. If we could approximate (3.96) by a finite series of $p + 1$ terms, the determinant would become soluble yielding $(p + 1)$ values of Y_n and $p + 1$ sets of constants A_s^n i.e. $(p + 1)$ eigenvectors each identifying a separate mode in the cavity.

The termination of the series in (3.96) is only possible if the original expansion of the kernel Eqn (3.89) can also be terminated i.e.

$$J_n(Nxy) = \sum_{s=0}^{p} 2(n + 2s + 1)(-1)^s \frac{J_{n+2s+1}(Ny)}{Ny} R_{n+2}^n(x), \tag{3.97}$$

and thus

$$T_{nm} = \mathrm{e}^{-in\phi} \sum_{s=0}^{p} A_s^{nm} R_{n+2s}^n(x)(n + 2s + 1)^{\frac{1}{2}}. \tag{3.98}$$

Using the relation in Eqn (9) (Appendix II)

$$\int_0^1 J_n(Nxy)R_{n+2s}^n(x)x \, dx = (-1)^s \frac{J_{n+2s+1}(Ny)}{Ny}.$$

It can be shown that T_{nm} is expressible as

$$e^{-in\phi}C_{nm} \sum_{s=0}^{p} A_s^{nm} \frac{J_{n+2s+1}(Nx)}{x} (-1)^s(n + 2s + 1)^{\frac{1}{2}}, \tag{3.99}$$

where C_{nm} is a constant and with normalization becomes equal to Y_{nm}. It is to be noted that the coefficients A_s^{nm} have the same value in Eqns (3.98) and (3.99). The losses (per iteration) are then

$$1 - |Y_{nm}|^{-2}. \tag{3.100}$$

It can be noted that when

$$N \to 0, \qquad A_s^{nm} \to 0 \qquad \text{and} \qquad A_n^{nm} \to 1$$

that is when N tends to zero the mode solution becomes

$$T_{nm} = e^{-in\phi}R_{n+2m}^n(x)(n + 2m + 1)^{\frac{1}{2}}.$$

Since $N = 2\pi a^2/b\lambda$ it can be seen that when dealing with cavities of the dimensions used at microwave frequencies, such as beam waveguides, N will in general have small value and the series of circle polynomials exhibits rapid and monotonic convergence, contrasting with the oscillatory convergence of the other series used. This enables a high degree of accuracy to be obtained in defining the modes of oscillation in the cavity. It is also found that this rapid convergence is maintained with higher values of the mode numbers m and n.

The mode of greatest interest is the dominant T_{00} mode for which calculation shows that for $N = 2\pi$, only six terms are required for an accurate mode field pattern with an accuracy of 1 in 10^5 at $x = 0$ to 1 in 10^3 at $x = 1$.

This incorporation of the circle polynomials can then be extended to include aberration effects which for the resonant cavity will include such things as irregular mirror surfaces, spherical mirrors and displaced, tilted or defocused mirrors. Bridges[29], who performed the above analysis, shows that the high rate of convergence is retained in these cases.

The other basic functions used in optical cavity mode designation are the Gauss-Laguerre functions[30] and the hyperspheroidal functions.[31] The former modifies the Laguerre function by multiplication with an exponential factor to provide rapid decay in the radial direction away from the axis. The orthogonality range of the Laguerre functions is from zero to infinity but the exponential factor can make the contribution outside a finite range as small as required. The Laguerre–Gaussian functions have been used for the radiation in the near field of a corrugated conical horn.[32]

The hyperspheroidal functions are a generalization of the prolate spheroidal functions which themselves are used in the *exact* analysis of the problem of diffraction by a circular disc. They have been derived in the reference as the general solution of the integral equation in Eqn (3.88).

This is obtained by forming the self-adjoint differential operator which commutes with this integral operator over a closed interval (normalized to $[0, 1]$) and from this, deriving the differential equation which these functions must satisfy. (This equation incidentally also incorporates the spheroidal and Lamé functions by definition of its possible singularities).

The solutions are parametrized by a number c which turns out to be the number of Fresnel zones seen by one mirror from the centre of the other. It then corresponds to N in the analysis given here. Heurtley then goes on to demonstrate that in the limit $c \rightarrow 0$ the hyperspheroidal functions degenerate to the circle polynomials and that for large c they expand asymptotically as a series of Gauss-Laguerre polynomials. It is probable therefore that a similar treatment to that given in this chapter, using hyperspheroidal functions in place of the circle polynomials, could be carried through. It is unlikely that the simplicity of the latter, and its dependence on familiar and tabulated functions will be superseded.

In essence, these functions rely on the modal properties of the radiating aperture or cavity which of course the scalar diffraction theory does not (but conceivably should). In view of their relevance, a number of the basic papers on the subject are listed in the bibliography. In at least one of these (Slepian) the extension to spaces of higher dimension is brought into consideration (see p. 383). This as we shall see is an area that holds great promise for the study of microwave antenna theory in the future.

REFERENCES

1. W. V. T. Rusch and P. D. Potter. "Analysis of Reflector Antennas", Academic Press, New York and London, 1970.
2. J. B. Keller. Diffraction by an aperture, *Jour. App. Phys.* **28** (1957), 426.
3. L. B. Felsen and N. Marcuvitz. "Radiation and Scattering of Waves", Prentice Hall, 1973.
4. C. J. Bouwkamp. "Diffraction theory. *In* "Reports on Progress in Physics", Vol XVII, 1954, p. 35.
5. S. Silver. "Microwave Antenna Theory and Design", M.I.T. Series, Vol. 12, McGraw Hill, 1949, p. 192.
6. G. N. Watson, "Bessel Functions", Cambridge University Press, 1948.
7. A. Erdelyi. *et al,* "Tables of Integral Transforms", Vol. 2, McGraw Hill, 1954.

8. S. Cornbleet. The diffraction fields of a non-uniform circular aperture, *In* "Symposium on E.M. Theory and Antennas" (E. C. Jordan, Ed.), Pergamon Press, 1959, p. 157.
   G. Lansraux, Conditions functionelles de la diffraction instrumentalle cas particulier des zeros d'amplitude de figure de diffraction de revolution, *Cahier de Physique*, September 1953, No 45, p. 29.
9. E. Pinney, Laguerre functions in the mathematical foundations of the electromagnetic theory of the paraboloidal reflector, *Jour. Maths. and Phys.* vol **25** (1946) 49, and Vol. **26** (1947) also M. S. Afifi. Radiation from a paraboloid of revolution, Electromagnetic wave theory, Part 2 (J. Brown, Ed.), Delft Symposium 1965, Pergamon Press, 1967, p. 669.
10. E. Jahnke and F. Emde. "Tables of Functions", Dover, 1945, p. 180.
11. G. C. McCormick. "McGill Symposium on Microwave Optics", Part 2, 1959, p. 363.
12. F. Zernike. Beugungstheorie des Schneidenvefahrens, *Physcia*, Vol. 1 (1934), 689.
13. N. Chako. Characteristic curves in image space. *In* "McGill Symposium on Microwave Optics", 1959, Part 1, p. 67.
14. S. H. Moss. Lommel transforms in diffraction theory, *Trans I.E.E.E.* **AP12** (1964), 777.
15. S. Cornbleet. Circular aperture pattern with ultra-low side-lobes, *Electronics Letters,* **2**(2) (1966), p. 79.
16. E. H. Linfoot. "Recent Advances in Optics", Oxford University Press, 1955, p. 51.
    M. Born and E. Wolf. Principles of Optics", Pergamon Press, 1959, p. 436.
17. H. C. Minnett and B. M. Thomas. Fields in the image space of symmetrical focusing antennas, *Proc. I.E.E.* **115** (10) (1968), 1419.
18. P. A. Matthews and A. L. Cullen. "A study of the Field Distribution at an Axial Focus of a Square Microwave Lens", I.E.E. monograph No. 186R, July 1956.
19. S. Cornbleet. Feed arrangement for the axis definition of a paraboloid reflector, *Electronics Letters,* **9** (3) (1973), 66.
20. R. C. Hansen and I. L. Bailin. A new method of near field analysis, *Trans I.R.E.* **AP7** (1960), Special Supplement p S458.
21. Ming Kwei Hu. Fresnel region field distributions of circular aperture antennas, *Trans. I.R.E.* **AP8** (1960), 344; and *Jour. Res. Nat. Bur. Stand. Sect. D.* **65** (2) (1961), 137.
22. L. K. de Size. "Uniform, Cosine and Cosine Squared Illumination with a Curved Phase Front", A.I.L. Report, No. 3585-4, December 1957.
23. B. R. A. Nijboer. The diffraction theory of optical aberrations, *Physica,* **13** (10) (1947), 605.
24. S. Cornbleet. Asymmetric phase effects in the circular aperture. *In* "Symposium on Quasi-Optics", Polytechnic Institute of Brooklyn, 1964, p 487.
25. R. M. McElvery and J. E. Smerczynski. The gain of a defocused circular aperture, *Quart. Jour. Appl Maths.* **29** (1964–5), 319.
26. J. H. McLeod. The axicon: a new type of optical element, *Jour. Opt. Soc. Amer.* **44** (1954), 592.
    J. H. McLeod. Axicons and their uses, *Jour. Opt. Soc. Amer.* **50** (2) (1960), 166.
27. S. Cornbleet. Superdirective property of the microwave axicon, *Proc. I.E.E.* **117** (5) (1970), 869.
28. A. G. Fox and T. Li. Resonant modes in an optical maser, *Bell System Tech. Jour.* **40** (1961), 453.

29. C. A. Bridges. "General Analysis of the Open Circular Resonator", Ph.D Thesis University of Surrey, 1969.
    C. A. Bridges and S. Cornbleet. Modes in confocal parabolic mirror cavities. Presented at U.S.R.I. Symposium on E.M. waves, Stresa, June 1968 (unpublished).
30. H. Kogelnik. Coupling and conversion coefficients for optical modes. *In* "Symposium on quasi-optics", Microwave Research, Vol. XIV, Brooklyn Polytechnic Press, 1964, p. 333.
31. J. C. Heurtley. Hyperspheroidal functions—optical resonators with circular mirrors. *In* "Symposium on quasi-optics", Microwave Research, Vol. XIV, Brooklyn Polytechnic Press, 1964, p. 367.
    W. Streifer and H. Gamo. On the Schmidt expansion for optical resonator modes. *In* "Symposium on quasi-optics", Microwave Research, Vol. XIV, Brooklyn Polytechnic Press, 1964, p. 351.
32. C. Aubry and D. Bitter, Radiation pattern of a corrugated conical horn in terms of Laguerre-Gaussian functions, *Electronics Letters*, **11** (7) (1975), 154.

BIBLIOGRAPHY

DIFFRACTION THEORY AND TRANSFORMS

C. J. Bouwkamp. Note on diffraction by a circular aperture, *Acta Physica Polonica* **27** (1965), 37.

C. J. Bouwkamp. Theoretical and Numerical treatment of diffraction through a circular aperture, *Trans. I.E.E.E.* **AP18** (2) (1970), 152.

A. Ishimaru and G. Held. Analysis and synthesis of radiation patterns from circular apertures, Chapter 4 ref. 5.

Y. Itoh. Evaluation of aberrations using the generalized prolate spheroidal wavefunctions, *Jour. Opt. Soc. Amer.* **60** (1) (1970), 10.

B. Karczewski. Fraunhofer diffraction of an electromagnetic wave, *Jour. Opt. Soc. Amer.* **51** (1961), 1055.

J. Kormska, Fraunhofer diffraction at apertures in the form of regular polygons, *Optica Acta,* **19** (10) (1972), 807 and **20** (7) (1973), 549.

Y. T. Lo and H. C. Hsuan. An equivalence between elliptical and circular arrays, *Trans. I.E.E.E.* **AP13** (1965), 247.

H. Osterberg. Rayleigh's Integral in the near Fresnel region, *Jour. Opt. Soc. Amer.* **55** (11) (1965), 1467.

A. Papoulis. Optical systems, singularity Functions complex Hankel transforms, *Jour. Opt. Soc. Amer.* **57** (2) (1967), 207.

A. C. Schell. The diffraction theory of large-aperture spherical reflector antennas, *Trans. I.E.E.E.* **AP11** (1963), 428.

S. Silver. Microwave aperture antennas and diffraction theory, *Jour. Opt. Soc. Amer.* **52** (2) (1962), p. 131.

J. Sinnott. Patterns for out-of-phase Taylor semicircular apertures, *Trans. I.E.E.E.* **AP14** (1956), 390.

H. Slevogt. Ein Vorschlag zur Darstellung des Lichtgebinges, Optik Band 22 Heft 6, 1965, 391.

J. P. Wild. Circular aerial arrays for radio astronomy, *Proc. Roy. Soc.* **262** (1961), 84.

PATTERN SQUINT

J. W. Duncan. Asymmetric phase error in circular apertures.

Y. T. Lo. On the beam deviation factor of a parabolic reflector, *Trans. I.R.E.* **AP8** (1960), 347.

A. W. Rudge. Multiple beam antennas: offset reflectors with offset feeds, *Trans. I.E.E.E.* **AP23** (3) (1975), 317.

J. Ruze. Lateral feed displacement in a paraboloid, *Trans. I.E.E.E.* **AP13** (1965), 660.

S. S. Sandler. Paraboloid reflector patterns for off axis feed, *Trans. I.R.E.* **AP** (1960), 368.

APPLICATIONS

S. Cornbleet. Determination of the aperture field of an antenna by a beam displacement method, *Proc. I.E.E.* **115** (10) (1968), 1398.

S. Cornbleet. Radiation patterns of circular apertures with structural shadows, *Proc. I.E.E.* **117** (8) (1970), 1620.

K. K. Dey. Microwave aerial measurements at reduced range by on-axis defocus of the feed, *Indian Jour. Pure and Applied Physics,* (9) (3) (1971), 179.

R. C. Hansen. Near field determination of antenna difference patterns. *In* "URSI Conference on E.M. Wave Theory", I.E.E. Conference Publication, No. 114, 1974, 179.

R. C. Johnson. H. A. Ecker and J. S. Hollis. Determination of far-field patterns from near field measurements, *Proc. I.E.E.E.* **61** (12) (1973), 1668.

APERTURE DISTRIBUTIONS

P. Jacquinot and B. Roizen-Dossier. Apodisation *In* "Progress in Optics", Vol. 3 (E. Wolf, Ed.), North Holland Press, 1964.

J. F. Kauffman, W. F. Croswell and L. J. Jowers. Analysis of the radiation patterns of reflector antennas, *Trans. I.E.E.E.* **AP24** (1) (1976), p. 53.

G. Lansraux and A. Boivin, Maximum factor of the encircled energy, *Can. Jour. Phys.* January 1961.

G. O. Olaofe. Diffraction by Gaussian apertures, *Jour. Opt. Soc. Amer.* **61** (12) (1971), 1654.

D. R. Rhodes. On the aperture and pattern space factors for rectangular and circular apertures, *Trans. I.E.E.E.* **AP19** (6) (1971), 763.

D. R. Rhodes. On the Taylor distribution, *Trans. I.E.E.E.* **AP20** (2) (1972), 143.

R. S. Richardson. A new family of illumination functions, *Trans. I.E.E.* **AP18** (1970), 284.

R. C. Schell and G. Tyras. Irradiance from an aperture with a truncated Gaussian field distribution, *Jour. Opt. Soc. Amer.* **61** (1) (1971), 31.

A. F. Sciambi. The effect of aperture illumination on circular aperture pattern characteristics, *Microwave Journal,* (August 1965), 79.

H. H. Snyder. On certain wave transmission coefficients for elliptical and rectangular apertures, *Trans. I.E.E.E.* **AP17** (1969), 107.

T. T. Taylor, Design of circular apertures for narrow beam-width and low side lobes *Trans. I.R.E.* **AP8** (1960), 17.

ABERRATIONS—NEAR FIELD

H. Arsenault and A. Boivin. Optical filter synthesis by holographic methods, *Jour. Opt. Soc. Amer.* **48** (11) (1968), 1490.

M. P. Bachynski and G. Bekefi. Aberrations in circularly symmetric microwave lenses, *Trans. I.R.E.* **AP4** (1956) 412.

D. J. Bem. Electric field distribution in the focal region of an offset paraboloid, *Proc. I.E.E.* **116** (5) (1969), 79.

J. C. Bennett, A. P. Anderson, P. A. McInnes and A. J. T. Whittaker. Investigation of the characteristics of a large reflector antenna using microwave holography, 1973 I.E.E.E. P.G. A-P Symposium Digest, p. 298.

S. C. Biswas and A. Boivin. Formal expansion of the diffraction integrals in the caustic region of an obliquely illuminated wide-angle spherical mirror, *Jour. Opt. Soc. Amer.* **63** (10) (1973), 1284.

T. E. Cherot Jnr. Calculation of the near field of Circular aperture antenna using the geometrical theory of diffraction, *Trans. I.E.E.E.* **EMC13** (2) (1971), 29.

T. S. Chu. A note on simulating Fraunhofer radiation patterns in the Fresnel region, *Trans. I.E.E.E.* **AP19** (1971), 691.

R. C. Hansen. Minimum spot size of focused antennas. *In* "Electromagnetic Wave Theory", Pt 2 (J. Brown, Ed.), Delft Symposium, 1965, Pergamon, 1967, p. 661.

G. Hyde. Studies of the focal region of a spherical reflector: stationary phase evaluation, *Trans. I.E.E.E.* **AP16** (6) (1968), 646; also with R. C. Spencer, *Polarization Effects*, **16** (1968), 399.

E. M. Kennaugh and R. H. Ott. Fields in the focal region of a parabolic receiving antenna, *Trans. I.E.E.E.* **AP12** (1964), 376.

M. Landry and Y. Chasse. Measurement of the electromagnetic field intensity in the focal region of a wide-angle paraboloid reflector, *Trans. I.E.E.E.* **AP19** (4) (1971), 539.

M. Novotny. Foci of axially symmetrical filters, *Optica Acta* **20** (3) (1973), 217.

B. Richards and E. Wolf. Electromagnetic diffraction in optical systems II: structure of the image field, *Proc. Roy. Soc. A.* **253** (1959), 358.

A. W. Rudge. Focal plane field distribution of parabolic reflectors, *Electronics Letters*, **5** (21) (1969), 510.

W. V. T. Rusch. The Physical optics of focused scatterers using source multiple expansion, *Trans. I.E.E.E.* **AP22** (2) (1974), 236.

J. J. Stengel and W. M. Yarnell. Pattern characteristics of an antenna focused in the Fresnel region, Convention Record of the I.R.E. pt. 1, 1962, p. 3.

W. H. Watson. The field distribution in the focal plane of a paraboloidal reflector, *Trans. I.E.E.E.* **AP12** (1964), 561.

B. G. Whitford and T. J. F. Pavlasek. Focal region fields of annular and sectoral microwave apertures, *Jour. Opt. Soc. Amer.* **58** (12) (1968), 1591.

CAVITY RESONATORS AND BEAM WAVEGUIDES

G. D. Boyd and J. P. Gordon. Confocal multimode resonator for millimeter through optical wavelength masers, *Bell Syst. Tech. Jour.* **40** (1961). 489.

G. D. Boyd and H. Kogelnik. Generalized confocal resonator theory, *Bell Syst. Tech. Jour.* **41** (1962), 1347.

G. Goubau and F. Schwering. On the guided propagation of electromagnetic wave beams, *Trans. I.R.E.* **AP9** (1961), 248.

J. C. Heurtley and W. Streifer. Optical resonator modes—Circular reflectors of spherical curvature, *Jour. Opt. Soc. Amer.* **55** (11) (1965), 1472.

H. J. Landau and H. O. Pollack. Prolate spheroidal wave functions–Fourier analysis, Pt II, *Bell Syst. Tech. Jour.* **40** (1961), 65.

D. M. McCumber. Eigenmodes of a symmetric cylindrical confocal laser resonator, *Bell Syst. Tech. Jour.* **44** (1965), 333.

J. R. Pierce. Modes in sequences of lenses, *Proc. Nat. Acad. Sci.* **47** (1961), 1808.

Tables of angular spheroidal wave functions, Vol 1 Prolate $m = 0$, Vol II Oblate $m = 0$ (inc 156 refs.), Naval Res. Lab. Washington D.C., June 1975.

D. Slepian and E. Sonnenblick. Eigenvalues associated with Prolate spheroidal wave functions of zero order, *Bell Syst. Tech. Jour.* **44** (1965), 1745.

D. Slepian and H. O. Pollack. Prolate spheroidal wave functions—Fourier analysis and uncertainty, Pt. I, *Bell Syst. Tech. Jour.* **40** (1961), 43.

D. Slepian. Prolate spheroidal wave functions—Fourier analysis and uncertainty, Pt IV, *Bell Syst. Tech. Jour.* **43** (1964), 3009.

DIFFRACTION THEORY TABLE

(for completeness some references are repeated in this section)

W. Adrejewski. Rigorous theory of diffraction of plane E.M. waves at a perfectly conducting circular disc and a circular aperture in a perfectly conducting plane screen, *Naturwissenschaften,* **38** (1951), 406.

C. L. Andrews. "The optics of the Electromagnetic Spectrum", Prentice Hall, 1960.

B. B. Baker and E. T. Copson. "The Mathematical Theory of Huygen's Principle", Oxford University Press, 1939.

H. G. Booker and P. C. Clemmow. The concept of an angular spectrum of plane waves and its relation to that of polar diagram and aperture distribution, *Proc. I.E.E.* **97** (3) (1950), 11.

M. Born and E. Wolf. "Principles of Optics", p. 374. Chapter 8 Section 3, Pergamon Press.

C. J. Bouwkamp. Diffraction theory. *In* "Reports on Progress in Physics", Vol. XVII, 1954, p. 35. (Including 500 refs).

W. Franz, ref in Bouwkamp.

J. C. Heurtley. Scalar Rayleigh–Sommerfeld and Kirchhoff diffraction integrals. A comparison of exact evaluations at axial points, *Jour. Opt. Soc. Amer.* **63** (8) (1973), 1003.

J. D. Jackson. "Classical Electrodynamics" John Wiley, 1962.

J. B. Keller. Diffraction by an aperture, *Jour. App. Phys.* **28** (1957), 426.

M. Kline and I. W. Kay. "Electromagnetic Theory and Geometrical Optics", Interscience, 1965.

F. Kottler. *Ann Phys.* **71** (1923), 457.

F. Kottler, Diffraction at a black screen. *In* "Progress in Optics", vol. VI, (1967), 333.

H. Levine and J. Schwinger. On the theory of electromagnetic wave diffraction by an aperture in an infinite conducting screen. *In* "Theory of E. M. Waves", Washington Square Symposium, Interscience, 1951, 1.

E. W. Marchand and E. Wolf. Boundary diffraction wave in the domain of the Rayleigh-Kirchhoff diffraction theory, *Jour. Opt. Soc. Amer.* **52** (1962), 76.

E. W. Marchand and E. Wolf. Transmission cross section for small apertures in black screens. *Jour. Opt. Soc. Amer.* **60** (11) (1970), 1501.

K. Miyamoto and E. Wolf. Generalization of the Maggi–Rubinowicz theory of the boundary diffraction wave, *Jour. Opt. Soc. Amer.* **52** (1962), 615.

C. H. Papas, "The Theory of Electromagnetic Wave Propagation", McGraw Hill, 1965.

A. Rubinowicz. The Miyamoto–Wolf diffraction wave. *In* "Progress in Optics" (E. Wolf, Ed.) vol. IV. 1965 p. 201, North Holland.

W. V. T. Rusch. Scattering from a hyperboloid reflector in a Cassegrain feed system, *Trans. I.E.E.E.* **AP11** (1963), 414.

M. I. Sancer. An analysis of the vector Kirchhoff equations and the associated boundary line charge, *Radio Science* **3** (2) (new series) (1968), 141.

S. A. Schelkunoff. Kirchhoff's formula, its vector analogue and other field equivalence theorems. *In* "Theory of E.M. Waves", Washington Square Symposium, Interscience, 1951, 107.

H. Severin. Methods of light optics for the calculation of the diffraction phenomena within the range of centimeter waves. Supplemento al vol IX serie IX *Nuovo Cimento* no 3, 1952, p. 381.

A. B. Shafer. Hamilton's mixed and angle characteristic functions and diffraction aberration theory. *Jour. Opt. Soc. Amer.* **57** (5) (1967), 630,

G. C. Sherman, Diffracted wave fields expressible by plane-wave expansions containing only homogeneous waves., *Jour. Opt. Soc. Amer.* **59** (6) (1969), 697; also Integral transform formulation of diffraction theory, *Jour. Opt. Soc. Amer.* **57** (12) (1967), 1490.

S. Silver. "Microwave Antenna Theory and Design", M.I.T. Radiation Laboratory Series, Vol. 12, 1949.

A. Sommerfeld. "Optics", Academic Press, 1964, Chapter V, p. 179.

J. A. Stratton and L. J. Chu. Diffraction theory of electromagnetic waves, *Physics Review,* **56** (1939), 99.

J. A. Stratton, "Electromagnetic Theory", McGraw Hill, 1941.

C. T. Tai. Kirchhoff theory: scalar, vector or dyadic?, *Trans. I.E.E.E.* **AP20** (1972), 114.

J. P. Vasseur. Diffraction of electromagnetic waves by apertures in a plane conducting screen, *L'Onde Electrique.* **32** (1952), 3, 55, 97.

DIFFRACTION PATTERNS

A. I. Mahan, C. V. Bitterli and S. M. Cannon. Far field diffraction patterns of single and multiple apertures bounded by arcs and radii of concentric circles, *Jour. Opt. Soc. Amer.* **54** (6) (1964), 721.

D. Carter. Wide-angle radiation in pencil beam antennas, *Jour. App. Phys.* **26** (6) (1975) 645.

W. H. Ierley and H. Zucker. A stationary phase method for the computation of the far field of open Cassegrain antennas, *Bell Syst. Tech. Jour.* (March 1970), 431.

H. H. Snyder. On certain wave transmission coefficients for elliptical and rectangular apertures, *Trans. I.E.E.E.* **AP17** (1969) 107.

4

The Synthesis of Far-field
Radiation Patterns

One of the attractions of the Neumann series description of the far-field radiation pattern is the insight it gives into the problem of constructing an aperture illumination function that will create a required or pre-determined radiation pattern. If, as in Popovkin[1] we regard the fundamental equation for radiation patterns, Eqn (3.1) as the operational relation

$$Af = g, \tag{4.1}$$

where g is the given function to be derived by the operation of A on the unknown distribution f, then we require for the synthesis problem the inverse operator A^{-1}. This, as it is stated in the reference, exists but is not continuous and then additional restrictions have to be made on the distribution $f(r)$. This function then becomes unstable and difficult of realization. In the classical sense the problem of antenna pattern synthesis is "incorrect". These aspects can all be demonstrated by the results of the method that has to be adopted in order to invert the series form of Eqn (3.28). This will come as no surprise to experienced workers on antennas who find that even large variations in the amplitude distribution has little effect on the fundamental *shape* of a radiated beam, such as, for example, turning it from a pencil beam into a conical beam. The major effects are on the scale of the pattern, that is the beam width, and on the side lobe configuration. Thus, to synthesize a completely arbitrary beam shape, even a fairly simple one, turns out to be a very difficult proposition. This is basically so since we have to make *a priori* assumptions regarding the phase distribution in the aperture. Without these however the problem as a whole is indeterminate.

4.1 SYMMETRICAL PATTERNS—METHOD OF WEBB KAPTEYN

For symmetrical patterns and with a uniform phase distribution in the aperture, the operation required for the inversion in Eqn (4.1) is the inverse of the finite Hankel transform of zero order.

Using the series description (Eqn (3.28))

$$g(u) = \sum_{s=0}^{\infty} \alpha_{2s+1} \frac{J_{2s+1}(u)}{u},$$

then we only require to find the coefficients α_{2s+1} in order to obtain the aperture distribution from

$$f(r) = \sum_{s=0}^{\infty} (-1)^s \alpha_{2s+1} R_{2s}^0(r)$$

To do this we apply the theory[2] of Webb–Kapteyn which is based on the result

$$\int_0^{\infty} J_{2m+1}(t) J_{2n+1}(t) \frac{dt}{t} = \frac{\delta_{m,n}}{2(2n+1)} \tag{4.2}$$

Then if an *odd* function $F(x)$ admits an expansion of the type

$$F(x) = \sum_{n=0}^{\infty} \alpha_{2n+1} J_{2n+1}(x) \tag{4.3}$$

the coefficients required are given by

$$\alpha_{2n+1} = (4n+2) \int_0^{\infty} F(t) J_{2n+1}(t) \frac{dt}{t} \tag{4.4}$$

Unfortunately, the validity of this Fourier-like expansion is restrained by severe restrictions on the functions $F(x)$ to which it can be applied. These are (Ref. 2, p. 535)

(a) the integral $\int_0^\infty F(t)\,dt$ exists and is absolutely convergent
(b) $F(t)$ has a continuous differential coefficient for all positive values of the variable which do not exceed x
(c) the function satisfies the equation

$$F'(t) = \tfrac{1}{2} \int_0^\infty \frac{J_1(v)}{v} [F(v + t) + F(v - t)]\,dv \qquad (4.5)$$

when $t \leqslant x$.

Applying this to the *even* function

$$g(u) = \Sigma\, \alpha_{2s+1} \frac{J_{2s+1}(u)}{u}$$

we have the more stringent conditions

(a) the integral $\int_0^\infty u\,g(u)\,du$ exists and is absolutely convergent (4.6a)
(b) $g(u)$ has a continuous differential coefficient, and (4.6b)
(c) the function $g(v)$ satisfies the relation

$$t\,g'(t) + g(t) = \tfrac{1}{2} \int_0^\infty \frac{J_1(v)}{v} \{(v + t)g(v + t) + (v - t)\,g(v - t)\}\,dv \qquad (4.6c)$$

for all t in the range $[0, u]$.

As stated in the reference, no simple criteria has been established for functions which satisfy Eqns (4.5) and (4.6).

We find that the ability to satisfy the latter is intimately bound up with the physical realizability of the pattern chosen. There are known functions with Neumann series expansions and there are other means of deriving Neumann series for arbitrary functions. But in every case not found to satisfy (4.6) the introduction of the α coefficients so obtained into the aperture distribution series (equation 3.28) leads to the divergence of that series. In the absence of a general proof to this effect we give some illustrations

$$(1) \quad g(u) = \cos u = 2\left\{1^2\frac{J_1(u)}{u} - 3^2\frac{J_3(u)}{u} + 5^2\frac{J_5(u)}{u}\dots\right\}$$

then with these coefficients the summation rule (Chapter 3, Eqn (3.44)) for coefficients gives

$$f(1) = \sum_{s=0}^{\infty} (2s + 1)^2$$

which is clearly divergent.

$$(2) \quad g(u) = \frac{\sin au}{au}$$

substitution into Eqn (4.6) requires

$$\cos au = \frac{1}{2a} \int_0^\infty \frac{J_1(t)}{t} \sin at \cos au \, dt,$$

which is true for $a \leqslant 1$ only. Therefore $(\sin au)/au$ satisfies relation 4.6. but not condition a on the absolute convergence of $\int_0^\infty \sin(au) \, du$. Hence the coefficients, obtained from Eqn (4.4) namely

$$\alpha_{2s+1} = 2(2n + 1) \int_0^\infty \frac{\sin au}{au} J_{2s+1}(u) \, du$$

$$= \frac{2}{a} \sin\left[(2s + 1) \sin^{-1} a \right] = \frac{2}{a} T_s(a),$$

are identical with those obtained in chapter 3 equation 3. and lead to an amplitude distribution with an infinite singularity at $r = a$ as shown in Fig. 4.1.

(3) $g(u) = J_{2n+1}(au)/au$ gives equality when introduced into the relation (4.6c) provided $a \leqslant 1$ and satisfies the two other criteria and is thus a viable function. From (4.4) the coefficients are then

$$\alpha_{2s+1} = 2(2s + 1) \int_0^\infty J_{2n+1}(au) J_{2s+1}(u) \frac{du}{au}$$

$$= \frac{1}{2} \sum_{m=0}^{s-n} \frac{(-1)^m a^{2n+2m}(s + m + n)!}{m!(2n + m + 1)! (s - n - m)!}, \quad s \geqslant n \qquad (4.7)$$

Comparing this with the alternative derivations of these coefficients gives us the two results

$$\frac{1}{2} \sum_{s \geqslant n}^\infty (-1)^s \sum_{m=0}^{s-n} (-1)^m \frac{a^{2n+2m}}{m!} \frac{(s + m + n)}{(2^n + m + 1)! (s - n - m)!} R_{2s}^0(r)$$

$$= (-1)^n R_{2n}^0 \left(\frac{r}{a} \right) \qquad 0 \leqslant r \leqslant a$$

$$= 0 \qquad a < r \leqslant 1, \qquad (4.8)$$

and

$$\frac{1}{2} \sum_{s \geqslant n}^\infty \sum_{m=0}^{s-n} (-1)^m \frac{a^{2n+2m}}{m!(2n + m + 1)!} \frac{(s + m + n)!}{(s - n - m)!} \frac{J_{2s+1}(u)}{u} = \frac{J_{2n+1}(au)}{au},$$

$$(4.9)$$

neither of which appear to be in the literature to date. The second of these is a particularly useful multiplication theorem for Bessel functions.

This last example gives rise to the concept of "approximately valid functions". Reference to the process of deriving the relation (4.5) shows that it is a requirement only in so far as its cancellation allows the series description of $F(x)$ and the original function to agree exactly. If it were not exact then, the series and the function would no longer agree, but the degree of approximation may be acceptable in a practical situation.

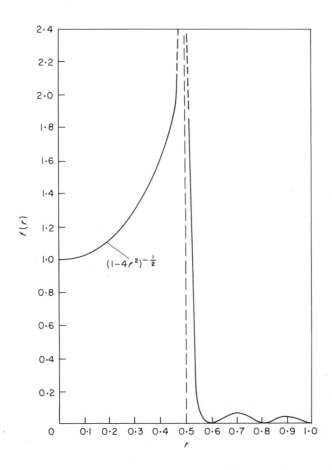

FIG. 4.1. *Series solution for aperture distribution require for a pattern* $g(u) = (\sin u/2)/u$;

Exact solution $(1 - 4r^2)^{-\frac{1}{2}}$

Series solution $\displaystyle\sum_{s=0}^{\infty} 4T_{2s+1}(\tfrac{1}{2})R_{2s}^0(r)$.

Thus, for example, we may choose to ignore the limitation $a \leqslant 1$ in the last expansion and endeavour to obtain an aperture function for the pattern

$$g(u) = J_1(2u)/u.$$

This, being half the beamwidth allowable by the aperture dimensions, would have the attributes of a "supergain" pattern.

The coefficients can still be derived from the series in Eqn (4.7) but when these are used to sum the expression

$$g(u) = \sum_{s=0}^{\infty} \alpha_{2s+1} J_{2s+1}(u)/u,$$

gives convergence and agreement with the required $J_1(2u)/u$ function up to values of u approximately equal to 8, after which the series becomes divergent. This is in precise agreement with the known physical result[3] that a finite aperture can give any degree of resolution, if, after a certain angular width (that is value of u) large amounts of energy are wasted in radiated side lobes.

It is apparent from this example that the magnitude of a, taken in the forbidden region larger than unity, creates this form of super resolution to an increasing degree, but if the high energy side lobes can be retained in the non radiating part of the u field, some degree of "supergain" would be achieved.

4.2 FUNCTIONS DESCRIBED BY THEIR MACLAURIN EXPANSION

Similar effects are observed when the Neumann series for a given radiation pattern function is derived from the Maclaurin expansion of the function. Incidentally however, many interesting Bessel function expansions can be obtained from the method independently of the practical consequences.

Neumann's own treatment for the expansion of an arbitrary function (Ref. 2, p. 523) can be summed up by the statement that:
if the Maclaurin expansion of $F(z)$ is

$$F(z) = \sum_{n=0}^{\infty} b_n z^n,$$

then

$$F(z) = \sum_{n=0}^{\infty} a_n J_n(z) \tag{4.10}$$

where a_n and b_n are related by

$$a_0 = b_0 : a_n = n \sum_{m=0}^{\leq n/2} 2^{n-2m} \frac{(n-m-1)!}{m!} b_{n-2m}. \tag{4.11}$$

The way in which the α coefficients of the Neumann series can be derived is obvious and hence the series description of the aperture distribution. No criteria has been established for the resultant convergence of these series and no success has been obtained in applying the method to such patterns as[4]

$$g(u) = \left[\frac{J_1(u)}{u} \right]^2 \quad \text{or} \quad g(u) = \frac{\cos(u^2 - A^2)^{\frac{1}{2}}}{\cosh A}$$

We are led therefore to make the conjecture that the only radiation patterns which can be synthesized from a circular aperture with a continuous inphase field distribution are those obeying the criteria of convergence, continuity and the relations given in Eqn (4.6).

Any attempt to obtain a meaningful physical understanding of Eqn (4.6c) must likewise fall into the category of pure conjecture. Integrating Eqn (4.6c) between the limits $[0, u]$ with respect to t results in the integral equation (after reversing the order of integration on the right hand side)

$$ug(u) = \frac{1}{2} \int_0^\infty \frac{J_1(v)}{v} \, dv \int_0^u \{(v + t) g(v + t) + (v - t) g(v - t)\} \, dt \tag{4.12}$$

and the two cases, cases 2 and 3 given above are the simplest solutions for which the inner integral on the right hand side gives a separable solution. That is

$$\int_0^u \{(v + t) g(v + t) + (v - t) g(v - t)\} \, dt = F_1(v) F_2(u)$$

and then $F_1(v)$ must equal a constant $= b$ and so

$$F_2(u) = \frac{1}{b} ug(u)$$

Equation (4.12) can be said to have the appearance of a far-field Huygens principle in which the kernel function is the uniform circular aperture function $J_1(v)/v$ in place of the free space function e^{ikr}/r and the inner integral is a symmetrized aperture weighting function, with all the appearance of being the average of a retarded and an advanced form. But beyond that our speculation should not go.

4.3 APERTURE EXPANSION AS A BESSEL SERIES

Other methods of pattern synthesis based on the scalar Kirchhoff formulation have been presented.[5] The method is based upon a different expansion for the radial amplitude distribution $f(r)$, this time in terms of scaled Bessel functions of order zero. That is

$$f(r) = \sum_{n=0}^{N} a_n J_0(u_n r) \qquad 0 < r < 1$$
$$= 0 \qquad\qquad\qquad r > 1. \qquad (4.13)$$

Substitution into the scalar integral for the condition of circular symmetry and uniform phase, require the application of Lommel's formula

$$\int J_0(u_n r) J_0(ur) r \, dr = \frac{r}{u_n^2 - u^2} \left[u J_0(u_n r) J_0'(ur) - u_n J_0'(u_n r) J_0(ur) \right], \quad (4.14)$$

giving

$$g(u) = \sum_{n=0}^{N} \frac{a_n}{u_n^2 - u^2} \left[u J_0(u_n) J_0'(u) - u_n J_0'(u_n) J_0(u) \right]. \qquad (4.15)$$

The right hand side can be simplified through the choice of N values of u_n to satisfy

$$J_0(u_n) \text{ or } J_0'(u_n) = \text{zero},$$

or more generally

$$u_n J_0'(u_n) + h J_0(u_n) = 0, \qquad (4.16)$$

a form which we shall be meeting again later.

With $g(u)$ so fitted at these N values of u_n the coefficients a_n become

$$a_n = \frac{2u_n^2}{(h^2 + u_n^2)} J_0^2(u_n) g(u_n). \qquad (4.17)$$

The first reference applies the theory to the positioning and the magnitudes of the side lobe zero and maxima, and thus is not strictly a beam shaping application. Ruze applies the same analysis to the production of a flat topped pattern. It is shown that the resultant series of Eqn (4.15) does represent such a pattern but the aperture distribution arising from the substitution of the same coefficients into the series in equation 4.13 is not illustrated. Asymmetric patterns can be derived by the application of the cyclic phase factor $e^{-ip\phi'}$ (Eqn (3.10)) for $p = 1$ only. Higher values of p would give patterns with higher periodicity in the ϕ coordinate and which have not had a practical application as yet. The analysis can then be repeated, but with first order Bessel function (since $p = 1$) and gives basically a pencil beam shape as before, but linearly shifted along the u axis. Thus again general pattern shaping cannot be said to have been achieved even though we have entered the area of non-uniform phases.

4.4 AN EXTENSION TO THE METHOD OF WEBB–KAPTEYN

We conclude with a final example which acts as an indication of the way that is needed to proceed in order to obtain general shaped patterns.

We use instead of the usual Neumann series, a series of scaled functions, each necessarily wider than its unscaled form. That is we assume

$$g(u) = \sum_{n=0}^{N} a_n \frac{J_1(\omega_n u)}{u} + b_n \frac{J_3(\omega_n u)}{u}, \qquad \omega_n < 1 \qquad (4.18)$$

All these functions, and the higher order ones which have not been included satisfy the criteria for physical realization which we have conjectured. The variety of shaped patterns that can be constructed from a few terms of this series is illustrated in Fig. 4.2. A method is available for obtaining the coefficients a_n and b_n (Ref. 2, p. 396) using the Schaftheitlin integrals

$$\int_0^\infty J_\mu(t)J_\nu(t)\, dt/t^{\mu+\nu} = \frac{\Gamma(\mu + \nu)\Gamma(\tfrac{1}{2})}{2^{\mu+\nu}\Gamma(\mu + \nu + \tfrac{1}{2})}\Gamma(\nu + \tfrac{1}{2})\Gamma(\mu + \tfrac{1}{2}). \qquad (4.19)$$

$$\mu + \nu > 0$$

The resulting integrations on the left hand side of equation 4.18 are of the form

$$\int_0^\infty u\, g(u)J_2(\omega_n u)\, du,$$

and may require special consideration or computation. However, this only results in N equations for the $2N$ unknown coefficients. A further condition can be applied by specifying the value of $g(u)$ at the origin, that is

$$g(0) = \lim_{u \to 0} g(u) = \tfrac{1}{2} \sum_{n=1}^{N} \frac{n a_n}{N}. \qquad (4.20)$$

Hence the expansion in 4.18 must be confined to three terms only for a completely determined solution.

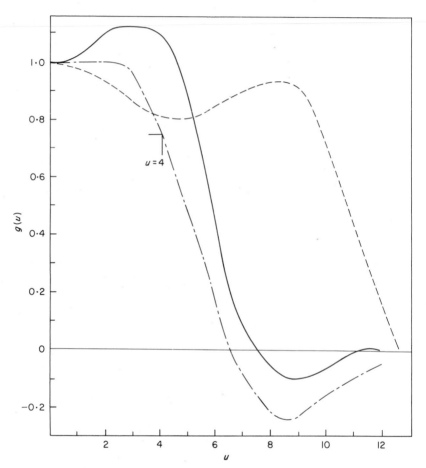

FIG. 4.2. *Patterns constructed from the functions* $g(u) = J_n(\omega u)/u$.

$$\left\{ \begin{array}{l} \underline{\hspace{2cm}} \quad \dfrac{2J_1(u)}{u} + \dfrac{2J_3(u)}{u} + \dfrac{J_3(\frac{1}{2}u)}{u} \\[3mm] \text{-----} \quad \dfrac{J_1(\frac{3}{4}u)}{u} + \dfrac{7J_3(\frac{1}{2}u)}{u} \text{ (Normalized)} \end{array} \right.$$

$—\cdot—\cdot—$ *Three term solution of equation (4.18) with* $\omega_2 = 0$
(for square topped pattern).

Now each term in the series is of the form $J_{2s+1}(au)/au$ for $a \leqslant 1$ and is therefore derived from an incomplete aperture distribution in which $f(r)$ is zero outside the value a (Chapter 3, Eqn (3.45)). Thus unless the appropriate condition is applied, the aperture distribution required to produce the pattern of Eqn (4.18), will be *discontinuous* at each value of ω_n used. It is soon to be found that when this further condition is applied and continuity of the aperture distribution (but not necessarily of its derivative) is insisted upon, then the patterns no longer are able to develop the shapes that the completely free choice of these coefficients gave.

Consequently, the next consideration must be that of the aperture distribution with deliberately placed discontinuities in the illumination function.

4.5 THE ZONED CIRCULAR APERTURE

Many practical microwave antennas and optical devices are known in which the aperture is divided into discrete zones. This includes all the zoned mirrors of the parageometrical optics design and the Fresnel zone plates discussed in chapter one. Antennas constructed from axial annuli, each with its own source and thus with independent control of amplitude and phase, have been designed by Koch[6] (Fig. 4.3(a)). By an ingenious coupling method between the coaxial elements he arranged the correct modal system in each annulus to give a manifestly uniform amplitude distribution in that region. This is a combination of the TE_{11} and TM_{11} modes for the central region, and of the TE_{11} and TE_{12} modes for each annular region is shown in Fig. 4.3(b). Much of the current work on antenna feeds or horns which by themselves create shaped beam patterns now centres on the creation of these modes by mode transformers at the throat of the horn[7] or by the addition of hybrid modes such as would be created by impedance surfaces, corrugations or dielectric layers in the interior of the flare of the horn. The essence of the design is to create the correct mode in the correct amplitude and phase to create what in practice becomes a zoned distribution. It can be noted from Fig. 4.3 that it is possible to insert an intermediate diametral conductor to divide each zone into two semi-circular halves without upsetting the mode pattern, and that subsequently each semi-circular zone can be considered to have independence of amplitude and phase. This degree of freedom, we shall see, can be of great assistance in the creation of asymmetric patterns.

Larger antennas, if not constructed by the zonal method already noted in Chapter 1, can be constructed from individual circular arrays of discrete elements, although the construction itself is intricate and would only be used

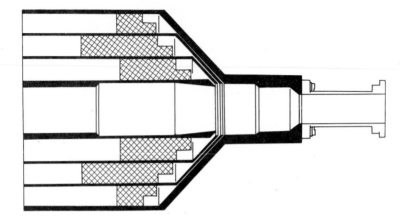

FIG. 4.3(a). *Cross section of multi-zone antenna (from Ref. 6).*

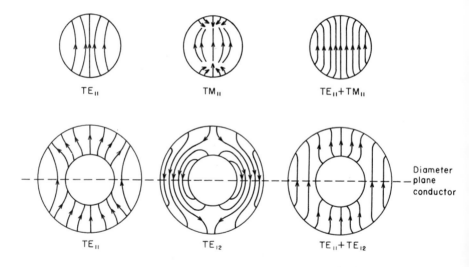

FIG. 4.3(b). *Mode summation to give zones of constant amplitude*

for antennas of a few wavelengths in diameter (Ref. 8, Fig. 4.4). The particular antenna illustrated, has the additional capability of being fed directly from a planar hybrid combination in a way that enables monopulse operation among the four quadrants of the antenna.

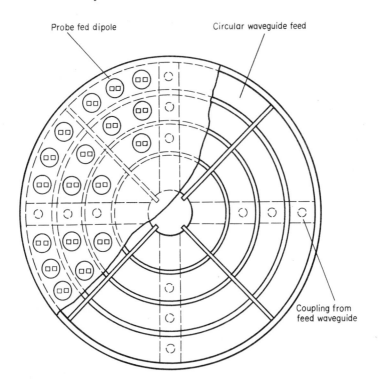

Probe fed dipole   Circular waveguide feed

Coupling from
feed waveguide

FIG. 4.4. Zoned array of discrete elements.

In the study of optics the Fresnel zone plate and phase zone plate are known, and the analysis encompasses these designs. Similar devices have possible applications in microwave antenna designs. The difference as before is the additional phase factor e^{-ibr^2} in the integrand for rays focusing on the axis at a local point. With its omission for the far field pattern of a microwave zoned aperture, the analysis becomes considerably simpler, but leads to many interesting effects applicable to these apertures and to multimode horns. The radii of the zones are arbitrary in the analysis of the free aperture. In the case of the horns these radii have to be chosen to be compatible with the various TE or TM mode combinations which create the zones.

4.5.1 Circular symmetric patterns: Fourier–Bessel and Dini Series

Applying the scalar integral to the construction of the radiation pattern of a single zone with inner radius r_{n-1} and outer radius r_n with a constant complex illumination of magnitude a_{n+1}, gives

$$g_n(u) = \int_{r_{n-1}}^{r_n} a_{n+1} J_0(ur) r\, dr = a_n \left\{ r_n \frac{J_1(ur_n)}{u} - r_{n-1} \frac{J_1(ur_{n-1})}{u} \right\}. \qquad (4.21)$$

Summing these for an aperture of unit radius divided into N zones gives

$$g(u) = \sum_{n=1}^{N} r_n \frac{J_1(ur_n)}{u} (a_n - a_{n+1}), \qquad (4.22a)$$

where $a_{N+1} = 0$,

Putting $A_n = a_n - a_{n+1}$ and r_n the outer radius of the nth zone equal to $\lambda_n r_1$ where $r_1 = 1/m$ and $r_N = 1$

$$g(u) = \sum_{n=1}^{N} \frac{A_n \lambda_n J_1(\lambda_n u/m)}{um}. \qquad (4.22b)$$

Finally putting $u = mv$

$$m^2 v g(mv) = \sum_{n=1}^{N} A_n \lambda_n J_1(\lambda_n v). \qquad (4.23)$$

This equation can be taken to be the starting point for the synthesis of radiation patterns given by the associated function, derived from the given function $g(u)$,

$$F(v) = m^2 v g(mv), \qquad (4.24)$$

$F(v)$ then has to be a generally well behaved function and $\lim_{v \to 0}(F(v)/v)$ must exist, or in terms of Eqn (4.22)

$$g(0) = \sum_{n=1}^{N} \frac{A_n \lambda_n^2}{2m}. \qquad (4.25)$$

This is a basic condition applicable to the amplitude coefficients $A_n = a_n - a_{n+1}$ by whatever means they are to be determined.

The procedure is to compare the series of Eqn (4.23) with the infinite series of the same kind

$$F(v) = \sum_{n=1}^{\infty} b_n J_\nu(\lambda_n v) \qquad (4.26)$$

there being several ways in which this latter expression can be used for the description of quite general functions $F(v)$. The choice depends entirely upon the radii of zones decided upon and hence upon the parameters λ_n.

The first choice is to take the $\lambda_n = \lambda_1, \lambda_2, \lambda_3 \ldots$ proportional to the positive zeros in ascending order of magnitude of the function previously met in Eqn (4.16) (Ref. 2, p. 577), namely

$$x^{-\nu}\{xJ'_\nu(x) + hJ_\nu(x)\}, \tag{4.27}$$

where in the present case $\nu = 1$, h is any given constant. The choice $h = 0$ or $h = 1$ then makes λ_n the zeros of $J'_1(x)$ and $J_0(x)$ respectively. The choice of infinite h gives the zeros of $J_1(x)$. These are all in accord with the multi-mode requirements for the correct boundary conditions for TE or TM propagation in circular horns. The coefficients of the series in Eqn (4.26) are then

$$b_n = \frac{2\lambda_n^2}{\{(\lambda_n^2 - 1)J_1^2(\lambda_n) + \lambda^2 J_1'(\lambda_n)\}} \int_0^1 \upsilon F(\upsilon) J_1(\lambda_n \upsilon) \, d\upsilon, \tag{4.28}$$

and hence for the finite series the amplitude coefficients are given by

$$A_n = \frac{2m^2 \lambda_n}{\{(\lambda_n^2 - 1)J_1^2(\lambda_n) + \lambda^2 J_1'(\lambda_n)\}} \int_0^1 \upsilon^2 g(m\upsilon) J_1(\lambda_n \upsilon) \, d\upsilon \tag{4.29}$$

The procedure being a Fourier–Bessel expansion it can be shown that with these coefficients in the *finite* series the resultant patterns will be the best mean-square fit to the required function $g(u)$.

The integrals on the right hand side of Eqn (4.29) can now be evaluated by the circle polynomial method of the previous chapter. That is

$$\int_0^1 \upsilon^2 g(m\upsilon) J_1(\lambda_n \upsilon) \, d\upsilon = \sum_{s=0}^\infty \alpha_{2s+1}(-1)^s \frac{J_{2s+2}(\lambda_n)}{\lambda_n}, \tag{4.30}$$

where

$$\alpha_{2s+1} = 2(2s + 2) \int_0^1 \upsilon^2 g(m\upsilon) R_{2s+1}^1(\upsilon) \, d\upsilon, \tag{4.31}$$

whence

$$A_n = \frac{2m^2}{\{(\lambda_n^2 - 1)J_1^2(\lambda_n) + \lambda_n^2 J_1'(\lambda_n)\}} \sum_{s=0}^\infty \alpha_{2s+1}(-1)^s J_{2s+2}(\lambda_n). \tag{4.32}$$

The three cases of major interest then give the following results

(a) h infinite; λ_n are proportional to the zeros of $J_1(x)$ and the series is the Fourier–Bessel series

$$A_n = \frac{2m^2}{\lambda_n [J_2(\lambda_n)]^2} \sum_{s=0}^\infty \alpha_{2s+1}(-1)^s J_{2s+2}(\lambda_n). \tag{4.33}$$

(b) $h = 0$; λ_n are proportional to the roots of $J'_1(x)$, the series is a Dini series (Ref. 2, p. 577)

$$A_n = \frac{2m^2}{(\lambda_n^2 - 1)[J_1(\lambda_n)]^2} \sum_{s=0}^{\infty} \alpha_{2s+1}(-1)^s J_{2s+2}(\lambda_n). \tag{4.34}$$

(c) $h = 1$; λ_n are proportional to the roots of $J_0(x)$

$$A_n = \frac{2m^2}{(\lambda_n^2 - 1)[J_1(\lambda_n)]^2 + \frac{1}{4}\lambda_n^2[J_2(\lambda_n)]^2} \sum_{s=0}^{\infty} \alpha_{2s+1}(-1)^s J_{2s+2}(\lambda_n) \tag{4.35}$$

All the solutions are subject to the summation condition of Eqn (4.25). If, for a given $g(u)$ pattern this summation is carried out there results an oscillatory function which therefore only crosses the prescribed value $g(0)$ at a discrete set of values. This limits the total freedom with which $g(u)$ can be selected in dimension but not in shape.

The effect can best be illustrated by an example and the synthesis of a flat topped pattern from an aperture of eight zones will be given.[9] The pattern is to be described by

$$g(u) = 1 \quad \text{for} \quad -u_0 < u < u_0$$

$$= 0 \qquad u_0 < |u| < N\pi$$

where N = number of zones in this case 8. That is

$$m^2 v g(mv) = m^2 v \quad |v| < v_0$$

$$= 0 \qquad |v| > v_0, \qquad v_0 < 1. \tag{4.36}$$

From Eqn (4.29) or (4.33)

$$A_n \lambda_n = \frac{2}{[J_2(\lambda_n)]^2} \int_0^{v_0} m^2 v^2 J_1(\lambda_n v) \, dv \tag{4.37}$$

and $m = \lambda_N$, where the integrals are

$$I = \int_0^{v_0} v^2 J_1(\lambda_n v) \, dv = v_0^3 \int_0^1 t^2 J_1(\lambda_n v_0 t) \, dt$$

$$= v_0^3 \frac{J_2(\lambda_n v_0)}{\lambda_n v_0},$$

either by the method of circle polynomials or Sneddon.[10] Hence

$$A_n \lambda_n = \frac{2u_0^2 \lambda_1^2}{[J_2(\lambda_n)]^2} \frac{J_2(\lambda_n u/\lambda_N)}{\lambda_n} \tag{4.38}$$

These coefficients have first to be substituted into the summation condition

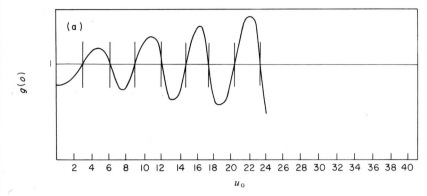

FIG. 4.5(a). Summation function for proper beam widths.

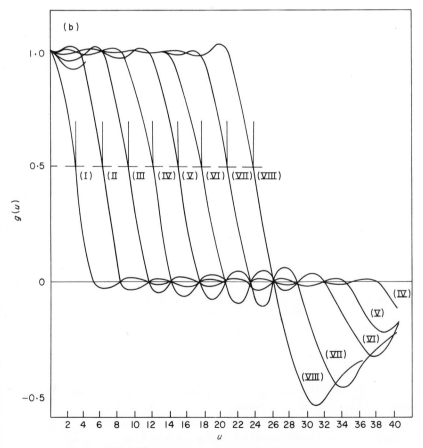

FIG. 4.5(b). Sector patterns—Fourier–Bessel solution.

of Eqn (4.25). This gives a curve of

$$\sum_{n=0}^{N} \frac{A_n \lambda_n^2}{2\lambda_N}$$

as a function of u_0 as shown in Fig. 4.5(a).

Only those values of u_0 which equal the required value of $g(0)$ can then give patterns with correct value at the origin. Other selections of u_0 would have other values of $g(0)$ and hence different shaped patterns would result. These discrete crossings show the relationship between the summation function and the proper set of radiation patterns as illustrated in the figure. The narrowest pattern is that of an ordinary pencil beam with a monotonically decreasing in-phase amplitude distribution among the zones. The widest fills the entire region for which $u = 2a \sin \theta/\lambda$ has real values, that is $u < 2\pi a/\lambda$ and corresponds to a beam filling the complete hemisphere. The required amplitudes for the zones is given in the accompanying table (table 4.1). Of

Table 4.1. Sector Patterns. Zone illumination amplitudes for sector patterns

	I	II	III	IV	V	VI	VII	VIII
a_1	1·0	1·0	1·0	1·0	1·0	1·0	1·0	1·0
a_2	0·95	0·84	0·605	0·402	0·15	−0·05	−0·257	−0·345
a_3	0·86	0·60	0·186	−0·08	−0·20	−0·11	0·017	0·162
a_4	0·74	0·335	−0·102	−0·142	0·004	0·114	0·022	−0·10
a_5	0·67	0·07	−0·156	0·018	0·078	−0·025	−0·08	0·051
a_6	0·423	−0·057	−0·057	0·083	−0·04	−0·015	0·039	−0·038
a_7	0·285	−0·10	0·044	0·004	−0·033	0·054	−0·058	0·012
a_8	0·136	−0·067	0·057	−0·044	0·036	−0·025	0·02	−0·006

course, the physical realization of such wide patterns is mainly hypothetical since it is based upon the original scalar integral for the small angle approximation (Eqn (3.1)) and omits the obliquity factor of $(1 + \cos \theta)$ (Appendix III). For a closer approximation this factor could be taken into account in the manner of the following example.

A further interesting result manifests itself if we continue the range of u in the computation into the complex region for which $\sin \theta > 1$. A far out side lobe appears for each solution which moves in toward the widening patterns

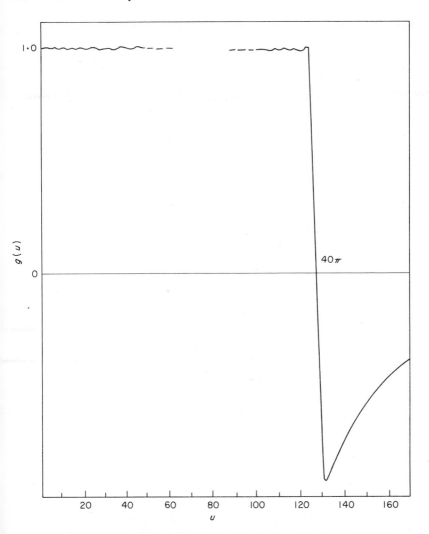

FIG. 4.6. Pattern from aperture of 40 zones.

as can be seen in Fig. 4.5(b). With the widest pattern the side lobe is greatest and contiguous with the pattern itself. Any attempt to create a super-wide pattern, the reverse of super directive in principle, is then defeated by the mutual destruction of the main pattern and the out-of-phase lobe. Mathematically it has the appearance of the Gibbs overshoot of linear Fourier series. Physically, as is known, "radiation patterns" in the complex region represent stored inductive energy in the form of trapped surface waves. This

K

would be created if there were interaction effects between the zones. This concept is in accord with the amplitude distributions shown in the table, where for the widest pattern, that with the highest lobe and therefore greatest stored energy, the zones are alternating in phase, a condition which would be expected to give rise to such inductive coupling between the zones. This is remarkable in so far as it has been produced by a purely scalar theory, yet has the final result that would be expected to arise from a full field and mode expansion. Other effects of interest to be observed from the patterns shown are the half amplitude points which occur at values of u corresponding to the crossing values of the summation curve. This is analogous to the geometrical shadow of an edge which crosses the diffraction shadow at the same level. The common set of zeros for all the patterns occurs in similar solutions and these zeros occur at the turning values of the summation function.

To investigate the overshoot lobe the result of the analysis for a 40 zone aperture is shown in Fig. (4.6). The cut off for the beam is now much sharper and the resultant overshoot much greater. It would appear that in the limit of an exactly square pattern the overshoot would have the same magnitude as the pattern itself.

In an identical manner the derivation of a shaped symmetrical pattern of an increasingly useful kind is shown in the next example. These are for "parabolic" patterns of the generic type

$$g(u) = 1 + p^2 u^2 \tag{4.39}$$

where p can be chosen to accommodate such patterns (or their binomial approximations) as $g(\theta) = \sec^2 \theta, \sec^2 \theta/2$ or to take into account the obliquity factor $(1 + \cos \theta)$.

Such patterns have increasing amplitude with increasing u and are generally desirable when the *uniform* illumination of a curved surface is required. These encompass the illumination of the surface of the earth from a stationary satellite antenna, and the more efficient illumination of a paraboloid from a source at its focus,[11] as observed previously (p. 94).

The patterns have similar properties to the square topped beam and one of the set is shown in Fig. 4.7. The zone illumination amplitudes are obtained from the result[12]

$$A_n \lambda_n = \frac{2}{\lambda_n \{J_2(\lambda_n)\}^2} \left\{ \left(u_0^2 + \tfrac{2}{3} p^2 u_0^4 \right) \frac{J_2(\lambda_n u/\lambda_N)}{\lambda_n} - \frac{p^2 u_0^4}{3} \frac{J_4(\lambda_n u)/\lambda_N)}{\lambda_n} \right\} \tag{4.40}$$

A consequence of this analysis allows the generation of patterns with a central hollow of any given depth, including conical beam shapes with a central null of any width. More than one method is available for this derivation.

(a) Superposition

If any two of the square topped patterns are superimposed in anti-phase, and with due regard to their absolute gains, a pattern with a central null

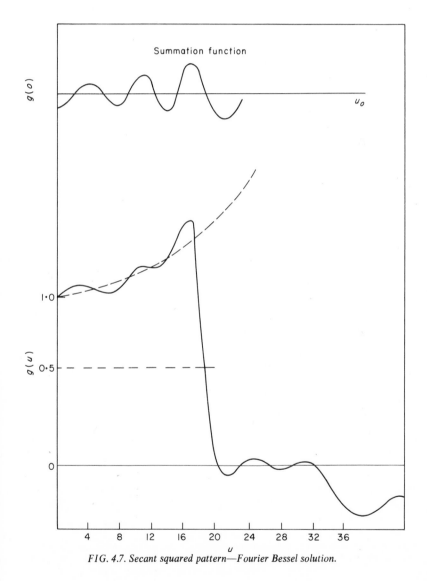

FIG. 4.7. *Secant squared pattern—Fourier Bessel solution.*

of the width of the smaller of the two patterns can be obtained. The required
zone amplitudes are then the appropriate subtraction of the zone amplitudes
of the two patterns concerned (Fig. 4.8).

(*b*) *By direct application to the function*

$$g(u) = 1 \qquad u_0 < u < u_1$$

$$\quad\;\; = 0 \qquad \text{all other values of } u.$$

Then

$$A_n \lambda_n = \frac{2m^2}{\{J_2(\lambda_n)\}^2} \int_{v_0}^{v_1} v^2 \, J_1(\lambda_n v) \, dv$$

$$= \frac{2}{\lambda_n \{J_2(\lambda_n)\}^2} \left\{ u_1^2 J_2\left(\frac{\lambda_n u_1}{\lambda_N}\right) - u_0^2 J_2\left(\frac{\lambda_n u_0}{\lambda_N}\right) \right\} \tag{4.41}$$

for those values of u_0 and u_1 which make $g(0) = 0$.

(*c*) By the choice of a value of u for which the summation function Eqn (4.25)
gives the value $g(0) = 0$ instead of its "proper" value $g(0) = 1$.

   This procedure compels the pattern to start off at a zero value after which
it recovers to give the remaining flat topped portion. A pattern based on
this process is illustrated in Fig. 4.9.

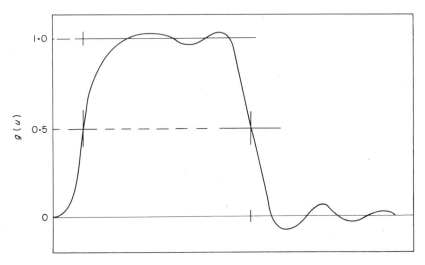

FIG. 4.8. *Conical pattern by superposition. Pattern created by the subtraction of pattern No. 1
from pattern No. V in Fig. 4.5(b).*

The patterns derived by using a Dini series instead of the Fourier–Bessel series have only two fundamental differences from the patterns so far described. Comparing the patterns for a flat topped beam from an aperture of eight zones by each method, shows that apart from the central pencil

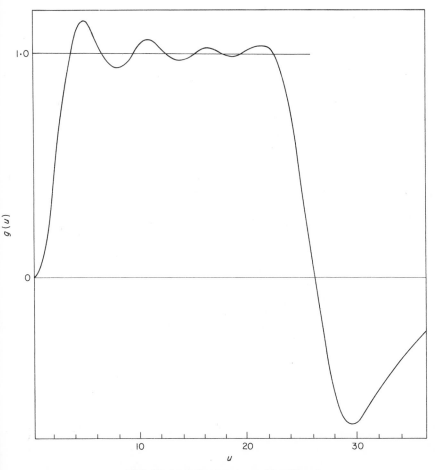

FIG. 4.9. Conical pattern created by $g(0) = 0$.

beam pattern, the approximation to a flat topped beam in the other members of the set is not as good as are the Fourier–Bessel patterns. This is to be expected since the latter is theoretically the best mean square approximation. The result is higher side-lobe levels for the Dini patterns and a less uniform level over the flat region of the pattern. However, the really important effect occurs for the widest patterns. These are all virtually identical except

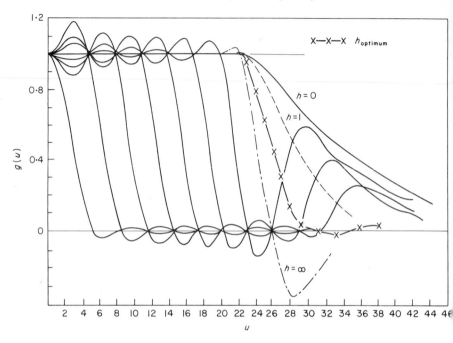

FIG. 4.10. Sector patterns. Dini series solution.

for the cut-off of the square beam. The effect is shown in Fig. 4.10. The value
of h gives the kind of Dini series from the expressions in Eqns (4.33)–(4.35).
As shown the effect of increasing h from zero to infinity is to increase the
sharpness of the cut-off from a gradual slope to the maximum at which level
the overshoot lobe is at its greatest. Obviously, a choice of value for h can be
made which optimizes the slope of the cut-off of the beam and the level of the
overshoot. One such possibility is guessed at in the figure.

4.5.2 Circularly symmetric patterns—Schlömilch series

An entirely different form of series results with the choice of zone radii
to give zones all of equal radial width.

In this case $r_n = nr_1$ and the substitution $u = Nv$ results in Eqn (4.22)
having the form

$$F(v) = N^2 vg(Nv) = \sum_{n=1}^{N} nA_n J_1(nv), \qquad (4.42)$$

where there are assumed to be N zones and whence $r_1 = 1/N$. This series is to be compared with the infinite series of the similar kind

$$S(v) = \sum_{n=1}^{\infty} A_n J_0(nv) \quad \text{the Schlömilch series.}$$

Assuming, as does Schlömilch himself, the admissibility of differentiating this series term by term we obtain

$$S'(v) = -\sum_{n=1}^{\infty} n A_n J_1(nv). \tag{4.43}$$

And hence comparing with equation 4.42

$$S(v) = -\int^{v} N^2 v g(Nv) \, dv. \tag{4.44}$$

Then (Ref. 2, p. 630) the coefficients A_n of Eqn (4.42) are given by

$$A_n = \frac{1}{\pi} \int_{-\pi}^{\pi} \int_0^{\pi/2} \sec\phi \, \frac{d}{d\phi}[S(v \sin\phi)] \cos nv \, d\phi \, dv, \tag{4.45}$$

where again the form of Eqn (4.42) insists that

$$g(0) = \sum_{n=1}^{N} \frac{n^2}{N^2} \frac{A_n}{2}. \tag{4.46}$$

This replaces the summation condition of the previous section. For a flat topped beam with

$$g(u) = 1 \qquad u < u_0,$$

$$g(u) = 0 \qquad u_0 < u < N\pi,$$

we find

$$A_n = \frac{-2N^2}{\pi} \left\{ \frac{u_0^2}{N^2 n} \sin\left(\frac{n u_0}{N}\right) + \frac{2u_0}{N n^2} \cos\left(\frac{n u_0}{N}\right) - \frac{2}{n^3} \sin\left(\frac{n u_0}{N}\right) \right\}, \tag{4.47}$$

the only complication being integration by parts. The widest of these forms is a good approximation to the flat topped beam but with a much slower cut-off at the end. There is also a narrowest pencil beam derived from monotonic aperture distribution. The intermediate patterns are very poor approximations to the required flat-topped pattern but improve slightly with the width of the pattern (Fig. 4.11).

A better flat topped pattern from this series can be obtained by using a remarkable property of the Schlömilch series pointed out by Watson (Ref. 2, p. 636) that the series

$$\frac{1}{2\Gamma(v+1)} + \sum_{m=1}^{\infty} (-1)^m \frac{J_v(mx)}{(mx/2)^v} = 0, \tag{4.48}$$

for x in the *open* range $-\pi < x < \pi$ oscillating when $x = 0$ and diverging when $x = \pi$. As stated in the reference, this theorem has no analogue in the theory of Fourier series and there is some conjecture whether the result is valid outside the range $-\frac{1}{2} < v < \frac{1}{2}$. Such conjectures are an open temptation to the computer, so putting $v = 1$, $x = v$ and $|v| < \pi$ in the above we obtain

$$2 \sum_{m=0}^{\infty} (-1)^{m-1} \frac{J_1(mv)}{mv} = \frac{1}{2}. \qquad (4.49)$$

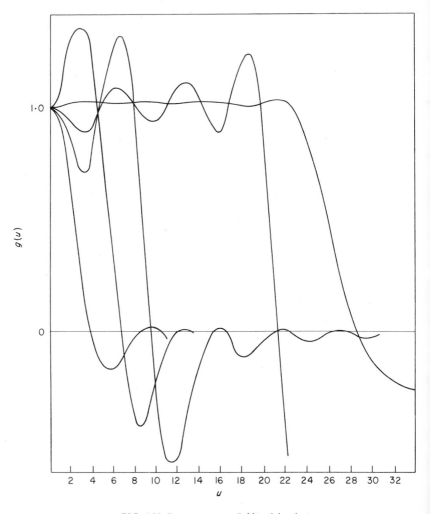

FIG. 4.11. Sector patterns. Schlömilch solution.

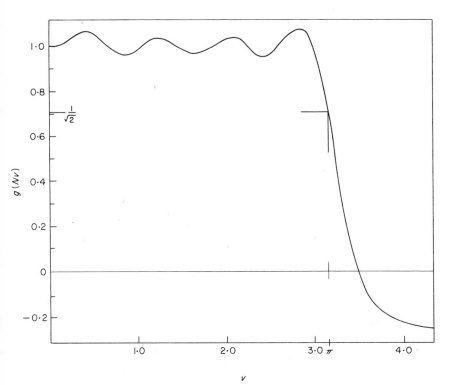

FIG. 4.12. *Sector pattern—null function solution.*

At the origin however, as we were warned, the series oscillates between $\pm\frac{1}{2}$. We find that this alternation of sign depends upon whether an odd or an even number of terms has been taken. So the oscillation can be effectively cancelled by superposing two series, one with an even and one with an odd number of terms as follows

$$\sum_{m=1}^{8} 2(-1)^{m-1}\frac{J_1(mv)}{mv} + \sum_{m=1}^{7} 2(-1)^{m-1}\frac{J_1(mv)}{mv} = 1. \qquad (4.50)$$

That is

$$\sum_{m=1}^{7} 4(-1)^{m-1}\frac{J_1(mv)}{mv} - \frac{J_1(8v)}{4v} = 1 \text{ for } 0 < v < \pi, \qquad (4.51)$$

that is for $u < N\pi$ with in this case $N = 8$.

The A_n coefficients and hence the zone amplitudes are already written in

Eqn (4.51) that is

$$\frac{nA_n}{N^2} = \frac{4(-1)^{n-1}}{n}, \qquad 1 < n < 7,$$

and

$$\frac{A_8}{N} = -\tfrac{1}{4}.$$

The resultant pattern is as shown in Fig. 4.12.

4.6. NON-SYMMETRICAL PATTERNS

Any general pattern $g(u)$ can be separated into its symmetrical and asymmetrical parts by the usual method

$$g(u)_{\text{sym}} = \tfrac{1}{2}\{g(u) + g(-u)\},$$
$$g(u)_{\text{asym}} = \tfrac{1}{2}\{g(u) - g(-u)\}.$$

The symmetrical part can be synthesized by the methods of the previous section and the asymmetrical part then added by the principle of super-position. In the following section we refer entirely to the asymmetrical patterns.

4.6.1 Cyclic phase variation

As noted in Chapter 3, Eqns (3.10) to (3.12), a phase variation of the form $e^{ip\phi'}$ in the angle coordinate of the aperture produces an asymmetric pattern for odd values of p. Little can be done with values greater than unity since patterns would have to be taken in a number of ϕ planes in order to observe the effect fully. For p equal to unity the major effect is observed in the plane $\phi = \pi/2$ and a series definition for patterns in that plane can be obtained from zoned apertures also.

Using the relation for the integration with respect to the ϕ' coordinate

$$\int_0^{2\pi} \exp\left[i(ur \sin \phi' - \phi')\right] d\phi' = J_1(ur). \tag{4.52}$$

We obtain for a single zone with outer radius r_n, uniform amplitude a_n and phase $e^{i\phi'}$

$$g_n(u) = \int_{r_{n-1}}^{r_n} a_n J_1(ur) r \, dr. \tag{4.53}$$

Invoking the method of circle polynomials to evaluate this integral as shown in Chapter 3

$$\int_0^1 J_1(ur) r \, dr = \sum_{s=1,3,5} \frac{2(s+1)}{s(s+2)} \frac{J_{s+1}(u)}{u}, \tag{4.54}$$

then for a single zone

$$g_n(u) = \sum_{s=1,3,5} \frac{2(s+1)}{s(s+2)} \left\{ r_n^2 \frac{J_{s+1}(r_n u)}{r_n u} - r_{n-1}^2 \frac{J_{s+1}(r_{n-1} u)}{r_{n-1} u} \right\}. \tag{4.55}$$

Summing as before for N zones gives

$$g(u) = \sum_{n=1}^{N} \frac{A_n r_n}{u} \left\{ \tfrac{4}{3} J_2(ur_n) + \tfrac{8}{15} J_4(ur_n) + \tfrac{12}{35} J_6(ur_n) + \ldots \right\}, \tag{4.56}$$

where $A_n = a_n - a_{n+1}$ and $a_{N+1} = 0$ as before.
With the same transformations as Eqn (4.22) and (4.23)

$$m^2 v g(mv) = \sum_{n=1}^{N} A_n \lambda_n \{ \tfrac{4}{3} J_2(\lambda_n v) + \tfrac{8}{15} J_4(\lambda_n v) + \ldots \}. \tag{4.57}$$

This double series does not lend itself in any obvious way to the synthesis process but could of course be applied to a trial and error procedure. Equation (4.56) being a completely odd function of u provides the asymmetrical part of the pattern as required.

4.6.2 The generalized Schlömilch series

Consider the aperture of N zones to be divided by a diameter, and that the excitation phase of the opposite halves of each zone be $\pm \alpha_n$ radians with respect to a nominal zero of phase which is the same for all zones (Fig. 4.13). The real part of the amplitude of each zone a_n is considered to be constant. Then from the relation given in Chapter 3, Eqn (3.9), for a complete aperture divided in this way, we have for a *single* zone

$$g_n\left(u, \pm \frac{\pi}{2} \right) = a_n \left\{ \cos \alpha_n \left(r_n^2 \frac{J_1(ur_n)}{ur_n} - r_{n-1}^2 \frac{J_1(ur_{n-1})}{ur_{n-1}} \right) \right.$$
$$\left. \pm \sin \alpha_n \left(r_n^2 \frac{H_1(ur_n)}{ur_n} - r_{n-1}^2 \frac{H_1(ur_{n-1})}{ur_{n-1}} \right) \right\}, \tag{4.58}$$

where $H_1(x)$ is the Struve function.

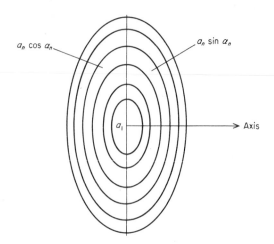

FIG. 4.13. *Aperture of phased semi-circular zones.*

Summing a series of such terms for an aperture of N zones, and choosing the zone radii to give zones of equal width† we obtain a generalization of the Schlömilch series given in Section 4.5.2.

$$g\left(u, \pm \frac{\pi}{2}\right) = \sum_{n=1}^{N} \frac{n}{uN} \left\{A_n J_1\left(\frac{un}{N}\right) \pm B_n H_1\left(\frac{un}{N}\right)\right\}, \qquad (4.59)$$

where we now have

$$A_n = a_n \cos \alpha_n - a_{n+1} \cos \alpha_{n+1},$$

$$B_n = a_n \sin \alpha_n - a_{n+1} \sin \alpha_{n+1},$$

and

$$a_{N+1} = 0. \qquad (4.60)$$

One limiting condition remains as for the ordinary series (Eqn (4.46)) since $H(x)/x \to 0$ as $x \to 0$

$$g(0) = \sum_{n=1}^{N} \frac{n^2}{N^2} \frac{A_n}{2}.$$

Putting $u = Nv$, $-\pi < v < \pi$, we obtain the associated function

$$N^2 v g(Nv) = \sum_{n=1}^{N} n A_n J_1(nv) \pm n B_n H_1(nv). \qquad (4.61)$$

† This condition is essential since $H_1(x)$ has no roots outside $x = 0$ and hence a Fourier–Struve series would not be possible for the odd functional part of this expansion.

In this relation, the first series represents the symmetrical part of the pattern and the second series the anti-symmetrical part.

We compare the series given in Eqn (4.61) with the complete Schlömilch series (Ref. 2, Chapter XIX)

$$S(v) + R = \sum_{n=1}^{\infty} A_n J_0(nv) + B_n H_0(nv), \qquad (4.62)$$

where R is a constant to be determined.

Integrating Eqn (4.61) we have

$$-\int^v N^2 v g(Nv)\, dv = \sum_{n=1}^{N} A_n J_0(nv) + B_n H_0(nv) - \frac{2B_n}{\pi}. \qquad (4.63)$$

Then letting

$$S(v) = -\int^v N^2 v g(Nv)\, dv,$$

and

$$R = \sum_{n=1}^{N} \frac{2B_n}{\pi}, \qquad (4.64)$$

the finite series can be compared with an infinite series to which it forms an approximation. Then, provided $\sum_{n=1}^{\infty} B_n$ is convergent as a precautionary measure, we have from Ref. 2, p. 630

$$A_n = \frac{1}{\pi} \int_{-\pi}^{\pi} \int_{0}^{2\pi} \sec \phi \, \frac{d}{d\phi} [S(v \sin \phi)] \cos nv \, d\phi \, dv,$$

and

$$B_n = \frac{1}{\pi} \int_{-\pi}^{\pi} \int_{0}^{2\pi} \sec \phi \, \frac{d}{d\phi} [S(v \sin \phi)] \sin nv \, d\phi \, dv, \qquad (4.65)$$

where the constant R disappears in the internal differentiations.

From A_n and B_n we now obtain the amplitudes and phases of the illumination functions of the individual zones from Eqn (4.60).

For illustration we can generate the often required cosecant^2 pattern[13] given by

$$|g(u)|^2 = \operatorname{cosec}^2 \theta$$

for the range $\theta_1 < \theta < \theta_2$ not containing the aixs $\theta = 0$ (Fig. 4.14). This, for $u = 2\pi a \sin \theta / \lambda$ becomes

$$g(u) = \frac{1}{u} \quad \text{for} \quad u_1 < u < u_2,$$

and hence

$$g(Nv) = \frac{1}{Nv} \quad \text{for} \quad v_1 < v < v_2.$$

We have directly from Eqn (4.65)

$$A_n = -\frac{N}{2} \int_{v_1}^{v_2} v \cos nv \, dv,$$

$$B_n = -\frac{N}{2} \int_{v_1}^{v_2} v \sin nv \, dv, \qquad (4.66)$$

where "permissible" values of v_1 and v_2 are obtained from the summation condition

$$\sum_{n=1}^{N} \frac{n^2}{N^2} \frac{A_n}{2} = g(0).$$

It can now be shown that the limitation on the range of θ can be varied by

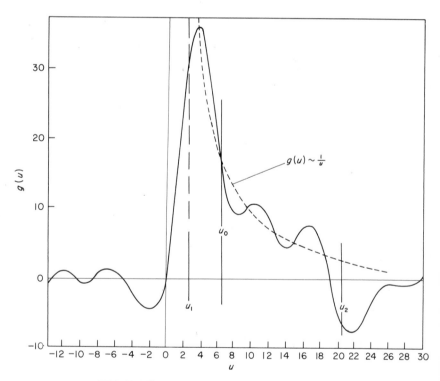

FIG. 4.14. *Theoretical cosecant pattern Schlömilch series.*

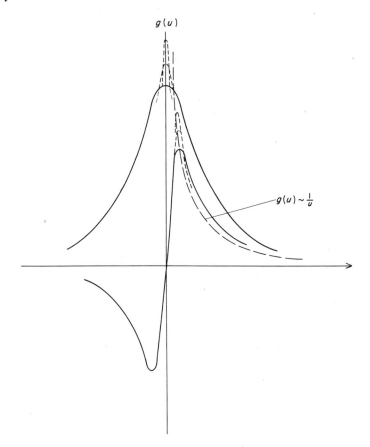

FIG. 4.15. *Odd and even Schlömilch series approximations to the infinite* $1/u$ *pattern.*

an axis shift to a mean value of the pattern at u_0 say (Fig. 4.14) corresponding to θ_0 in the original pattern. The pattern then, given by

$$g(u) = \frac{1}{u - u_0}$$

results in the *same* relations for A_n and B_n but now of course v_1 and v_2 can be on opposite sides of the *new* axis and the value of $g(0)$ in the summation condition becomes

$$g(0) = \frac{-1}{u_0}$$

Some computational procedure is required to obtain the values of v_1 and v_2 and the results for one such pattern are shown in Fig. 4.14.

For the complete cosecant pattern in the range

$$-N\pi < u < N\pi \quad \text{or} \quad -\pi < v < \pi,$$

the relations in (4.66) take on the particularly simple forms

$$A_n = \frac{-2N}{n^2}\{(-1)^n - 1\},$$

$$B_n = \frac{\pi N}{n}(-1)^n.$$

Thus $g(0) = 1 + (1/N)$ and $\sum_{n=1}^{\infty} B_n$ converges.

The symmetrical and anti-symmetrical parts then take on increasingly close approximations to the even and odd forms of $g(u) = 1/u$ as shown in Fig. 4.15. As would be more expected from a Fourier representation, to which the Schlomilch expansion is allied (Ref. 2, p. 620) the odd function takes on the mean value at a discontinuity and thus retains a value of zero at the origin. This makes it possible to form empirical summations of the odd and even forms of the pattern[9] to create asymmetric patterns of the cosecant squared type.

4.7 FRESNEL ZONE PLATES

4.7.1 Amplitude zone plates

One obvious application of the zone theory is in the construction and application of the Fresnel zone plate. These as used in optics are essentially for the focusing of rays at a local point on the axis and thus the optical theory invariably incorporates the focusing factor e^{ibr^2}. The theory given above is essentially simplified by the omission of such a factor for the far field of a microwave antenna. Unfortunately, most of the published optical results do not allow the simple limit $b \to 0$ to be made which would allow for a transition to be made between the two theoretical approaches.

This problem does not arise if the zone plate is considered to be an *addition* to an already collimated beam of rays. In practical terms this implies the addition of the zone plate to the existing aperture of, for example, a focused paraboloid. For narrow enough zones (the minimum being $\lambda/2$) the amplitude distribution can be considered to be constant over the zone and the foregoing analysis applied directly.

As a collimating device in its own right the zone plate has major drawbacks mainly in relation its aperture efficiency. When illuminated by a point source, the amplitude at each zone could still be assumed to be constant, but an addition "spherical" phase term has to be incorporated making the a_n coefficients complex.

The simplicity of the series resulting in the relation in Eqn (4.22) was the objective for its use in a pattern synthesis procedure. It, of course, is directly applicable to the more simple problem of presenting the major aspects of a radiated pattern from a *known* distribution of zone amplitudes and phases, and hence to the Fresnel zone plate whether used as a collimating device or as a pattern shaping device added to an already collimated pattern.

Several quite interesting results can be achieved by inserting quite arbitrary distributions of zone radii, amplitudes and phases, into the coefficients and obtaining resultant radiation patterns.

In a standard form of Fresnel zone plate the zones are alternately transparent and opaque and the radii are proportional to the square root of the zone number.[14] That is with

$$r_n = \beta n^{\frac{1}{2}}, \qquad \beta = (\lambda f)^{\frac{1}{2}},$$

f = focal distance

$$g(u) = \sum_{n=1}^{N} \beta A_n n^{\frac{1}{2}} \frac{J_1(\beta u n^{\frac{1}{2}})}{u}, \qquad (4.67)$$

where A_n are alternately positive and negative. The actual form of A_n depends on the illumination function creating the appropriate a_n and upon whether the central zone is transparent or opaque.

4.7.2 Phase zone plates

Zone plates which are totally transparent do not have the drawback of inefficiency associated with the alternate opaque zones of the amplitude zone plate. Such a plate relies for its action in creating phase shift zones, and has both a collimating and beam shaping action as does the amplitude zone plate.

Its action can again be directly evaluated from the series form of Eqn (4.22), when added to an already collimated aperture. In a practical form this would appear as a sheet of dielectric material with zones of various depths acting as phase shifters over those regions as shown in Fig. 4.16. It is a close approximation to a stepped lens,[15] but with an additional degree of freedom not available to a simple collimating device. It, in fact, provides the additional surface, which as was shown in Chapter 1, could be applied to any one optical advantage. The design procedure given above shows how the pattern from an antenna incorporating such a plate may be evaluated.

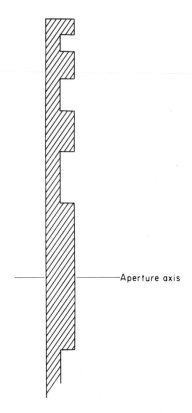

FIG. 4.16. *Microwave phase zone plate.*

REFERENCES

1. V. I. Popovkin, G. I. Shcherbakov and V. I. Yelumeyev. Optimum solutions of problems in antenna synthesis, *Rad. Eng. and Electronic Phys.* **14** (7) (1969), p. 1025.
2. G. N. Watson, "The Theory of Bessel Functions", Cambridge University Press, 1944, p. 533.
3. G. Toraldo di Francia. Super gain antennas and optical resolving power, *Supplement Nuovo Cimento* **1X** Series IX (3) (1952), p. 426.
4. R. C. Hansen. "Microwave Scanning Antennas", Academic Press, London and New York, 1964, Vol. 1, Chap. I, p. 67.
   T. T. Taylor. Design of circular apertures for narrow beam width and low side lobes, *Trans. I.R.E.* **AP8** (1) (1960), p. 17.
   J. Sinnott. Patterns for out of phase semi-circular apertures, *Trans. I.E.E.E.* **AP14** (3) (1966), p. 390.

5. A. Ishimaru and G. Held. Analysis and synthesis of radiation patterns from circular apertures, *Can. J. Phys.* **39** (1960), p. 78.
   J. Ruze. Circular aperture synthesis, *Trans. I.E.E.E.* **AP12** (6) (1964), p. 691.
6. G. F. Koch. Koaxialstrahler als Erreger fur rauscharme Parabolantenn, F.T.Z., January, 1968.
   W. Rebhan. "Theoretical Analysis of Antennas with Sector Shaped Radiation Patterns for Communications Satellites", Zentral Laboratorium fur Nachrichtentechnik Siemens and Halske A.G. Munich.
7. P. D. Potter. A new horn antenna with suppressed sidelobes and equal beamwidths, *Microwave Journal,* (6) (1963), p. 71.
8. S. Cornbleet and J. Brown. Circular array antenna, Brit. Patent no. 1234751, June 9, 1971.
   S. Cornbleet. Monopulse combining network, *Electronics Letters,* **1** (6) (1965), p. 158.
9. S. Cornbleet. Beam shaping for optimum illumination patterns, United Nations Conference on the Peaceful Uses of Outer Space, Vienna, 1968, reprinted in *Electronics Letters,* **4** (23) 1968.
   S. Cornbleet. Pattern synthesis from zoned circular apertures, *Trans. I.E.E.E.* **AP14** (1966), p. 646.
10. I. N. Sneddon. "Mixed Boundary Value Problems in Potential Theory", North Holland Publishing Co., 1966, p. 27.
11. J. S. Ajioka. Shaped beam antenna for earth coverage from a stabilized satellite, *Trans. I.E.E.E.* **AP18** (3) (1970), p. 323.
12. O. Schlömilch. On Bessel's function, *Zeitschrift für Math and Phys.* **2** (1857), p. 155.
13. S. Silver, Microwave antenna theory and design, M.I.T. Series, Vol. 12, 1949, p. 469.
14. L. F. van Buskirk and C. E. Hendrix. The zone plate as a radio frequency focusing element, *Trans. I.R.E.* **AP9** (1961), p. 319.
15. F. Sobel, F. L. Wentworth and J. C. Wiltse. Quasi-optical surface wave-guide and other components for 100–300 GHz region, *Trans. I.R.E.* **MTT9** (1961), p. 512.

BIBLIOGRAPHY

ZONED APERTURES

M. Bottema. Fresnel zone plate diffraction patterns, *Jour. Opt. Soc. Amer.* **59** (12) (1969), p. 1632.
J. Dyson. Circular and spiral diffraction gratings, *Proc. Roy. Soc.* **248A** (1958), p. 93.
S. Fujiwara. Fresnel conic mirror, *Jour. Opt. Soc. Amer.* **51** (11) (1961), p. 1305.
B. A. Lippmann. Exact calculation of the field due to a single Fresnel zone, *Jour. Opt. Soc. Amer.* **55** (4) (1965), p. 360.
M. V. R. K. Murty. Spherical zone plate diffraction grating, *Jour. Opt. Soc. Amer.* **50** (1960), p. 923.
B. J. Thompson. Diffraction by semi transparent and phase annuli, *Jour. Opt. Soc. Amer.* **55** (2) (1965), p. 145.
G. S. Waldman. Variations on the Fresnel zone plate, *Jour. Opt. Soc. Amer.* **56** (2) (1966), p. 215.

M. Young. Zone plates and their aberrations, *Jour. Opt. Soc. Amer.* **62** (8) (1972), p. 972.

ARRAYS OF RINGS

A. Boivin. On the theory of diffraction by concentric arrays of ring shaped apertures, *Jour. Opt. Soc. Amer.* **42** (1) (1952), p. 60.

D. K. Cheng and F. I. Tseng. Maximisation of directive gain for circular and elliptical arrays, *Proc. I.E.E.* **114** (5) (1967), p. 589.

R. Das, Concentric ring array, *Trans. I.E.E.E.* **AP14** (3) (1966), p. 398.

C. O. Stearns and A. C. Stewart. An investigation of concentric ring antennas with low side lobes, *Trans. I.E.E.E.* **AP13** (1965), p. 856.

H. Thielen. Ein Mehrmoden–Koaxialreger fur Parabolantennen, *N.T.Z.* **6** (1971), p. 307.

D. Tichenor and R. N. Bracewell. Fraunhofer diffraction of concentric annular slits, *Jour. Opt. Soc. Amer.* **63** (12) (1973).

J. P. Wild. Circular aerial arrays for radioastronomy, *Proc. Roy. Soc.* **262A** (1961), p. 214.

SYNTHESIS THEORY

L. D. Bakhrakh and A. P. Kurochkin. Application of the convolution transform in optical modelling of microwave antennas and in construction of radiation patterns and the scattering characteristic, *Radio Eng. and Electronic Phys.* **14** (6) (1969), p. 953.

R. G. Cooke. Gibbs' phenomenon in Fourier–Bessel series and integrals, *Proc. London Math. Soc.* Vol. 27 (1928), p. 171. Also: On the summability of Schlömilch series, *Proc. London Math. Soc.* Vol. 41 (1935), p. 176.

G. A. Deschamps and H. S. Cabayan. Antenna synthesis and solution of inverse problems by regularization methods, *Trans. I.E.E.E.* **AP20** (3) (1972), p. 268.

R. L. Fante. Optimum distribution over a circular aperture for best mean-square approximation to a given radiation pattern, *Trans. I.E.E.E.* **AP18** (2) (1970), p. 177.

N. Feld and L. D. Bakhrakh. Present state of antenna synthesis theory, *Radio Eng. and Electron Phys.* **8** (2) (1963), p. 163.

L. P. Grabar. Stabilization and numerical solution of unstable problems encountered in the theory of antenna synthesis, *Radio Eng. and Electronic Phys.* **14** (9) (1969), p. 1352.

J. H. Harris and A. E. Shanks. A method for synthesis of optimum directional patterns from non-planar apertures, *Trans. I.E.E.E.* **AP10** (1962), p. 228.

R. Holland. Optimization criterion for illuminating circular antenna apertures, *Trans. I.E.E.E.* **AP19** (1971), p. 436.

I. G. Klyatskin and V. B. Zhukov. Method of advanced potentials and antenna synthesis, *Telecomm. Radio Engineering,* **23** (5) (1968), p. 76.

S. D. Kremenetsky and V. A. Skachkov. Synthesis of a plane curvilinear radiator, *Radio Eng. and Electronic Phys.* **14** (10) (1969), p. 1611.

A. C. Ludwig. Radiation pattern synthesis for circular aperture horn antennas, *Trans. I.E.E.E.* **AP14** (4) (1966), p. 434.

A. Papoulis. Optical systems, singularity functions, complex Hankel transforms, *Jour. Opt. Soc. Amer.* **57** (2) (1967), p. 207.

C. E. Petersen. Aperture synthesis beam shapes, *Proc. Inst. Rad. Elect. Eng. (Australia)* **31** (10) (1970), p. 361.

D. R. Rhodes. On a class of optimum aperture distributions for pattern shaping, *Trans. I.E.E.E.* **AP20** (3) (1972), p. 262.

P. M. Woodward. A method of calculating the field over a plane aperture to produce a given polar diagram, *Proc. I.E.E.* **93** pt III (1956), p. 1554.

5

Polarization

One of the subjects that clearly demonstrates the unity of physics is the polarization property of the electromagnetic field. Most of the fundamental properties of the polarization of light or of radio waves are long established and can be found in the classical literature. In recent years studies being made of the polarization of light, have produced a representation which shows the deeper connection between the common description of polarization as an elliptically rotating field vector and the spin properties of particles and the entire group theoretic approach to theoretical physics.

Many more applications of this representation have been used in optics than in microwave studies, with the result that the beauty of this particular unifying theory has not been appreciated by antenna designers, who may be able to apply these concepts to problems of immediate practical concern.

The first part of the chapter is concerned with the complete definition of the polarization ellipse. In this we derive in full some of the fundamental relations often merely quoted in the standard literature, but which are required in subsequent analysis. The newer optical representation is introduced and applied to some problems of interest to antenna designers and microwave engineers. In the final part, an indication is given of some of the intimate connections which exist between polarization and general theoretical physics with the expectation that some concepts in the latter could be introduced to assist in problems of polarized antennas.

5.1 THE POLARIZATION ELLIPSE

We can express the total electric field E in terms of its vector components along fixed x and y directions denoted by the unit vectors $\hat{e}_x$ and $\hat{e}_y$. Propagation is assumed to be in the positive z direction. Then

$$\mathbf{E}_x = E(x)\hat{e}_x \exp i\{\omega t - kz\},$$
$$\mathbf{E}_y = E(y)\hat{e}_y \exp i\{\omega t - kz\}, \tag{5.1}$$

$E(x)$ and $E(y)$ being complex and $k = 2\pi/\lambda$.

Putting $E(x) = a_x e^{i\delta x}$ and $E(y) = a_y e^{i\delta y}$, then, at any fixed plane, say $z = 0$, the *real* part of the amplitudes are

$$E_x = a_x \cos(\omega t + \delta_x),$$
$$E_y = a_y \cos(\omega t + \delta_y). \tag{5.2}$$

Elimination of the ωt factor results in the relation

$$\frac{E_x^2}{a_x^2} + \frac{E_y^2}{a_y^2} - \frac{2E_x E_y}{a_x a_y} \cos\delta = \sin^2\delta, \tag{5.3}$$

where $\delta = \delta_y - \delta_x$. This is the equation of an ellipse which, because of the presence of the cross product term, has not been referred to its principal axes. We assume that the major axis of the ellipse makes an angle θ with the x axis. An ellipse with major axis $2a$ and minor axis $2b$ with centre at the origin and whose major axis makes angle θ with the x axis (Fig. 5.1) has the equation (obtained by rotating the ellipse $(x^2/a^2) + (y^2/b^2) = 1$ through a positive angle θ)

$$x^2\left(\frac{\cos^2\theta}{a^2} + \frac{\sin^2\theta}{b^2}\right) + y^2\left(\frac{\sin^2\theta}{a^2} + \frac{\cos^2\theta}{b^2}\right) - 2xy\left(\frac{1}{b^2} - \frac{1}{a^2}\right)\sin\theta \cos\theta = 1 \tag{5.4}$$

Taking lines parallel to the x and y axes such that their intersections with Eqn (5.4) have equal roots we find that this ellipse fits into the rectangle

$$x = \pm\{a^2 \cos^2\theta + b^2 \sin^2\theta\}^{\frac{1}{2}} = \pm c,$$
$$y = \pm\{a^2 \sin^2\theta + b^2 \cos^2\theta\}^{\frac{1}{2}} = \pm d. \tag{5.5}$$

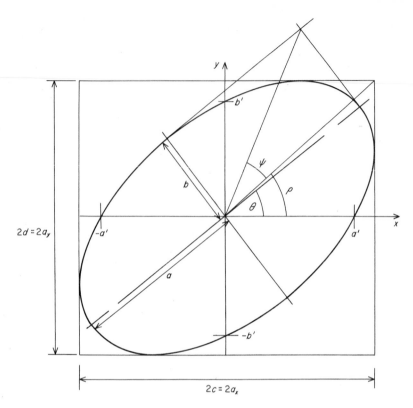

FIG. 5.1. Geometry of polarization ellipse.

The ellipse of Eqn (5.4) intersects the axes at

$$x = \pm a' \qquad \text{where} \qquad a'^2 = \frac{a^2 b^2}{b^2 \cos^2 \theta + a^2 \sin^2 \theta},$$

$$y = \pm b' \qquad \text{where} \qquad b'^2 = \frac{a^2 b^2}{b^2 \sin^2 \theta + a^2 \cos^2 \theta}.$$

(5.6a)

Solving these we find

$$a^2 = \frac{a'^2 b'^2 \cos 2\theta}{b'^2 \cos^2 \theta - a'^2 \sin^2 \theta}; \qquad b^2 = \frac{a'^2 b'^2 \cos^2 \theta}{a'^2 \cos^2 \theta - b'^2 \sin^2 \theta},$$

(5.6b)

and thus

$$\frac{1}{a'^2} + \frac{1}{b'^2} = \frac{1}{a^2} + \frac{1}{b^2},$$

and

$$\frac{1}{b^2} - \frac{1}{a^2} = \frac{1}{\cos 2\theta}\left(\frac{1}{b'^2} - \frac{1}{a'^2}\right). \tag{5.7}$$

The ellipse of Eqn (5.4) with this substitution becomes

$$\frac{x^2}{a'^2} + \frac{y^2}{b'^2} - \frac{2xy \sin\theta \cos\theta}{\cos 2\theta}\left(\frac{1}{b'^2} - \frac{1}{a'^2}\right) = 1. \tag{5.8}$$

Comparing this with Eqn (5.3) in the form

$$\frac{E_x^2}{(a_x \sin\delta)^2} + \frac{E_y^2}{(a_y \sin\delta)^2} - \frac{2E_x E_y \cos\delta}{a_x a_y \sin^2\delta} = 1 \tag{5.3}$$

we find that the tip of the E vector describes the same ellipse, in a rectangle $\pm a_x$ and $\pm a_y$ in magnitude, if

$$a' = a_x \sin\delta; \qquad b' = a_y \sin\delta,$$

and

$$(a_x^2 - a_y^2)\tan 2\theta = 2a_x a_y \cos\delta. \tag{5.9}$$

From the enclosing rectangle given by (5.5) we have

$$c^2 + d^2 = a^2 + b^2 = a_x^2 + a_y^2,$$

and from (5.6)

$$a' = ab/d.$$

Therefore

$$a' = ab/a_y \qquad \text{or} \qquad ab = a_x a_y \sin\delta. \tag{5.10}$$

Hence

$$\frac{2ab}{a^2 + b^2} = \frac{2a_x a_y \sin\delta}{a_x^2 + a_y^2}. \tag{5.11}$$

The ellipticity is defined by

$$\varepsilon = 1 - (b/a)$$

Putting $b/a = \tan\psi$ Eqns (5.9) and (5.11) become

$$\tan 2\theta = \frac{2a_x a_y \cos\delta}{a_x^2 - a_y^2}, \qquad \sin 2\psi = \frac{2a_x a_y \sin\delta}{a_x^2 + a_y^2}. \tag{5.12}$$

The numerical value of $\tan \psi$ gives the axial ratio b/a and the sign differentiates the two senses for the description of the ellipse by the convention

$$0 < \delta < \pi \begin{cases} \text{the polarization is right-handed and} \\ \tan \psi = +b/a \end{cases}$$

$$-\pi < \delta < 0 \begin{cases} \text{polarization is left-handed and} \\ \tan \psi = -b/a \end{cases} \tag{5.13}$$

5.2 STOKES PARAMETERS

Four new parameters may be derived from Eqn (5.12). These are usually denoted by the symbols

$$I = a_x^2 + a_y^2 = |E(x)|^2 + |E(y)|^2,$$

$$Q = a_x^2 - a_y^2 = |E(x)|^2 - |E(y)|^2,$$

$$U = 2a_x a_y \cos \delta = -2\text{Re}\{E(x)E(y)^*\}, \tag{5.14}$$

$$V = 2a_x a_y \sin \delta = -2\text{Im}\{E(x)E(y)^*\}$$

We note that I is the "intensity" in optics or the "power" in electromagnetic field theory. Only three of the four parameters are independent since they are related by

$$I^2 = Q^2 + U^2 + V^2. \tag{5.15}$$

We denote the set given by Eqn (5.14) as the Stokes set $\{I, Q, U, V\}_{\text{lin}}$ and then

$$\tan 2\theta = \left(\frac{U}{Q}\right)_{\text{lin}}, \qquad \sin 2\psi = \left(\frac{V}{I}\right)_{\text{lin}} \tag{5.16}$$

This set of parameters is capable of an interesting physical interpretation.[1] Consider a set of hypothetical filters $F_1 F_2 F_3$ and F_4 with the following properties

Each has a transmittance of 50%.
Each is normal to the incident beam.
F_1 is independent of the incident polarization.
F_2 only transmits horizontally polarized radiation (a vertical wire grating).
F_3 only transmits 45° linearly polarized radiation.
F_4 only transmits right-handed circular polarization.

A power detector is assumed which is independent of the incident polarization. Then if each of the filters is placed in turn between the source and the detector and the readings multiplied by 2 are $V_1 V_2 V_3$ and V_4 then

$$I \equiv V_1, \qquad Q \equiv V_2 - V_1, \qquad U \equiv V_3 - V_1, \qquad V \equiv V_4 - V_1$$

will give the Stokes parameters of the radiation.

Elliptical polarization can also be defined in terms of right and left-handed circularly polarized waves travelling in the same (positive z) direction. The basis vectors are then

$$\hat{\alpha}_r = \frac{1}{\sqrt{2}}(\hat{e}_x - i\hat{e}_y) \quad \text{and} \quad \hat{\alpha}_l = \frac{1}{\sqrt{2}}(\hat{e}_x + i\hat{e}_y), \qquad (5.17)$$

then if

$$\mathbf{E} = E(r)\hat{\alpha}_r + E(l)\hat{\alpha}_l$$

the Stokes parameters are

$$\begin{aligned}
I &= |E(r)|^2 + |E(l)|^2, \\
Q &= -2\text{Re}\,\{E(l)E(r)^*\}, \\
U &= 2\text{Im}\,\{E(l)E(r)^*\}, \\
V &= |E(r)|^2 - |E(l)|^2.
\end{aligned} \qquad (5.18)$$

This set will be denoted by $\{I, Q, U, V\}_{\text{circ}}$.

Putting $\mathbf{E}$ of Eqn (5.1) in the form

$$\begin{aligned}
\mathbf{E} &= E(x)\hat{e}_x + E(y)\hat{e}_y \\
&= \tfrac{1}{2}[E(x) + iE(y)][\hat{e}_x - i\hat{e}_y] + \tfrac{1}{2}[E(x) - iE(y)][\hat{e}_x + i\hat{e}_y] \qquad (5.19)
\end{aligned}$$

and comparing with (5.16) and (5.17) gives

$$E(r) = \frac{1}{\sqrt{2}}(E(x) + iE(y)),$$

$$E(l) = \frac{1}{\sqrt{2}}(E(x) - iE(y)). \qquad (5.20)$$

Equation (5.17) can be put in the equivalent form

$$\begin{pmatrix} \hat{\alpha}_r \\ \hat{\alpha}_l \end{pmatrix} = \frac{1}{\sqrt{2}}\begin{pmatrix} 1 & -i \\ 1 & i \end{pmatrix}\begin{pmatrix} \hat{e}_x \\ \hat{e}_y \end{pmatrix}. \qquad (5.21)$$

If $E(r) = a_r e^{i\delta_r}$ and $E(l) = a_l e^{i\delta_l}$ Eqn (5.20) is

$$
\begin{pmatrix} a_r \exp(i\delta_r) \\ a_l \exp(i\delta_l) \end{pmatrix} = \frac{1}{\sqrt{2}} \begin{pmatrix} 1 & i \\ 1 & -i \end{pmatrix} \begin{pmatrix} a_x \exp(i\delta_x) \\ a_y \exp(i\delta_y) \end{pmatrix} \tag{5.22}
$$

and its inverse is

$$
\begin{pmatrix} a_x \exp(i\delta_x) \\ a_y \exp(i\delta_y) \end{pmatrix} = \frac{1}{\sqrt{2}} \begin{pmatrix} 1 & 1 \\ -i & i \end{pmatrix} \begin{pmatrix} a_r \exp(i\delta_r) \\ a_l \exp(i\delta_l) \end{pmatrix} \tag{5.23}
$$

Rotation of reference frame

If the (x, y) frame is rotated through an angle γ in a right handed screw sense with respect to the propagation vector **k**, linear fields are changed by a rotation matrix but the circularly polarized fields are only changed in phase. If primes denote the rotated system we have for the linear system

$$
\begin{pmatrix} a'_x \exp(i\delta'_x) \\ a'_y \exp(i\delta'_y) \end{pmatrix} = \begin{pmatrix} \cos\gamma & -\sin\gamma \\ \sin\gamma & \cos\gamma \end{pmatrix} \begin{pmatrix} a_x \exp(i\delta_x) \\ a_y \exp(i\delta_y) \end{pmatrix} \tag{5.24}
$$

and for the circular system

$$
\begin{pmatrix} a'_r \exp(i\delta'_r) \\ a'_l \exp(i\delta'_l) \end{pmatrix} = \begin{pmatrix} \exp(i\gamma) & 0 \\ 0 & \exp(-i\gamma) \end{pmatrix} \begin{pmatrix} a_r \exp(i\delta_r) \\ a_l \exp(i\delta_l) \end{pmatrix} \tag{5.25}
$$

It can be simply shown that equations 5.22 and 5.23 transform by these rules.

The column vectors (I, Q, U, V) for the same rotation have the same transformation law in both the linear and the circular representations. This is[1]

$$
\begin{pmatrix} I' \\ Q' \\ U' \\ V' \end{pmatrix} = \begin{pmatrix} 1 & 0 & 0 & 0 \\ 0 & \cos 2\gamma & -\sin 2\gamma & 0 \\ 0 & \sin 2\gamma & \cos 2\gamma & 0 \\ 0 & 0 & 0 & 1 \end{pmatrix} \begin{pmatrix} I \\ Q \\ U \\ V \end{pmatrix} \tag{5.26}
$$

5.3 JONES' VECTORS

The parameters I, Q, U and V all have the dimensions of power and require a 4×4 matrix for transformation an illustration of which is Eqn (5.26). Such matrices are called Mueller matrices and the transformation laws the

Mueller calculus. On the other hand Eqns (5.24) and (5.25) represent the polarization by a two (complex) element column vector and their transformation by 2×2 matrices. This method was developed by Jones.[2] There are nine major differences between the Mueller calculus and the Jones calculus listed in Shurcliff (Ref. 1, p. 122). The outstanding one, from the microwave point of view, is that the Jones calculus retains the phase information of the individual components. The Mueller calculus is able to deal with a completely depolarized wave which of course, is not normally encountered in microwave practice. Consequently Mueller matrices can be derived from Jones matrices in much the same way that power can be derived from complex field components, but the reverse is in general not possible. Such a derivation can be found in the work of Schmeider[3] which gives the method of obtaining a Mueller matrix from the equivalent Jones matrix. For ease of handling, Jones vectors can be "normalized" by the division by any appropriate factor factor which leaves the vector in its simplest normalized state.

The interpretation of a normalized Jones vector is then made as follows

(i) convert the vector to its full form $\begin{pmatrix} a_x \exp(i\delta_y) \\ a_y \exp(i\delta_x) \end{pmatrix}$

(ii) compute $\rho = |\tan^{-1}(a_y/a_x)|$ and $\delta = \delta_y - \delta_x$

(iii) the azimuth of the ellipse is then given by $\tan 2\theta = \tan 2\rho \cos \delta$ (see Eqn (5.39))

TABLE 5.1 *Jones' Vectors.*

Horizontal polarization	$\begin{pmatrix} a_x \exp(i\delta_x) \\ 0 \end{pmatrix}$	$\begin{pmatrix} 1 \\ 0 \end{pmatrix}$
Vertical polarization	$\begin{pmatrix} 0 \\ a_y \exp(i\delta_y) \end{pmatrix}$	$\begin{pmatrix} 0 \\ 1 \end{pmatrix}$
$\pm 45°$ linear polarization	$\begin{pmatrix} a_x \exp(i\delta_x) \\ \pm a_x \exp(i\delta_x) \end{pmatrix}$	$\frac{1}{\sqrt{2}}\begin{pmatrix} 1 \\ \pm 1 \end{pmatrix}$
Linear polarization at angle $\rho = \|\tan^{-1}(a_y/a_x)\|$	$\begin{pmatrix} a_x \exp(i\delta_x) \\ \pm a_y \exp(i\delta_x) \end{pmatrix}$	$\begin{pmatrix} \cos \rho \\ \pm \sin \rho \end{pmatrix}$
R.H. Circular polarization	$\begin{pmatrix} a_x \exp(i\delta_x) \\ a_x \exp[i(\delta_x + \frac{1}{2}\pi)] \end{pmatrix}$	$\frac{1}{\sqrt{2}}\begin{pmatrix} -i \\ 1 \end{pmatrix}$
L.H. Circular polarization	$\begin{pmatrix} a_x \exp(i\delta_x) \\ a_x \exp[i(\delta_x - \frac{1}{2}\pi)] \end{pmatrix}$	$\frac{1}{\sqrt{2}}\begin{pmatrix} i \\ 1 \end{pmatrix}$
Ellipitical polarization $\delta = \delta_y - \delta_x$	$\begin{pmatrix} a_x \exp(i\delta_x) \\ a_y \exp(i\delta_y) \end{pmatrix}$	$\begin{pmatrix} \cos \rho \exp(-\frac{1}{2}i\delta) \\ \sin \rho \exp(\frac{1}{2}i\delta) \end{pmatrix}$

TABLE 5.2 *Jones matrices.*

(1) Free Space	$\begin{pmatrix} 1 & 0 \\ 0 & 1 \end{pmatrix}$

(2) Ideal isotropic absorber with transmission (voltage) coefficient τ (This could be separated into different τ_x, τ_y values) $\begin{pmatrix} \tau & 0 \\ 0 & \tau \end{pmatrix}$

Linear polarizers

(3) Horizontal	$\begin{pmatrix} 1 & 0 \\ 0 & 0 \end{pmatrix}$
(4) Vertical	$\begin{pmatrix} 0 & 0 \\ 0 & 1 \end{pmatrix}$
(5) $\pm 45°$	$\frac{1}{2}\begin{pmatrix} 1 & \pm 1 \\ \pm 1 & 1 \end{pmatrix}$
(6) Polarization Rotator to angle α	$\begin{pmatrix} \cos^2\alpha & \cos\alpha \sin\alpha \\ \cos\alpha \sin\alpha & \sin^2\alpha \end{pmatrix}$
(7) Linear Phase Shifter, free space distance d	$\begin{pmatrix} \exp(i2\pi d/\lambda) & 0 \\ 0 & \exp(i2\pi d/\lambda) \end{pmatrix}$

Phase shift of 90°, azimuth of fast axis γ

(8) $\gamma = 0$	$\begin{pmatrix} \exp(i\pi/4) & 0 \\ 0 & \exp(-i\pi/4) \end{pmatrix}$
(9) $\gamma = 90°$	$\begin{pmatrix} \exp(-i\pi/4) & 0 \\ 0 & \exp(i\pi/4) \end{pmatrix}$
(10) $\gamma = \pm 45°$	$1/\sqrt{2}\begin{pmatrix} 1 & \pm i \\ \pm i & 1 \end{pmatrix}$
(11) γ general	$\begin{pmatrix} \cos^2\gamma \exp(i\pi/4 \\ + \sin^2\gamma \exp(-i\pi/4) & i\sqrt{2}\cos\gamma \sin\gamma \\ i\sqrt{2}\cos\gamma \sin\gamma & \cos^2\gamma \exp(-i\pi/4) \\ + \sin^2\gamma \exp(i\pi/4) \end{pmatrix}$
(12) Phase shift of 180°, fast axis at angle γ	$\begin{pmatrix} \cos 2\gamma & \sin 2\gamma \\ \sin 2\gamma & -\cos 2\gamma \end{pmatrix}$
(13) General phase shift ϕ, fast axis at angle γ	$\begin{pmatrix} \cos^2\gamma \exp(i\phi/2) \\ + \sin^2\gamma \exp(-i\phi/2) & 2i\cos\gamma \sin\gamma \sin(\phi/2) \\ 2i\cos\gamma \sin\gamma \sin(\phi/2) & \cos^2\gamma \exp(-i\phi/2) \\ + \sin^2\gamma \exp(i\phi/2) \end{pmatrix}$
(14) Transmitter of R.H. circular polarization only i.e. total reflection of L.H. circular polarization	$\frac{1}{2}\begin{pmatrix} 1 & -i \\ i & 1 \end{pmatrix}$
(15) Transmitter of L.H. circular polarization	$\frac{1}{2}\begin{pmatrix} 1 & i \\ -i & 1 \end{pmatrix}$

(iv) the axial ratio is given by $\tan \psi$ where $\sin \psi = \sin 2\rho |\sin \delta|$
(v) the handedness is given by $\operatorname{sgn}(\sin \delta)$ that is if

$$\sin \delta > 0 \text{ ellipse is right handed}$$

$$\sin \delta < 0 \text{ ellipse is left handed.}$$

In particular (see Section 5.4) if a normalized Jones vector is represented by $\begin{pmatrix} m \\ n \end{pmatrix}$ the Jones vector of the orthogonally polarized state is $\begin{pmatrix} -n^* \\ m^* \end{pmatrix}$ (*refers to complex conjugate).

The transformation of the Jones vector of a wave, as it is transmitted through a system of optical components, is made through the application of the Jones matrix appropriate to those components. These can be derived heuristically by analysing the properties of each component individually. The matrices in the table are those given in the original work of Jones (loc. cit.)

5.4 REFLECTED WAVES

As in most optical processes the analysis so far has been confined to a progressive wave through a system of optical components and the surface reflections at each component have been ignored. In microwave practice

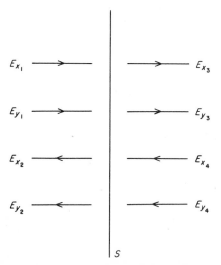

FIG. 5.2. *Intermediate surface: reflection and transmission components.*

considerable loss of power by internal reflections can occur and accumulate in a system which contains many internal surfaces. This is one of the facts limiting the number of such surfaces to be found in a microwave optical system. The reflected power, if not scattered into generally undesirable directions, will be returned to the source and give rise to impedance mismatch. This for any high power source is deleterious to performance. In communications systems, internal reflections can cause serious inter-channel interference and, in general, have to be kept to the minimum possible.

A necessary extension to the above theory therefore is to include in the matrix description the effects of reflected waves.

We can consider S, Fig. 5.2, as an intermediate plane surface in an optical system with a normally incident beam giving rise to a reflected wave in addition to the transmitted wave. The Jones vectors for waves in both directions can be written as a column vector for four components, the upper two for propagation in the positive z direction and the lower two for the negative z direction. These vectors are then related by a 4×4 matrix, not to be confused with the Mueller matrix, which for this problem would require doubling to an 8×8 matrix.

Considering both reflection and transmission the components of Fig. 5.2 are related by

$$
\begin{pmatrix} E_{x2} \\ E_{y2} \\ E_{x3} \\ E_{y3} \end{pmatrix} = \begin{pmatrix} R_{11} & R_{12} & T_{11} & T_{12} \\ R_{21} & R_{22} & T_{21} & T_{22} \\ T_{11} & T_{12} & R_{11} & R_{12} \\ T_{21} & T_{22} & R_{21} & R_{22} \end{pmatrix} \begin{pmatrix} E_{x1} \\ E_{y1} \\ E_{x4} \\ E_{y4} \end{pmatrix}, \tag{5.27}
$$

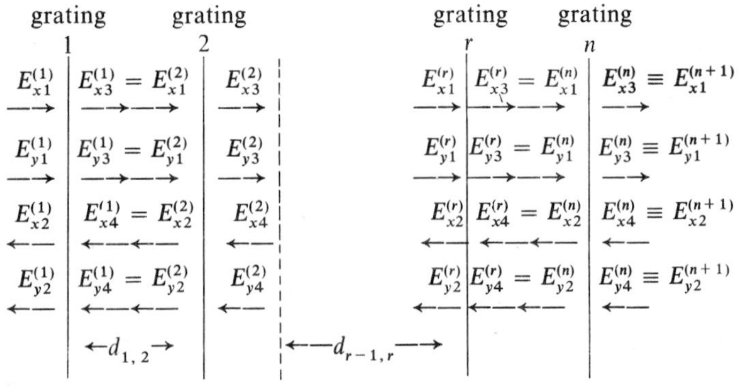

FIG. 5.3. Reflection and transmission components for a system of n surfaces.

E_x and E_y being, as before, complex field components parallel to fixed x and y axes.

The matrix in Eqn (5.27) can be rearranged to give a form applicable to a succession of surfaces, that is a cascade matrix,[4] so that components on the transmission side may be obtained from the components on the incident side. This results in

$$\begin{pmatrix} E_{x_3} \\ E_{y_3} \\ E_{x_4} \\ E_{y_4} \end{pmatrix} = A \begin{pmatrix} E_{x_1} \\ E_{y_1} \\ E_{x_2} \\ E_{y_2} \end{pmatrix}$$

where A has elements $A_{ij}/(T_{22}T_{11} - T_{12}T_{21})$ and where

$$A_{11} = \begin{Bmatrix} T_{11}(T_{11}T_{22} - T_{12}T_{21}) \\ + R_{11}(T_{12}R_{21} - T_{22}R_{11}) \\ + R_{12}(T_{21}R_{11} - T_{11}R_{21}) \end{Bmatrix} \qquad A_{21} = \begin{Bmatrix} T_{21}(T_{11}T_{22} - T_{12}T_{21}) \\ + R_{21}(T_{12}R_{21} - T_{22}R_{11}) \\ + R_{22}(T_{21}R_{11} - T_{11}R_{21}) \end{Bmatrix}$$

$$A_{12} = \begin{Bmatrix} T_{12}(T_{11}T_{22} - T_{12}T_{21}) \\ + R_{11}(T_{12}R_{22} - T_{22}R_{12}) \\ + R_{12}(T_{21}R_{12} - T_{11}R_{22}) \end{Bmatrix} \qquad A_{22} = \begin{Bmatrix} T_{22}(T_{11}T_{22} - T_{12}T_{21}) \\ + R_{21}(T_{12}R_{22} - T_{22}R_{12}) \\ + R_{22}(T_{21}R_{12} - T_{11}R_{22}) \end{Bmatrix}$$

$$A_{13} = (T_{22}R_{11} - T_{21}R_{12}) \qquad A_{23} = (T_{22}R_{21} - T_{21}R_{22})$$

$$A_{14} = (T_{11}T_{12} - T_{12}R_{11}) \qquad A_{24} = (T_{11}R_{22} - T_{12}R_{21})$$

$$A_{31} = (T_{12}R_{21} - T_{22}R_{11}) \qquad A_{41} = (T_{21}R_{11} - T_{11}R_{21})$$

$$A_{32} = (T_{12}R_{22} - T_{22}R_{12}) \qquad A_{42} = (T_{21}R_{12} - T_{11}R_{22})$$

$$A_{33} = T_{22} \qquad A_{43} = -T_{12}$$

$$A_{34} = -T_{12} \qquad A_{44} = T_{11}. \qquad (5.28)$$

The space between the surfaces is represented by the space matrix shown in the table for a linear phase shift in free space of magnitude $2\pi d/\lambda$, where d is the distance between the surfaces. For the forward travelling wave this phase shift is positive and for the reflected wave it is negative. The spacing matrix is therefore a diagonal matrix $D^{r,s}$ with elements

$$D_{11} = D_{22} = \exp\{i2\pi d^{r,s}/\lambda\},$$
$$D_{33} = D_{44} = \exp\{-i2\pi d^{r,s}/\lambda\}, \qquad (5.29)$$

where $d^{r,s}$ is the free space distance between the r and s surfaces. Since the transmitted fields at one surface are the incident fields upon the next as

L'

shown in Fig. 5.3, for an overall system of n surfaces therefore

$$
\begin{pmatrix} E_{x_1} \\ E_{y_1} \\ E_{x_2} \\ E_{y_2} \end{pmatrix}_{(n+1)} = A_n D^{n,\,n-1} A_{n-1} \ldots D^{2,\,1} A_1 \begin{pmatrix} E_{x_1} \\ E_{y_1} \\ E_{x_2} \\ E_{y_2} \end{pmatrix}_{incident} . \tag{5.30}
$$

This gives the transmitted field $\begin{pmatrix} E_{x_1}^{(n+1)} \\ E_{y_1}^{(n+1)} \end{pmatrix}$ and the reflected field $\begin{pmatrix} E_{x_2}^{inc} \\ E_{y_2}^{inc} \end{pmatrix}$ in terms of the incident field $\begin{pmatrix} E_{x_1}^{inc} \\ E_{y_1}^{inc} \end{pmatrix}$ with the field $\begin{pmatrix} E_{x_2}^{(n+1)} \\ E_{y_2}^{(n+1)} \end{pmatrix}$ identically zero, it being the reflection from the non-existent $(n + 1)$th surface. In those cases where the medium between the surfaces is not free space, the phase factor in Eqn (5.29) can be modified by the inclusion of the appropriate refractive index. This, for normal incidence, oblique parallel incidence or oblique perpendicular incidence, are given in Chapter 2, Section 10.

5.5 TRANSMISSION AND RECEPTION OF ELLIPTICALLY POLARIZED WAVES

We consider the power received by an antenna whose elliptically polarized transmitted radiation is the ellipse

$$
\begin{pmatrix} a_x \exp(i\delta_x) \\ a_y \exp(i\delta_y) \end{pmatrix}
$$

with major axis $2a$ and minor axis $2b$ and whose major axis makes angle θ with the x axis. The relations of Section 5.1 Eqns (5.1) to (5.12) therefore apply to this ellipse.

A second elliptically polarized wave is incident upon this antenna, and this incident ellipse can, without loss of generality be referred to the principal x and y axes.

If we designate the properties of the incident elliptical wave by capital letters we therefore have $A_x = A$, $A_y = B$ and the axial ratio is given by $\tan \psi_i = B/A$.

The field received by the first antenna is then obtained from the product

(note the transposition)

$$E = (A_x \quad iA_y)\begin{pmatrix} a_y \exp(-i\delta_y) \\ a_x \exp(-i\delta_x) \end{pmatrix},$$ (5.31)

and hence the received power

$$P = EE^* = A_x^2 a_y^2 + A_y^2 a_x^2 + 2A_x A_y a_x a_y \sin \delta.$$ (5.32)

Use of the relations (Equations (5.10), (5.6a) and (5.6b))

$$a_x a_y \sin \delta = ab,$$

$$a'b' = a_x a_y \sin^2 \delta,$$

allows for the substitution for $a_x a_y$ in terms of a and b in Eqn (5.32) to give the result

$$P = EE^* = (A^2 a^2 + B^2 b^2) \cos^2 \theta + (B^2 a^2 + A^2 b^2) \sin^2 \theta \pm 2ABab,$$ (5.33)

where the positive or negative sign is taken when the ellipses have the same handedness or opposed handedness.

The same result is obtained by rotating the receiving ellipse to the axes of the incident ellipse. That is if

$$E = (A \quad iB)\begin{pmatrix} \cos \theta & -\sin \theta \\ \sin \theta & \cos \theta \end{pmatrix}\begin{pmatrix} a \\ \pm ib \end{pmatrix},$$ (5.34)

then $P = EE^*$ gives the result of Eqn (5.33).

The power in the incident field is $P_i = A^2 + B^2$ and the power of the receiving antenna is $P_r = a^2 + b^2$.

Equation (5.33) can then be put in the form

$$P = P_i P_r \eta,$$ (5.35)

where η is an efficiency factor. Putting $B/A = r_i$, $b/a = r_r$, the axial ratio of the respective ellipses

$$\eta = \frac{[(1 + r_i^2)(1 + r_r^2) + (1 - r_i^2)(1 - r_r^2) \cos 2\theta \pm 4r_i r_r]}{2(1 + r_i^2)(1 + r_r^2)}$$

the same result as Kales.[5]

5.6 THE POINCARÉ SPHERE

From the definitions and analysis of section 5.1, Eqns (5.9) to (5.12) we can derive

$$\tan \psi = \frac{b}{a}, \qquad \frac{a_y}{a_x} = \tan \rho,$$

$$a_x^2 + a_y^2 = a^2 + b^2,$$

$$a_x^2 - a_y^2 = (a^2 - b^2) \cos 2\theta, \qquad \text{from Eqn (5.7)}$$

$$a_x a_y \sin \delta = \pm ab,$$

$$2a_x a_y \cos \delta = (a^2 - b^2) \sin 2\theta, \qquad (5.37)$$

and hence

$$\cos 2\psi = \cos 2\rho \cos 2\theta + \sin 2\rho \sin 2\theta \cos \delta. \qquad (5.38)$$

Equation (5.38) immediately suggests a spherical triangle, one internal angle being $\pi/2$ and another δ.

Other relations which can be derived from the set in Eqn (5.37) and the spherical triangle are then

$$\pm \sin 2\psi = \sin 2\rho \sin \delta,$$

$$\tan 2\theta = \tan 2\rho \cos \delta,$$

$$\cos 2\rho = \cos 2\psi \cos 2\theta, \qquad (5.39)$$

$$\pm \tan 2\psi = \sin 2\theta \tan \delta.$$

These are identical (with change in notation only) to those given in Rumsey.[5]

If we now put

$$\frac{a_y e^{i\delta_x}}{a_x e^{i\delta_y}} = \frac{a_y}{a_x} e^{i\delta} \equiv u + iv$$

then

$$u^2 + v^2 + 2u \cot 2\theta - 1 = 0 \qquad (5.40)$$

and

$$u^2 + v^2 - 2v \operatorname{cosec} 2\psi + 1 = 0 \qquad (5.41)$$

Constant values of θ in Eqn (5.40) gives circles of radius $\operatorname{cosec} 2\theta$ centred on the point $(-\cot 2\theta, 0)$ in the u, v plane.

In Eqn (5.41) constant values of ψ gives circles of radius $\cot 2\psi$ centred at $(0, \operatorname{cosec} 2\psi)$. These circles are orthogonal in the u, v plane and their point of intersection has polar angle δ.

Equations (5.40) and (5.41) can be recognized as analogous to equations of constant resistance and reactance that comprise the system of orthogonal circles in the Cartesian form of the impedance chart.[5] The connection between Eqns (5.38), (5.40) and (5.41) shows that the impedance chart is a stereographic projection of the Poincaré sphere.

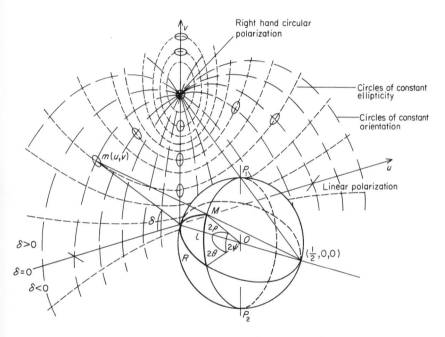

FIG. 5.4. Stereographic projection of the Poincaré sphere.

We consider a sphere of radius $\frac{1}{2}$ centred at the origin (Fig. 5.4), which is tangential to the u, v plane at its origin, the point $(-\frac{1}{2}, 0, 0)$. The point m in the plane is the stereographic projection of the point M on the sphere from the point $(\frac{1}{2}, 0, 0)$ the opposite end of the diameter perpendicular to the u, v plane at its origin. If l is the latitude and k the longitude of M with respect to the polar axis $P_1 P_2$ as shown then

(i) the equation of the sphere is $x^2 + y^2 + z^2 = \frac{1}{4}$
(ii) M has coordinates (x, y, z) and m has coordinates $(-\frac{1}{2}, \tan \rho \cos \delta, \tan \rho \sin \delta)$.

which are connected by

$$2x = \cos 2\psi \, \cos 2\theta,$$

$$2y = \cos 2\psi \, \sin 2\theta, \qquad\qquad (5.42)$$

$$2z = \sin 2\psi.$$

Since $2z = \sin l$, $l = 2\psi$, and since $\tan k = y/x = \tan 2\theta$; $k = 2\theta$.

From the equation for the spherical triangle given by (5.38) the final side of the triangle is the angle 2ρ as shown.

The upper half of the projection being for $\delta > 0$ is for right handed elliptical polarizations and the lower half for left handed. The projection of the poles are then the points corresponding to right-handed circular polarization in the upper half plane and left handed circular polarization in the lower half.

In order to use the Poincaré sphere the effects of the various devices listed

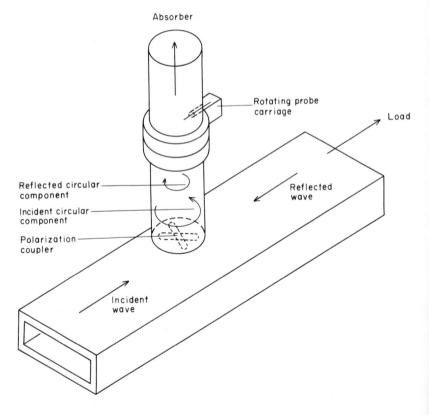

FIG. 5.5. Polarization impedance meter.

in the table for Jones matrices have to be converted to rotations on the sphere. Other projections of the sphere could be considered, for example orthographic projection, which would give rise to other forms of impedance chart. This has been considered fully by Deschamps.[5] Conformal mappings of the impedance chart, for example the Smith chart, could by stereographic projection give other forms of the polarization sphere. Rotations or other automorphisms which map the sphere onto itself are then representations of polarization changes of state and continuous automorphisms represent a continuous change in polarization state as a wave progresses through an optically active medium. This analogy between polarization and impedance has, as a practical consequence, the application of a circular polarizer as a standing wave impedance indicator. If as shown in Fig. 5.5 the coupling from rectangular waveguide to circular waveguide is made through a correctly positioned slot in the broad wall of the waveguide, positively travelling and negatively travelling waves in the rectangular waveguide will induce right-handed and left-handed circularly polarized components in the circular guide. The amplitudes of these will be proportional to the amplitudes of the waves in the rectangular waveguide and thus the elliptical wave which they combine to produce, will be of the axial ratio equal to the ratio of the two waves in the rectangular waveguide and hence to the standing wave ratio. A measurement of this ellipticity can be made by rotating a probe in the circular waveguide. The position of this on the polarization chart is then identical to its position on the equivalent impedance chart and the orientation of the ellipse determines the phase.

5.7 APPLICATIONS TO MICROWAVE ANTENNAS

5.7.1 Circular polarization

A standard form of microwave circular polarizer for incident linear polarization consists of a set of thin parallel plates (Fig. 5.6) set at an azimuth of 45°. Propagation through the plate medium consists of two modes, the TEM at a wavelength equal to the free space wavelength λ_0 for which the **E** field vector is perpendicular to the plates, and the TE_{01} at a wavelength of λ_g for which the **E** field vector is parallel to the plates. The two wavelengths are related by the waveguide relation

$$\lambda_g = \lambda_0 / \{1 - (\lambda_0/2a)^2\}^{\frac{1}{2}}$$

where a is the separation between the plates. In most practical applications the ratio λ_0/λ_g lies in the range 0·5 to 0·8. The two modes can then be separated by any prescribed phase value ϕ by adjusting the width d to suit the relation

$$\frac{2\pi d}{\lambda_0} - \frac{2\pi d}{\lambda_g} = \phi. \tag{5.43}$$

The direction parallel to the edges of the plate is then the azimuth γ of the "fast" axis in the terminology of Table 5.2. For a phase shift $\phi = \pi/2$ and

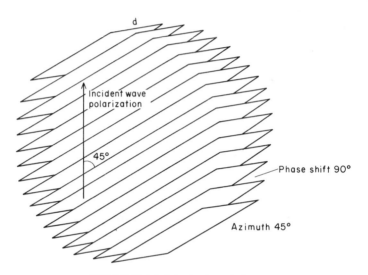

FIG. 5.6. Metal plate circular polarizer.

fast axis azimuth $\gamma = \pi/4$ the appropriate Jones matrix, No. 10 of Table 5.2, is $\frac{1}{\sqrt{2}}\begin{pmatrix} 1 & i \\ i & 1 \end{pmatrix}$. Applied to an incident x polarized wave this gives right-handed circular polarization through the relation

$$\frac{1}{\sqrt{2}}\begin{pmatrix} 1 & i \\ i & 1 \end{pmatrix}\begin{pmatrix} 1 \\ 0 \end{pmatrix} = \frac{1}{\sqrt{2}}\begin{pmatrix} 1 \\ i \end{pmatrix}.$$

With the more general form of phase shift matrix, No. 13 of Table 5.2, we can now extend this analysis to determine the effect of the same polarizer operating over a band of frequencies containing the frequency for which ϕ of Eqn (5.43) is exactly $\pi/2$. This gives the operating bandwidth of the polarizer, subject to prescribed limits on the permissible ellipticity of the transmitted wave.

The same matrix can be used to design a metal plate polarizer (and obtain the operating bandwidth) that will create any given ellipticity of polarization from an incident linearly polarized beam. Without loss of generality this can be taken to be x polarized. Then to create a given ellipticity defined by the Jones vector $\begin{pmatrix} me^{i\theta} \\ ne^{i\pi/2} \end{pmatrix}$ (since the absolute phase is arbitrary) we require the solution for γ and ϕ of

$$\begin{pmatrix} \cos^2 \gamma e^{i\phi/2} + \sin^2 \gamma e^{-i\phi/2} & 2i \cos\gamma \sin\gamma \sin\tfrac{1}{2}\phi \\ 2i \cos\gamma \sin\gamma \sin\tfrac{1}{2}\phi & \cos^2 \gamma e^{-i\phi/2} + \sin^2 \gamma e^{i\phi/2} \end{pmatrix} \begin{pmatrix} 1 \\ 0 \end{pmatrix} = \begin{pmatrix} me^{i\theta} \\ in \end{pmatrix}$$

This is readily found to be

$$\tan 2\gamma = \frac{n}{m \sin\theta}$$

$$\sin\tfrac{1}{2}\phi = m^2 \sin^2\theta + n^2$$

or $\qquad \cos\tfrac{1}{2}\phi = m \cos\theta$ since $m^2 + n^2 = 1$

5.7.2 Polarization rotators

The application of a series of phase shifters of the metal plate type illustrated in Fig. 5.6 can be shown to have the property of rotating the direction of polarization of any linearly polarized wave through a prescribed angle, which, for simplicity, we illustrate by a rotation through 90°.

We consider first two 180° phase shifters in series whose fast axes make an angle of 45°. We choose in the first instance the azimuths to be $22\tfrac{1}{2}°$ and $67\tfrac{1}{2}°$ and show later that this occasions no loss of generality.

The Jones matrices are then (from No. 12 of Table 5.2),

$$\begin{pmatrix} \cos 2\gamma & \sin 2\gamma \\ \sin 2\gamma & -\cos 2\gamma \end{pmatrix}$$

with $\gamma = 22\tfrac{1}{2}°$ for the first and $67\tfrac{1}{2}°$ for the second metal plate system.

Applying these to an incident linearly polarized wave polarized in the arbitrary direction α results in

$$\begin{pmatrix} -\dfrac{1}{\sqrt{2}} & \dfrac{1}{\sqrt{2}} \\ \dfrac{1}{\sqrt{2}} & \dfrac{1}{\sqrt{2}} \end{pmatrix} \begin{pmatrix} \dfrac{1}{\sqrt{2}} & \dfrac{1}{\sqrt{2}} \\ \dfrac{1}{\sqrt{2}} & -\dfrac{1}{\sqrt{2}} \end{pmatrix} \begin{pmatrix} \cos\alpha \\ \sin\alpha \end{pmatrix} = \begin{pmatrix} 0 & -1 \\ 1 & 0 \end{pmatrix} \begin{pmatrix} \cos\alpha \\ \sin\alpha \end{pmatrix} = \begin{pmatrix} -\sin\alpha \\ \cos\alpha \end{pmatrix}$$

That is the final polarization, $\begin{pmatrix} -\sin\alpha \\ \cos\alpha \end{pmatrix}$, is at right angles to the incident

polarization $\begin{pmatrix} \cos\alpha \\ \sin\alpha \end{pmatrix}$.

The independence of this result to the choice of azimuths can be seen from the arbitrary choice of the incident polarization direction α.

It can also be shown by rotating each of the phase shifters through an arbitrary azimuthal angle β while retaining the 45° separation. The Jones matrix for the combination becomes

$$\frac{1}{2}\begin{pmatrix} \cos\beta & \sin\beta \\ -\sin\beta & \cos\beta \end{pmatrix}\begin{pmatrix} -1 & 1 \\ 1 & 1 \end{pmatrix}\begin{pmatrix} \cos\beta & \sin\beta \\ -\sin\beta & \cos\beta \end{pmatrix}\begin{pmatrix} 1 & 1 \\ 1 & -1 \end{pmatrix} = \begin{pmatrix} 0 & -1 \\ 1 & 0 \end{pmatrix}$$

which rotates the polarization through 90° as before.

Application of the more general form of the matrix No. 12 now shows that the polarization rotation caused by a combination of phase shifters of this kind is simply twice the angle between the azimuths of the two individual elements.

Another series combination with the property of rotation of the polarization of a linearly polarized wave through 90°, is the sequence shown in

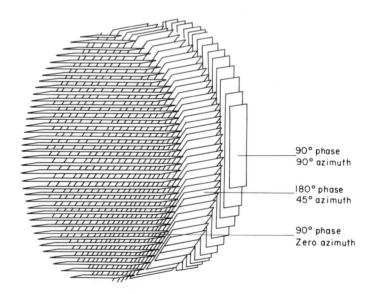

90° phase
90° azimuth

180° phase
45° azimuth

90° phase
Zero azimuth

FIG. 5.7. Polarization rotator 90°.

Fig. 5.7. This consists of

(a) a 90° phase shifter with azimuth at 0°.
(b) a 180° phase shifter with azimuth 45°,
(c) a 90° phase shifter with azimuth at 90°.

The result by the product of the respective matrices is

$$\frac{1}{2}\begin{pmatrix} 1 - i & 0 \\ 0 & 1 + i \end{pmatrix}\begin{pmatrix} 0 & 1 \\ 1 & 0 \end{pmatrix}\begin{pmatrix} 1 + i & 0 \\ 0 & 1 - i \end{pmatrix} = i\begin{pmatrix} 0 & -1 \\ 1 & 0 \end{pmatrix},$$

this being the matrix appropriate to a 90° rotation and with a shift in the phase of $\pi/2$.

This series is also capable of generalization and leads to the following combination

(a) a 90° phase shifter with azimuth γ,
(b) a 180° phase shifter with azimuth zero,
(c) a 90° phase shifter with azimuth $-\gamma$.

This rotates the angle of a linearly polarized incident wave through the angle $\pi - 2\gamma$. In its most symmetric form $\gamma = 60°$ and the polarization is rotated through 60°.

Although the devices illustrated here have been given in a free-space form, that is the effect upon an infinite normally incident plane wave, identical results can be obtained with analogues using the dominant mode in a circular waveguide. The differential phase required can be obtained by the insertion of a vane of dielectric along a diameter of the guide. The "fast" axis is the axis perpendicular to this vane and the length of the vane including its matching sections, determines the value of ϕ. The matrix theory can then be applied in full to these waveguide components. Similar electronically actuated devices using ferrite materials will have the same design principles.

5.8 MULTIPLE INCLINED GRATINGS

Major disadvantages of the polarization rotators discussed in the previous sections are the common microwave ones of large size and complexity in manufacture. A polarization rotator will now be described that will rotate linear polarization from any predetermined direction into any other fixed direction without loss. This is done by a series of closely spaced parallel wire gratings each of which has a slightly different azimuth from the preceding grating. The incident polarization is then perpendicular to the wires of the first grating and the transmitted polarization is perpendicular to the wires of the final grating. Analysis shows that this can be done for gratings which are as close as 1/16 of a wavelength and hence a rotation through as much as 90° can be made within a structure less than half a wavelength in depth. This compares very favourably with the metal plate structures. The analysis is performed for plane gratings and for normal incidence.

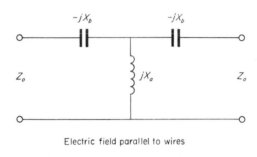

Electric field parallel to wires

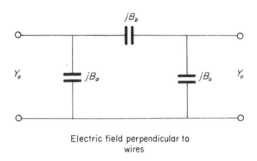

Electric field perpendicular to
wires

FIG. 5.8. Equivalent circuits for parallel wire gratings.

Each grating consists of parallel fine wires of circular cross-section, the spacing between the wires, the diameter of the wires and the separation between the gratings are all small fractions of the wavelength of the incident radiation.

The problem becomes one of determining how all the incremental reflected components which are parallel to the wires at each intermediate surface, combine through reflection, and re-reflection through the previous gratings in such a way that the overall reflection coefficient is minimized.

To do this we employ the cascade matrix set up in Section 5.4, Eqn (5.30).

For any individual grating, the complex transmission and reflection co-efficients parallel and perpendicular to the grating wires will be designated $T_\parallel T_\perp$, $R_\parallel$ and $R_\perp$.

Then the elements T_{ij} and R_{ij} that go to make up the cascade matrix A_{ij} in Eqn (5.28) are

$$T_{11} = T_\parallel \cos^2 \theta_r + T_\perp \sin^2 \theta_r,$$

$$T_{12} = T_{21} = \tfrac{1}{2}(T_\parallel - T_\perp) \sin 2\theta_r,$$

$$T_{22} = T_\perp \cos^2 \theta_r + T_\parallel \sin^2 \theta_r,$$

$$R_{11} = R_\parallel \cos^2 \theta_r + R_\perp \sin^2 \theta_r, \qquad (5.44)$$

$$R_{12} = R_{21} = \tfrac{1}{2}(R_\parallel - R_\perp) \sin 2\theta_r,$$

$$R_{22} = R_\perp \cos^2 \theta_r + R_\parallel \sin^2 \theta_r,$$

where the azimuth of the individual grating, is the (variable) angle θ_r made by the wires with respect to the fixed x axis.

For a close grating of wires with circular cross-section the coefficients $T_\parallel$, $T_\perp$, $R_\parallel$ and $R_\perp$ can be obtained from the equivalent circuits of Fig. 5.8 They are[6]

$$T_\parallel = \frac{2jX}{2Xx - x^2 + 1 + 2j(X - x)},$$

$$T_\perp = \frac{2jB}{-2Bd - b^2 + 1 + 2j(B + b)},$$

$$R_\parallel = \frac{2Xx - x^2 - 1}{2Xx - x^2 + 1 + 2j(X - x)}, \qquad (5.45)$$

$$R_\perp = \frac{2Bb + b^2 + 1}{-2Bb - b^2 + 1 + 2j(B + b)},$$

where

$$B_a = \frac{2\pi^2 a^2}{\lambda d}\left\{1 - \frac{\pi^2 a^2}{\lambda^2}\left[\frac{11}{2} + 2\log\left(\frac{2\pi a}{d}\right)\right] - \frac{\pi^2 a^2}{6d^2}\right\},$$

$$B_b = \frac{d}{\lambda}\left[\frac{3}{4} - \log\left(\frac{2\pi a}{d}\right)\right] + \frac{d\lambda}{2\pi^2 a^2},$$

$$X_a = \frac{d}{\lambda}\left\{\log\left(\frac{d}{2\pi a}\right) + \tfrac{1}{2}\sum_{m=-\infty}^{\infty}\left[\left(m^2 - \frac{d^2}{\lambda^2}\right)^{-\frac{1}{2}} - \frac{1}{|m|}\right]\right\},$$

$$X_b = \frac{d}{\lambda}\left(\frac{2\pi a}{d}\right)^2, \qquad (5.46)$$

and

$$X = X_a/Z_0, \qquad x = X_b/Z_0, \qquad B = B_b/Y_0 \qquad \text{and} \qquad b = B_a/Y_0. \qquad (5.47)$$

The analysis need by no means be confined to wires of circular cross-section. The reflection and transmission coefficients for gratings of elements with other cross-sections, including thin strips and elliptical cross-sections, can be found in the literature, in a form suitable for direct substitution into either the equivalent circuit Eqns (5.46) or Eqn (5.44).[7]

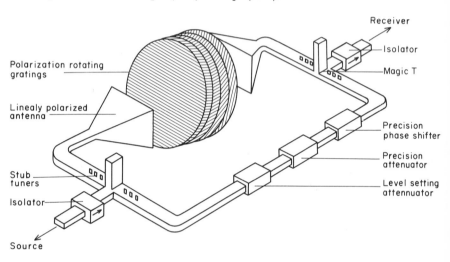

FIG. 5.9. *Experimental apparatus for polarizing grating measurement (Schematic).*

In the practical case where the gratings are only separated by a fraction ($\simeq \frac{1}{16}$) of a wavelength, the question of grating interaction has to be considered. The analysis has only considered plane TEM wave propagation and ignored evanescent or higher order modes. This has been investigated experimentally in a microwave bridge by the method shown diagrammatically in Fig. 5.9. A set of five fine wire gratings was used to rotate an incident linear polarization through 90°. The results agreed with the analysis for a range of spacings between the individual gratings which indicates a bandwidth of 3:1 for almost 100% transmission. In fact the arrangement can be used at any frequency for which the spacing between gratings is well below the critical value of $\lambda/2$. At this value, a resonant reflection effect occurs (Fig. 5.10) and this is found to increase in sharpness with increasing number of gratings. Such a reflection resonance indicates the possible use of this grating array as a polarization frequency filter.

One possible application for a polarization rotator of this type is to ajdust the polarization of the 360° azimuthal scanning antenna known as the "barrel" reflector. As discussed in Chapter 1, (p. 58) this relies for its operation upon the use of a linearly polarized feed and a grating system all at ±45° to the vertical. The "barrel" is a parabolic torus whose circular cross-section is of sufficiently large radius to give fairly precise focal properties on the

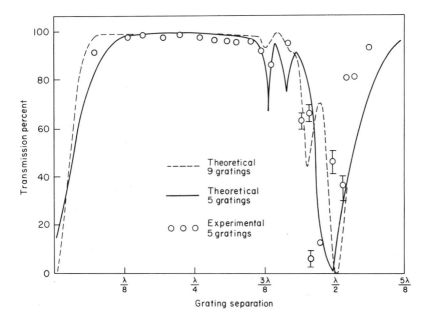

FIG. 5.10. *Polarization rotation through 90° with closely spaced gratings.*

same circle as the vertical parabolic cross-section. The reflector is constructed from a grating of wires at 45° and parallel to the polarization of the feed. After reflection the polarization is at right angles to the wire over the region of the reflector diametrically opposite the collimating area, and is subsequently transmitted. However, it remains polarized at 45° and this for many applications is undesirable. It can thus be corrected to the more normal vertical or horizontal polarizations by the addition of a few additional layers of wire gratings in the manner described above (Fig. 5.11).

This application of the polarization cascade matrix is only one of many that could be considered. It has been chosen to illustrate the applicability of the method under the extreme condition of closely packed elements. Other devices with wider element spacing and wire separation but with

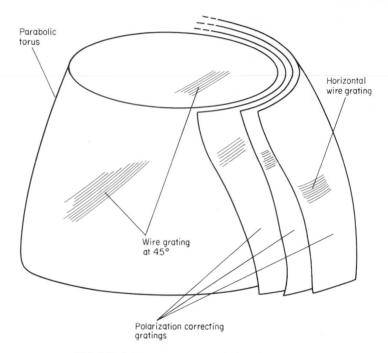

Parabolic
torus

Horizontal
wire grating

Wire grating
at 45°

Polarization correcting
gratings

FIG. 5.11. Polarization correction for barrel antenna.

more rigorous engineering tolerances can be designed by this process. This
would include systems such as circular polarizers and analysers. The method
can also be applied to the operation of any of these systems at other than
normal incidence by suitably modifying the grating coefficients in Eqns (5.44)
and (5.45) for oblique incidence and adjusting the transmission and spacing
matrices accordingly.

5.9 POLARIZATION TWIST REFLECTORS

Any transmission system which is *symmetrical* about a mid-plane can be
converted into a reflection system with identical properties by the insertion
of a conducting surface at the mid-plane. This is true in optical filter theory
as it is in electric circuit theory and thus applies to the previous examples of
this chapter.

In a most elementary application of this principle a polarization rotator which consists of a single stack of plates can be converted into a reflector with the same polarization rotation property by a conducting surface at the centre of the plates bisecting the stack. In the case of a 90° polarization rotator the resultant structure has the simple form of a set of parallel plane fins on a plane conducting packing surface, in which a differential phase of π radians is obtained between the TEM mode perpendicularly polarized to the fins and the TE_{01} mode of the parallel component. (Fig. 12(a)).

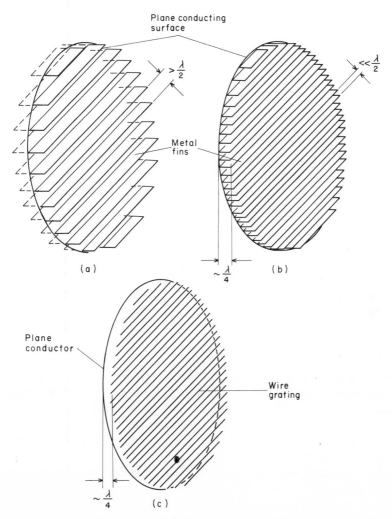

FIG. 5.12. *Polarization rotating reflectors.* (a) *Bisected polarization rotator,* (b) *and thinned version with closely spaced fins,* (c) *bisected grating polarization rotator.*

   Since the device is no longer acting as a transmitting system, propagation of both modes is no longer essential and the structure takes on the even more simple form whereby the fins are sufficiently close to completely reflect the parallel polarized component and the differential phase is obtained from the path length between the fins of the TEM component (Fig. 12(b)). In the main this depth of the fins is approximately a quarter wavelength, however the edges of the fins do not act as a precise short circuit to the parallel polarized component and introduce a slight additional phase shift. This manifests

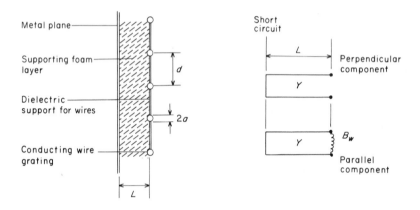

FIG. 5.13. Polarization twist reflector and equivalent circuits.

itself as a small shift of the exact position of the short circuit terminals for the parallel component and an even smaller shift of the open circuit terminals for the perpendicular component. For precision the exact amounts of these shifts can be obtained from the formulae presented in Ref. 6.

   In a similar way an even number of *parallel* wire gratings can be shown to introduce a polarization rotation of 90° by the matrix method of Section 3.3. For a simple structure (Fig. 5.12(c)) two gratings are sufficient and both theory and experiment have shown[7] that such a combination can act as a 90° polarization rotator. The action of the gratings in this case can be visualized as an inductance in the case of the parallel component and a (very small) capacitance to the perpendicular component.

   The resulting phase shift can be arranged to produce the required polarization rotation.

   In a practical structure the wires of each grating would be supported in a transparent dielectric medium and the two gratings kept separate and parallel by a light density dielectric medium such as rigid plastic foam. The addition of the effects of such refractive media into the matrix calculation, as has been shown, presents no difficulty.

The insertion of a metal wall into the mid-plane of this structure now creates a particularly effective form of polarization rotating reflector, or twist-reflector.

As a design procedure the matrix method is not wholly satisfactory as no obvious iterative procedure can be found to derive the matrix elements which alone define the system parameters. Hence a large number of *a priori* systems have to be analysed and an optimum obtained usually by graphical methods.

A method for the design of this type of twist reflector has been given by Hannan[8] using the circuit analogue, which has been found to produce accurate results for most applications.

If, as in Fig. 5.13, the wires are of radius a and have a grating spacing of d the approximate admittance from Eqn (5.46) is

$$\frac{1}{X_a} = B_w = \lambda/\{d \log (d/2\pi a)\} \text{ at normal incidence}$$

At oblique incidence at an angle θ, this becomes modified to

$$B_w = \lambda/\{d \cos \theta \log (d/2\pi a)\}$$

If the wires are spaced a distance L from the conducting surface, then, referred to a terminal plane at the grating, the perpendicular component sees a short circuited length of transmission line with admittance Y of length L and the parallel component sees a similar short circuited line but with a parallel admittance B_w at the terminal plane.

For operation as a 90° polarization rotator the phase differential is contained in the condition

$$Y_\perp = Y_\parallel^{-1}, \tag{5.48}$$

where

$$Y_\perp = i \cot (2\pi L \cos \theta/\lambda)$$

and

$$Y_\parallel = i \cot (2\pi L \cos \theta/\lambda) - iB_w$$

the solution of which gives

$$B_w = 2 \operatorname{cosec} (4\pi L \cos \theta/\lambda)$$

or

$$2 \operatorname{cosec} (4\pi L \cos \theta/\lambda) = \lambda/\{d \cos \theta \log (d/2\pi a)\}. \tag{5.49}$$

Optimum performance is obtained if this relation holds to the second order for changes in λ or θ. Differentiating therefore with respect to θ we obtain

$$(4\pi L \cos \theta)/\lambda = \tan\{(4\pi L \cos)/\lambda\}$$

and hence

$$(L \cos \theta)/\lambda = 0{\cdot}358. \tag{5.50}$$

Substitution into Eqn (5.49) then gives

$$\lambda/\{d \cos \theta \log(d/2\pi a)\} = 2{\cdot}046. \tag{5.51}$$

For any given angle of θ this can simply be solved by iteration. In a practical design θ would be taken to be a weighted mean of the angles of incidence the reflector is likely to intercept and the wire radius a chosen from the catalogue of available materials. The spacing d can then be determined.

Finally minor adjustments in the separation L can be made to account for the effects of the supporting dielectric materials.

Many applications exist for a reflector of this kind and are listed in Hansen.[9] In the optical sense the main interest lies in the many two reflector arrangements that become possible through its application as described in Chapter 1. If the first reflector is constructed as a grating of closely spaced wires which reflects the parallel polarized incident radiation entirely, the second reflector can be a twist reflector which turns the polarization through 90° and thus makes it pass through the first reflector without loss. This principle is illustrated in Fig. 55 of Chapter 1 and was first used in a paraboloid-plane combination as shown.[10] In this form it can scan a pencil beam over the entire forward hemisphere. As always other arrangements are possible which, while limiting the scan through the introduction of aberrations, make possible other desirable effects such as varying the illumination function.

The properties of these twist reflectors have been found to be constant over the wide range of incident angles that such systems and those illustrated in Chapter 1 require.

5.10 IMPOSSIBLE POLARIZERS

There are few polarizing systems whose application to microwaves would be highly desirable but whose design can be shown to be impossible by the nature of the Jones matrix required to describe them. For example, in nearly all open antenna systems and particularly in communications systems over long terrestrial paths or through the atmosphere to and from satellites, signal fading occurs at random. This can be caused by absorption and scattering mechanisms. Quite often however, and not always taken into consideration, it can be caused fully or in part by polarization rotation. Polarization diversity reception at the receiving antenna could be an unwarranted technical complication. Should, however, a device with total polarization agility be possible, the fade could be adjusted instantaneously, as opposed to having a form of servo-driven polarization seeker that would be otherwise required. Such a device if achievable, would be described by the matrix No. 6 in Table 5.2, which rotates any arbitrary incident linear polarization into a given prescribed direction, and without reflection loss of any kind. Such a device has not, so far, been shown possible of practical achievement.

Similarly we observe that polarizers corresponding to matrices numbered 14 and 15 in the table have as yet no physical counterparts. These refer to the total reflectors of one hand of circular polarization and hence total transmitters of the opposite hand.

The feature that distinguishes these matrices from the others in the table is that they are singular.

They cannot therefore be constructed from a series product of the other non-singular matrices each of which corresponds to a realizable practical device.

We are led therefore to the assumption that a lossless, which includes loss by reflection, polarizing device is only capable of practical realization if the Jones matrix representing it is non-singular.

5.11 BI-VECTORS AND QUATERNIONS IN POLARIZATION ANALYSIS

The theory so far given in this chapter has been concerned exclusively with wave propagation in the positive or negative z axis directions. For these cases the polarization state can be uniquely defined by the two component Jones vector or by a combination of the component circular polarizations to give the spinor representation of Eqn (5.22).

We now wish to describe the propagation of an elliptically polarized wave in any general direction in a three dimensional space, in terms of what may be called "basis polarizations" along orthogonal Cartesian axes.

Polarization studies in the field of optics and microwave antennas have involved concepts which, by themselves, would have led to many of the mathematical methods now applied to the study of particle spin. As long ago as 1843 Hamilton himself had found in his discovery of quaternions

"there seems to be something analogous to polarized intensity in the purely imaginary part and to the unpolarized energy (indifferent of direction) in the real part, of a quaternion, and thus we have some slight glimpse of a future calculus of polarities . . . " (letter to Graves, 17th October, 1843[11]).

In this discovery he had in fact anticipated the advent of the spin matrices by approximately a century. Spin matrix methods are now current in the study of optical polarization and concurrently quaternion methods are coming into regular usage.[12, 3]

The application in optics has been mainly to the fields of reflection polarimetry and ellipsometry.[13] These are studies of imperfect or irregular conducting surfaces or of the properties of dielectric materials in the form of thin films backed by perfect conductors through the measurement of the ellipticity of a reflected wave arising from an obliquely illuminating linearly polarized wave. This procedure could well be adapted to the measurement of the dielectric properties of materials at microwave frequencies, for example, such properties at elevated temperatures, or the thickness of transparent layers.

The method by which polarization studies as applied to microwave antennas derive all the properties attributed to quaternions is fascinating. In the classic set of papers published in 1951[5] dealing with the polarization of a generally directed plane wave Kales used the concept of a rotating vector in the plane normal to the propagation direction and the theory of bi-vectors to describe it. The results agree with the results obtained through the spinor method as we have shown in Eqns (5.31) and (5.36).

Both bi-vectors and quaternions have been used in the fundamental study of the relativistic invariance of Maxwell's equations.[14] Rumsey and later Crout[15] in considering the polarization of the radiation from n-arm spiral antennas introduce a bi-complex notation which only falls short of the quaternions themselves in a single particular. A similar consideration and the same result occurs in an entirely different context in a paper by Lewin[16] in which he discusses the wave equation in orthogonal curvilinear coordinates. In brief, the bi-vector notation in calling for a complex combination of the form $\mathbf{E} + i\mathbf{Z}\mathbf{H}$ of the electric and magnetic field vectors requires to distinguish between the complex $\sqrt{-1}$ as used there with the $\sqrt{-1}$ used in the angular frequency of the temporal component $e^{j\omega t}$. Such a bi-complex notation is the essence of the quaternion method. Commutation properties are required of course and these are given as $i^2 = j^2 = -1$ but the essential non-commutative property $ij = -ji$ that distinguishes the quaternions from all that had gone before, is not included.

The full relationship between bi-vectors and quaternions is given by Ehlers *et al.*[17] We shall demonstrate this by a physical quaternion description of a polarized wave.

We specify (Appendix II.4) three hypercomplex numbers α, β, γ with the algebraic properties

$$\alpha^2 = \beta^2 = \gamma^2 = -1,$$
$$\alpha\beta = \gamma, \qquad \beta\gamma = \alpha, \qquad \gamma\alpha = \beta,$$

from which we find fundamentally $\alpha\beta = -\beta\alpha$ (since for example $\alpha\beta\beta\gamma = \gamma\alpha$ but $= -\alpha\gamma$). This last commutation rule must be adhered to in all products of quaternions.

The "ordinary" complex number $e^{i\vartheta}$ now refers to a circular motion in the Argand plane that is to a rotation about the z axis. It is compounded of the two harmonic motions $\cos\vartheta$, $\sin\vartheta$, which are in phase quadrature and orthogonal in space. If we let each of the complex numbers α, β and γ refer similarly to rotations about the three orthogonal Cartesian axes then say

$e^{\alpha\vartheta}$ refers to a rotation about the z axis,

$e^{\beta\vartheta}$ refers to a rotation about the y axis,

and

$e^{\gamma\vartheta}$ refers to a rotation about the x axis.

This notation is the only variant between quaternion notation and the spin matrices.

The required circularity is a real circle $e^{i\psi}$ in a plane perpendicular to the direction (θ, ϕ) of wave propagation. That is ψ is a real rotation angle and i the ordinary $\sqrt{-1}$. This means in principle that $i = \sqrt{-1}$ has to be projected into the three basis coordinate directions and each component has to represent a basis circularity of the form $e^{\alpha\vartheta}$ . . . .

This can be done by the projection (Appendix II)

$$i = \alpha\cos\theta + \beta\sin\theta\sin\phi + \gamma\sin\theta\cos\phi. \qquad (5.52)$$

The validity of Eqn (5.52) can readily be tested by the direct square product of both sides which give -1 when the rules of quaternion multiplication are used. Then the rotation in the plane normal to the θ, ϕ direction becomes

$$\cos\psi + i\sin\psi = \cos\psi + \sin\psi\{\alpha\cos\theta + \beta\sin\theta\sin\phi + \gamma\sin\theta\cos\phi\},$$
$$(5.53)$$

which relation is established algebraically in the appendix.

This is then of the form $\cos\psi +$ (complex number) $\sin\psi$ for *any* choice of θ and ϕ. For example rotations about the x, y, and z axes are given by

$$x \text{ axis} \quad \theta = \frac{\pi}{2}, \quad \phi = 0 \quad e^{i\psi} = \cos\psi + \gamma \sin\psi$$

$$y \text{ axis} \quad \theta = \frac{\pi}{2} \quad \phi = \frac{\pi}{2}, \quad e^{i\psi} = \cos\psi + \beta \sin\psi \qquad (5.54)$$

$$z \text{ axis} \quad \theta = 0 \qquad\qquad e^{i\psi} = \cos\psi + \alpha \sin\psi$$

The connection with particle spin is now made through the fact that *complex* spin matrices have the same algebra as the complex α, β, γ given above and so we find[12]

$$\alpha = i\sigma_3, \qquad \beta = i\sigma_2, \qquad \gamma = i\sigma_1,$$

where

$$\sigma_1 = \begin{pmatrix} 0 & 1 \\ 1 & 0 \end{pmatrix} \quad \sigma_2 = \begin{pmatrix} 0 & -i \\ i & 0 \end{pmatrix} \quad \sigma_3 = \begin{pmatrix} 1 & 0 \\ 0 & -1 \end{pmatrix}$$

are the Pauli spin matrices.

A complete study of this aspect is not possible within the context of this volume although some of the benefits to be hoped for will be indicated in the final chapter.

The group theory of the quaternion algebra leads in its own way to representations of rotation in both three and four dimensions. The transformation of polarization as a rotation of the Poincaré sphere then becomes an operator transformation by a quaternion multiplier. In four dimensions the complex–complex form of the rotation gives Lorentz transformations and exemplifies the well known connection between spinors and operators of the Lorentz group.

The entire study of the rotation group therefore has a relevance to the study of polarization and to all other applications which can be described by a quaternion operation. A great deal of the recent group theoretic studies of the fundamental particles can thus be said to have an immediate relevance to optical and microwave polarization and allied problems.

REFERENCES

1. W. A. Shurcliff. "Polarized Light", Harvard University Press, 1962.
   R. C. Newton. "Scattering Theory of Waves and Particles", McGraw Hill, 1966, pp. 4–10.

P. Beckman. "The Depolarization of Electromagnetic Waves", Golem Press, 1968.

2. R. C. Jones. A new calculus for the treatment of optical systems, Parts I–VIII, *Jour Opt Soc. Amer.* **31** (1941), 488; **31** (1941), 493, with H. Hurwitz; **31** (1941), 500: **32** (1942), 486; **37** (1947), 107; **37** (1947), 110; **38** (1948), 671; **46** (1956), 126.

3. R. W. Schmeider. Stokes algebra formalism, *Jour. Opt. Soc. Amer.* **59** (1959), 297.

4. N. Hill and S. Cornbleet, Microwave transmission through a series of inclined gratings, *Proc. I.E.E.* **12** (4) (1973), 407.
W. E. Groves. Transmission of electromagnetic waves through pairs of parallel wire grids, *Jour. App. Phys.* **24** (1953), 845.

5. Techniques for handling elliptically polarized waves with special reference to antennas, *Proc. I.R.E.* **39** (1951), 533–556.
H. G. Booker. Introduction, p. 533;
V. H. Rumsey. Transmission between elliptically polarized antennae, p. 535.
G. A. Deschamps. Geometrical representation of the polarization of a plane electromagnetic wave, p. 540.
M. L. Kales. Elliptically polarized waves and antennas, p. 544.
J. I. Bohnert, Measurements of elliptically polarized waves, p. 549.
M. G. Morgan and W. R. Evans Jnr. Synthesis and analysis of elliptic polarization loci in terms of space-quadrature sinusoidal components, p. 552.
R. M. A. Azzam and N. M. Bishara. Polarization transfer function of an optical system as a bilinear transformation, *Jour. Opt. Soc. Amer.* **62** (7) (1972), p. 222.
L. J. Kaplan. Bilinear transformation of polarization, *Jour. Opt. Soc. Amer.* **62** (10) (1962), p. 1239.
H. M. Barlow and A. L. Cullen. "Microwave Measurements", Constable, 1950, Appendix II, p. 376.

6. N. Marcuvitz. "Waveguide Handbook", M.I.T. Series, Vol. 10, McGraw Hill, 1951.

7. J. K. Skwirzynski and J. C. Thackray. Transmission of electromagnetic waves through wire gratings, *Marconi Review* **22** (1959), p. 77.

8. P. W. Hannan. Microwave antennas derived from the Cassegrain telescope, *Trans. I.R.E.* **AP9** (1961), p. 140.

9. L. K. de Size and J. F. Ramsey. Reflecting systems. *In* "Microwave Scanning Antennas". (R. C. Hansen, Ed.) Vol. 1, Academic Press, London and New York, 1964, p. 128.

10. P. F. Mariner and C. A. Cochrane. High frequency radio aerials, British Patent no. 716939 August 1953.
P. F. Mariner. Microwave Aerials with full hemispherical scanning, *L'Onde Electrique,* Supplement August 1958, No. 376. Proceedings of the International Congress on Ultra high frequency circuits and antennas, Paris, 21–26 October, 1957, Vol. II, p. 767.

11. H. Halberstram and R. E. Ingram. "Collected Papers of W. R. Hamilton", Vol. 3, "Algebra", Cambridge University Press, 1967.

12. A. S. Marathay. Matrix operator description of the propagation of polarized light, *Jour. Opt. Soc. Amer.* **61** (10) (1971), p. 1363.
C. Whitney. Pauli-algebraic operators in polarization optics, *Jour. Opt. Soc. Amer.* **61** (9) (1971), p. 1207.
G. Eichmann. Complex polarization variable description of polarizing instruments, *Jour. Opt. Soc. Amer.* **61** (11) (1971), p. 1585.

13. R. M. A. Azzam and N. M. Bishara. Ellipsometric measurement of the polarization transfer function of an optical system, *Jour. Opt. Soc. Amer.* **62** (3) (1972), p. 336.
    M. Ghezzo. Thickness calculations for a transparent film from ellipsometric measurements, *Jour. Opt. Soc. Amer.* **58** (3) (1968), p. 368.
14. L. Silberstein. "The Theory of Relativity", MacMillan, 1924.
15. V. H. Rumsey. "Frequency Independent Antennas", Academic Press, New York and London, 1966.
    P. D. Crout, The determination of antenna patterns of *n*-arm antennas by means of bicomplex functions, *Trans. I.E.E.E.* **AP** (1970), p. 686.
16. L. Lewin, A decoupled formulation of the vector wave equation in orthogonal curvilinear coordinates, *Trans. I.E.E.E.* **MTT-20** (5) (1972), p. 339.
17. J. Ehlers, W. Rindler and I. Robinson. Quaternions bivectors and the Lorentz group. *In* "Perspectives in Geometry and Relativity". (B. Hoffmann, Ed.), Indiana University Press, 1966, p. 134.

BIBLIOGRAPHY

TRANSMISSION AND REFLECTION

C. Imbert. Calculation and experimental proof of the transverse shift induced by total internal reflection of a circularly polarized light beam, *Physical Review D* 3rd series **5** (4) (1972), p. 787.

V. A. Libin. Polarization analyzer, *Rad. Eng. and Electron Phys.* **6** (4), 1961, p. 289.

K. M. Mitzner. Change in polarization on reflection from a tilted plane, *Radio Science,* **1** (new series) (1) (1966), p. 27.

L. E. Raburn. The calculation of reflector antenna polarized radiation. *Trans. I.R.E.* Vol. AP8 No. 1 Jan. 1960, p. 43.

W. Swindell. Handedness of polarization after total reflection of linearly polarized light, *Jour. Opt. Soc. Amer.* **62** (2) (1972), p. 294.

P. A. Watson and S. I. Ghobrial. Cross polarizing effects of a water film on a parabolic reflector at microwave frequencies, *Trans. I.E.E.E.* **AP20** (1972), p. 668.

POLARIZATION MATRICES AND POINCARÉ SPHERE

E. Collett. Mueller Stokes formulation of Fresnel's equations, *Amer. Jour. Physics,* (39), (1971), p. 517.

C. D. Graves. Radar polarization power scattering matrix, *Proc. I.R.E.* **44** (1956), p. 248.

T. Hagfors. A study of the depolarization of lunar radar echoes, *Radio Sci.* **2** (5) (1967), p. 445.

G. H. Knittel. The polarization sphere as a graphical aid is determining the polarization of an antenna by amplitude measurement only, *Trans. I.E.E.E.* **AP15** (2) (1967), p. 217.

J. R. Priebe. Operational form of the Mueller Matrices, *Jour. Opt. Soc. Amer.* **59** (2) (1969), p. 176.

G. N. Ramachandran and S. Ramaseshan, Magneto optical rotation application of the Poincaré sphere, *Jour. Opt. Soc. Amer.* **52** (1) (1952), p. 49.

A. N. Sivov. The electrodynamic theory of the close-set plane grid composed of parallel conductors *Rad. Eng. and Electron Phys.* **6**(4) (1961) p. 1.

J. E. Vos and B. S. Blaise. A new way of representing the effect of optically anisotropic elements by rotation of a sphere, *Optica Acta* **17** (3) (1970), p. 197.

K. C. Westfold. New analysis of the polarization of radiation and the Faraday effect in terms of complex vectors, *Jour. Opt. Soc. Amer.* **49** (7) (1959), p. 717.

MEASUREMENTS

A. C. Ludwig, The definition of cross polarization, *Trans. I.E.E.E* **AP21** (1973), p. 116 also

G. H. Knittel, Comment on above *Trans. I.E.E.E.* **AP21** (1973) 917.

E. B. Joy and D. T. Paris. A practical method for determining the complex polarization ratio of arbitrary antenna, *Trans I.E.E.E.* **AP21** (4) (1973) 432.

A. C. Newell and D. M. Kearns. Determination of both polarization and power gain of antennas by a generalized 3-antenna measurement method. *Electronics Letters* **7** (3) (1971) 68.

WIRE GRATINGS

M. G. Andreason. Scattering from parallel metallic cylinders with arbitrary cross-section, *Trans. I.E.E.E.* **AP12** (1964), p. 746.

W. Franz. The transmission of electric waves through wire grids, *Z. Agnew Phys.* **1** (1949), p. 416.

D. S. Lerner. A wave polarization converter for circular polarization, *Trans. I.E.E.E.* **AP13** (1965), p. 3.

E. A. Lewis and J. P. Casey. Electromagnetic reflection and transmission by gratings or resistive wires, *Jour. App. Phys.* **23** (1952), p. 605.

G. G. Macfarlane. Surface impedence of an infinite parallel wire grid at oblique angles of incidence, *Jour. I.E.E.* **93** (3A) (1946), p. 1523.

V. Twersky. On the scattering of waves by an infinite grating, *Trans. I.R.E.* **AP4** (1956), p. 330. Also **AP10** (1962), p. 737.

L. A. Vainshtein. The diffraction of electromagnetic waves at a grating consisting of parallel conducting strips, *Zh. Tekh. Fiz.* **25** (1955), p. 847.

Yu. P. Vinichenko, L. N. Zakhar'yev and A. A. Lemansky, Diffraction of a plane wave by a double grating of thin circular cylinders, *Radio Eng. and Electron Phys.* **15** (12) (1970), p. 2196.

W. Wasylkiwskyj, On the transmission coefficient of an infinite grating of perfectly conducting circular cylinders, *Trans. I.E.E.E.* **AP19** (1971), p. 704.

6

Progress in Optics

The title of this concluding chapter is taken as that of the excellent series of volumes[1] that annually reviews the current state of all aspects of optical theory. Among these topics can be found any number of subjects relevant to microwave antenna design and which, in a more complete survey would have been included in a volume with this title. It is, however, the only title that is possible to give to a chapter that satisfactorily indicates that, for all its longevity, the subject of optics and particularly the generalization that constitutes the wide angle, non-homogeneous area of microwave optics, is still subject to progress, and to progress of a most exciting nature.

It is the intention of this chapter to show how some of the most recent studies in theoretical physics, as well as many almost forgotten ones, are relevant to the subject of microwave optics even in its most basic requirement, that of the design of practical optical antenna systems.

Many of the subjects related to the general survey are however likely to be unfamiliar to an antenna designing engineer, and a treatment of these at even an elementary level would go beyond the scope of the book. Furthermore, such digressions into definitions, basic details and rigorous derivation of results, although desirable, would detract from the objective, which is to illustrate the picture that is emerging of the central role being played by geometrical optics and its counterpart, particle mechanics, in the whole structure of modern physics. If we can draw on this for our practical purposes, then an advantage has been gained, and we can feel the subject to be more than engineering design but a true science in its own right.

The implication just given is that the subject to be discussed given some rigorous foundation, would in fact form a logical treatise. This is not at all the case. At times the connective material becomes extremely tenuous and the reader will understand that the limit of the author's understanding is being reached. At other times disparate subjects may be adjoined on the minimum of evidence because they are similar in the general tenor of their treatments. One's hope is that the story as it appears is of sufficient interest to make those antenna designers who wish, seek out the sources of the theories and use them to their own design purpose. As many of such sources as is known to the author will of course be indicated.

Those readers who may be much more familiar with the material than the author is himself may find that some of the conclusions drawn are common knowledge, but then one can say that they do not appear to have applied it to the field in which we are interested.

The charge could be made that much of this vagueness could be overcome by postponing this essay into the unknown until some more rigorous analysis can be applied, that actually gives results and precise details. The answer to that is that there is no certainty that this will happen and since one does not write many books per annum, it would be a waste of much of the material of the previous chapters, to have to repeat it merely to arrive at this point in the present context.

The essay that follows is thus a collection of loosely bound material, which indicates to this author at least an underlying field of pure theoretical physics, by no means beyond the understanding of a microwave antenna design engineer if some fundamental concepts are accepted, which we call geometrical optics or particle mechanics, which we feel contains a reservoir of analytical method which we can use to design optical devices or find out about fundamental particles by a projection, in the literal sense, into the field of reality.

It is, in the strict definition of the word, a conjecture and should be accepted as indicating possibilities and not defining certain processes.

6.1 OPTICAL TRANSFORMATION MATRICES

The position and direction of a ray at any point along its path through any optical system can be specified by four parameters which are the direction cosines of the tangent at a point on the ray and the coordinates of the intersection of the tangent with a given fixed plane, usually the plane $z = 0$, perpendicular to the axis of the system.

The transformation undergone by these parameters as the ray intersects a system of refracting surfaces which separate media of constant refractive index can be derived with the application of an invariant function discovered by Herzberger[2] in 1931. If the ray properties of position and direction, the original starting point and direction, and the properties of each refracting surface and medium are known, then the intersection of a ray with any surface can be obtained from its intersection with the preceding surface and the distances between these points, evaluated by the standard ray tracing technique. If we consider a two parameter system of rays then all of these properties can be put in terms of the parameters, which we call u and v. The ray tracing equation itself is a repeated application of the law of refraction, together with the necessary geometrical relations for the points of incidence on successive surfaces and the normals to the surfaces at these points. If a point on a ray has a position vector $\mathbf{r}$ and tangential direction $\hat{\mathbf{s}}$ with respect to a fixed origin, then Stavroudis[3] shows by a method of differentiating the ray tracing equation with respect to the parameters u and v that the function

$$\eta \left[\frac{\partial \mathbf{r}}{\partial u} \cdot \frac{\partial \hat{\mathbf{s}}}{\partial v} - \frac{\partial \mathbf{r}}{\partial v} \cdot \frac{\partial \hat{\mathbf{s}}}{\partial u} \right] \tag{6.1}$$

is invariant. That is it has the same value whether the point on the ray be taken in the object space, the image space or at any of the intermediate spaces between refracting surfaces. This function is termed the "fundamental optical invariant" and is closely related to the action of Fermat's principle for optical rays.

If the family of rays all issue from a single source point then at that point

$$\frac{\partial \mathbf{r}}{\partial u} = \frac{\partial \mathbf{r}}{\partial v} = 0,$$

and thus the invariant vanishes in the object space. Consequently it vanishes everywhere with the corollary that by a correct system of surfaces the rays can be brought to a point focus again if required. It also implies that *only* rays that have issued from a single point focus (possibly virtual) can be so brought together into a focus again. It follows that the rays form a normal congruence (or orthotomic system) if and only if the optical invariant is zero.

By a continuity process the discrete surfaces of this theory can be merged into a medium of continuously variable refractive index for which an invariant of this type can be found.

The scalar version of the optical invariant is derived in terms of the original four ray parameters the direction cosines of $\hat{s}$ and the coordinates of the intersection of $\hat{s}$ with the plane $z = 0$. The parameters of the invariant in the object space are then related to the parameters of the invariant in the image space by virtue of the equality of the invariant function.

This relation can be expressed as a 4×4 matrix and the resulting matrix equation is termed the "lens equation". It is admitted that a considerable algebraic effort has to go into obtaining the details of this matrix and its properties. It is found to be a Jacobian form, unitary and related to a second constant matrix by a similarity transformation. It is unnecessary to write down this matrix in this summary since definition of the symbols alone would take up considerable space. It is sufficient for our argument to know that such a matrix exists and for us to go into the consequences that such a relationship infers. For full details, of course, Ref. 3 should be consulted.

A certain simplification is then made to this lens equation by considering only the rotationally symmetric system. For this particular case a solution is obtained (Ref. 3, p. 251) and the substitution of this particular result back into the general equation leaves a system of "residual equations". These, again with considerable algebraic manipulation, are shown to be the product of matrix factors.

The particular solution for the rotationally symmetric system is then separated into a basic matrix product the individual factors of which are shown to represent the *transfer* from one refracting surface to the next and to the *refraction* at each surface. Each of the matrices concerned is Jacobian as is the product.

The development of a general matrix treatment of this kind leads on naturally to the concept of a "lens group" since nearly all the well-known groups have a matrix representation and the next stage is to determine the nature of the groups or sub-groups into which the lens group, the rotationally symmetric lens group, and the paraxial lens group can be put. The last of these is the, so far, *approximation* to the case where the rays are confined to small angles $\sin \theta \simeq \tan \theta$ near to the axis of the optical system.

Some elementary properties of groups will be taken as known at this point. We shall require a few names and definitions only so that we can recognise them when they appear in different contexts.

A group that has a one to one functional relationship or mapping, with a second group is *homomorphic* with that group. Two systems represented by homomorphic groups have a great deal in common for example their matrix representations can be similar. If the two groups are one and the same, the mapping is between elements of a single group and is termed *isomorphic*.

The groups are symbolized by letters and given a suffix which determines the order of the group and relates to the space dimensions. The work we shall be considering uses almost exclusively 4×4 matrices and hence groups will be appended with the suffix 4.

The general group is the group of all 4×4 non-singular matrices and is termed $GL(4)$. It is isomorphic with the group of linear transformations on a vector space of four dimensions.

Sub-groups can be defined by the invariance properties of certain quadratic forms under such a transformation.

The rotation group for example leaves the symmetric bi-linear form $x_1y_1 + x_2y_2 + x_3y_3 + x_4y_4$ invariant. This is the scalar product and the group leaving the scalar product invariant is a pure rotation, which must be an orthogonal transformation. This has the title $O(4)$.

The second group, and that most required for the continuation of the lens theory, is the group which leaves the skew symmetric bilinear form $x_1y_3 - x_3y_1 + x_2y_4 - x_4y_2$ invariant. Such a group, which must be of even order, is termed the symplectic group (a word meaning skew) and is known as $Sp(4)$. The matrix representation of this group consists of matrices M which satisfy

$$M^T J M = J$$

($M^T \equiv$ transpose of M) where J is the constant matrix

$$J = \begin{pmatrix} 0 & 0 & 1 & 0 \\ 0 & 0 & 0 & 1 \\ -1 & 0 & 0 & 0 \\ 0 & -1 & 0 & 0 \end{pmatrix}$$

J is a basis matrix of the quaternion group (Appendix II, p. 397). Groups with other titles, and their invariant forms and matrix algebra, are listed by Cartan[4] and cover the complete symmetry group.

In some cases groups or sub-groups can be constructed from a finite number of basic elements which are termed the generators of the (sub) groups. As shown in Ref. 3 the generators of the symplectic group are represented by

(i) translations
$$\begin{pmatrix} 1 & 0 & a & c \\ 0 & 1 & c & b \\ 0 & 0 & 1 & 0 \\ 0 & 0 & 0 & 1 \end{pmatrix}. \tag{6.2}$$

(ii) rotations

$$
\begin{pmatrix}
 & U & 0 & 0 \\
 &   & 0 & 0 \\
0 & 0 &   &   \\
0 & 0 & U^{T-1} &
\end{pmatrix},
\tag{6.3}
$$

where $|U| = \pm 1$, and

(iii) the semi-involutions

$$
\begin{pmatrix}
Q & I - Q \\
-I + Q & Q
\end{pmatrix}.
\tag{6.4}
$$

where Q is one of the matrices

$$
\begin{pmatrix} 0 & 0 \\ 0 & 0 \end{pmatrix}, \quad
\begin{pmatrix} 1 & 0 \\ 0 & 0 \end{pmatrix}, \quad
\begin{pmatrix} 0 & 0 \\ 0 & 1 \end{pmatrix} \quad \text{or} \quad
\begin{pmatrix} 1 & 0 \\ 0 & 1 \end{pmatrix}
$$

Stavroudis (Ref. 3, p. 293) then goes on to show that the transfer matrix between two refracting surfaces is identical in form to a translation matrix even to the symmetric property of the matrix $\begin{pmatrix} a & c \\ c & b \end{pmatrix}$ in (6.2). The refraction matrix is shown to be composed of a rotation, a translation and one of the four possible semi-involutions. Hence the group of rotationally symmetric lenses comprises a proper sub-group of $Sp4$. The significance of the remaining semi-involutions in this context has yet to be determined.

Finally, and most fundamentally, the paraxial case is considered. This is done by expanding the optical relations and discarding all but the first order linear terms. The resulting simplification is drastic, reducing the matrix order from four to two, and the resulting transformation becomes a proper sub-group of $Sp(4)$, homomorphic in fact with $Sp(2)$. That an *arbitrary* approximation should result in a *proper* sub-group is of some significance. It implies that the paraxial case is not simply the approximation it is considered to be, but results in a well defined particular form of lens structure.

It is impossible in this brief summary to convey the immensity of the calculations that have been performed or the elegance of the final result. We give here the words of the author himself (Ref. 3, p. 299) in reviewing this study: "... although the idea of optical image formation as a transformation has persisted for many years, the idea of the use of the Jacobian of such a transformation appears to have been missed. The key to its application is, of course, the fundamental optical invariant of Herzberger .... Notwithstanding the clarity of the goal of such an application the techniques that must be used all but obscure it in an avalanche of incredibly difficult calculations."

M

In view of this it may be permitted that we accept these results and allow ourselves some speculative propositions and questions based upon them.

One of the first implications concerns the matter first mentioned in Chapter 1. It was shown there (p. 32) that for (rotationally symmetric) microwave waveguide lenses at least, a single parameter could be used to describe an entire system of lenses, all with the same, single property of collimation of rays from a point source. The treatment given above has not as yet been applied to the constrained microwave case and will only be done so if the author's dire warning about the complexity of the calculation can be shown to be alleviated somewhat for this case. Under certain of the conditions applicable to the microwave case, a paraxial version of the optical case becomes feasible, and thus the group representation applies. In such a situation the *continuous* transformation of the lens shape by the variation of the *single* parameter α will become a *continuous* group transformation. This enables us to associate the operation known as "lens bending" (Refs in Chapter 1) with the continuous transformation group which is known as a Lie group.

A similar connection must be made with the rays in a non-uniform medium. This acts as a continuous lens and hence as a continuous lens transformation. As was shown the analogy between rays in a non-homogeneous medium and particles in a non-homogeneous potential is well catalogued and so it is of no surprise to find Lie groups occurring in the group theory of quantum mechanics.[5]

We also have had in Chapter 1 a different kind of optical transformation which is Damien's theorem. In this the transformation is between *surfaces*, that of refraction and that of the wave front; and so is very different in a fundamental aspect from the previous transformation which is between canonical *ray* coordinates. And yet there must be some way in which the refraction matrix combination will apply to both. We next make a similar survey of some other important transformations relating to optical systems.

6.2 TRANSFORMATIONS IN TWO DIMENSIONS

It has been shown by Luneburg[6] that generalizations of one lens design can be obtained by the conformal mapping of the ray paths with an associated transformation of the refractive index. This is applied to the perfect optical instrument which is the Maxwell fish-eye. The process itself is revealing. While not acting upon the three dimensional problem as a whole, the transformation is made upon a two-dimensional cross section and made into a practical lens by rotation about the axis. That is in essence the following procedure

(i) a projection onto a space of lesser dimension
(ii) a transformation in that space
(iii) a projection back into a space of higher dimension

and that the projections (i) and (iii) need not be inverses and that (ii) can be quite a general transformation. Luneburg uses the method to transform the complete space and hence transforms the refractive index everywhere in that space. The lens as such is delimited in each case by the curve $\eta \geqslant 1$ the limit of sensible refractive indices. The transformation itself is worth noting. The conformal mapping is the transformation $w = f(z)$

$$z = x + iy, \quad w = u + iv.$$

$$(6.5)$$

Then the line element in a space with refractive index $\eta(x, y)$ is

$$ds^2 = \eta^2(x, y)(dx^2 + dy^2), \qquad (6.6)$$

which by virtue of the Cauchy–Riemann conditions transforms into

$$ds^2 = \eta_1^2(u, v)|f'|^2 (du^2 + dv^2), \qquad (6.7)$$

$$|f'|^2 = \left(\frac{\partial u}{\partial x}\right)^2 + \left(\frac{\partial v}{\partial x}\right)^2 = \left(\frac{\partial u}{\partial y}\right)^2 + \left(\frac{\partial v}{\partial y}\right)^2$$

and this is equivalent to a transformation of refractive index

$$\eta_1 = \frac{\eta}{|f'|}. \qquad (6.8)$$

Applying this transformation to the refractive index law of the Maxwell fish-eye, Luneburg deduces that all perfect optical systems can be formed from media with refractive index laws, in the plane, given by

$$\eta = \frac{2|f'(z)|}{1 + |f^2(z)|}, \quad z = x + iy. \qquad (6.9)$$

The law of Eqn (6.8) is applied to other lenses by Zaborov,[7] in which it is termed the isometric transformation, and in particular to the cylindrical lens (meridional section), the short focus horn (Chapter 2, p. 116), with the refractive index $\eta = \text{sech } a\rho$.

This requirement for the refractive index to be transformed throughout all space, as well as the two-dimensionality of the study are limitations upon the solution to the more general problem. The former effect may be vitiated if it can be made sufficiently local, so that an approximate result, zero or small enough outside a certain region, can be used instead of the complete global result. The effect is more then of a perturbation theory than of a complete field transformation.

6.3 BATEMAN'S TRANSFORMATION OF OPTICAL SYSTEMS

The surprise then occurs to find that this problem was tackled and solved about 1910 in a series of papers by Bateman[8] based, as in his own title, on the conformal transformations of a space of *four* dimensions.

There appears to be very cogent reasons in the theory of the *algebra* we shall be using as to why this problem cannot be solved in three dimensions. Attempts to apply conformal mapping in three dimensions to the *statical* problem involve the use of multi-valued potentials (Jeans[9]) which gives the field distribution of a line source over a semi-infinite plane by a method of images in a Riemann surface. Sommerfeld (loc. cit., p. 283) continues the process for a point source over a semi-infinite plane by proposing a Riemann space having the same relation to real space that the Riemann surface has to the uncut complex plane. However, neither method appears to be applicable to wave solutions and are highly complex in the static case for other than the basic problems mentioned.

Bateman bases his study upon the requirement for transformations to leave the wave-equation invariant

$$dx^2 + dy^2 + dz^2 - dt^2 = 0. \tag{6.10}$$

A transformation that does this is extremely well known, namely the Lorentz transformation of the restricted theory of relativity. But even in that context it is to be applied in a free-space situation. To include the effects of refractive pondero-motive materials is still a highly complicated even questionable, procedure and not one that could be easily adapted to the shapes that constitute lens systems. However, Bateman[10] finds that many other transformations leave both the wave equation and the eikonal equation invariant and that the Lorentz transformation is only a particular version of these more general possibilities. Some in fact refer back to earlier work by Bianchi in differential geometry and of Maxwell himself.

He singles out for particular interest the two transformations[11]

$$X = \frac{x}{r^2 - c^2 t^2}, \qquad Y = \frac{y}{r^2 - c^2 t^2}, \qquad Z = \frac{z}{r^2 - c^2 t^2}, \qquad T = \frac{t}{r^2 - c^2 t^2},$$

$$\tag{6.11}$$

and

$$X = \frac{x}{z - ct}, \qquad Y = \frac{y}{z - ct}, \qquad Z = \frac{r^2 - 1}{2(z - ct)}, \qquad cT = \frac{r^2 + 1}{2(z - ct)}, \quad (6.12)$$

$$r^2 = x^2 + y^2 + z^2,$$

to which we shall refer later. In the references given the conformality of such transformations is discussed and Bateman (Ref. 10, p. 82) applies the transformation given in (6.11) directly to an optical lens (spherical surfaces paraxial case) to obtain a transformed lens.

The form of the transformations are inversions, but whereas the origin normally transforms by inversion into the point at infinity, in these cases the entire circle $r = ct$ (or the point $z = ct$) does. In a further discussion on more general transformations of the form

$$\lambda[dx^2 + dy^2 + dz^2 - dt^2] + \mu[(s_x\,dx + s_y\,dy + s_z\,dz)^2 - s^2\,dt^2]$$
$$= dx'^2 + dy'^2 + dz'^2 - dt'^2, \quad (6.13)$$

Bateman derives those transformations "that can be used to solve problems in the reflection or refraction of light when the orthotomic surfaces before or after incidence are known". As applied to the paraboloid, one example of the transformation is of the form

$$x' = a - r, \qquad t' = t, \qquad\qquad (6.14)$$

$$y' = \frac{ay}{r + z}, \qquad z' = \frac{az}{r + z}.$$

The fact that these are all inversions by nature illustrates a connection with the method of Damien in Chapter 1 (p. 25). Rather revealingly the reference concludes with a note much in accord with this author's own investigations: "It should be remarked that the present method cannot be used to solve any problem in reflection; so far transformations have only been found in cases where a single plane or spherical wave is transformed into a single plane or spherical wave . . . ." This is however, sufficient for our microwave antenna purpose as can now be illustrated.

A transformation that matches the inversion of Eqn (6.14) can be found that transforms the elliptical single refracting surface into the hyperbolic. We have (Chapter 1, Eqns (1.9) and 1.11)) for the elliptical profile

$$r = \frac{f(\eta - 1)}{\eta - \cos\theta},$$

to which we apply the transformation $\theta' = \theta$,

$$R = \frac{kr}{r + z}, \qquad \text{(from } y = r\sin\theta\text{)}, \qquad\qquad (6.15)$$

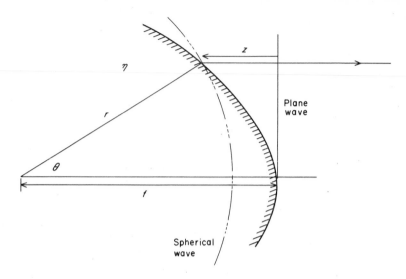

FIG. 6.1. Elliptical refractor.

making the radius of inversion k equal to the focal length f, we choose (fairly arbitrarily) z to be the path length along the axis between the point where the wave front is still spherical up to the point where it becomes plane (Fig. 6.1) that is $z = r \cos \theta - f$.

The transformation then becomes

$$R = \frac{fr}{r - f + r \cos \theta},$$

and its inverse by direct solution

$$r = \frac{fR}{R - f + R \cos \theta}.$$

Substitution of

$$r = \frac{f(\eta - 1)}{\eta - \cos \theta}$$

results in

$$R = \frac{f(\eta - 1)}{\eta \cos \theta - 1}.$$

and inversely.

That is the transformation converts the elliptical profile to the hyperbolic and vice versa. With $\eta = -1$ the transformation shows that the parabola $2f/(1 + \cos \theta)$ remains invariant. The plane $r = f \cos \theta$ is transformed into the circle $R = f$, this last being the clue to the operation of these surfaces in converting spherical to plane waves.

However, when one tries to enlarge upon this, for example, by transforming one Cartesian oval into another the procedure is found to be wanting.

One conclusion that is reached by this exercise is that the problem is of an essential four dimensional nature. We next look for a process that can perform these conformal mappings in a four dimensional space in the same way that functions of a complex variable perform them in two dimensions.

6.4 FUNCTIONS OF A HYPERCOMPLEX VARIABLE

The method of approach, in all logic, would be to use the four dimensional complex number for four dimensional mappings. These hypercomplex numbers must be the quaternions that we have already met in the study of polarization (Chapter 5, p. 323) and in fundamental geometrical optics (Appendix I, p. 365). The basic algebra is given in Appendix II where it can be seen that the jump to the four dimensional case quite definitely excludes the possibility of a similar structure being used in three dimensions.

A quaternion $Q = (a, \mathbf{A}) = (a, a_1, a_2, a_3)$ satisfies the relations (Appendix II, p. 391)

$$(Q - a)^2 = b^2, \qquad b^2 = \mathbf{A} \circ \mathbf{A},$$

where since

$$\bar{Q} = (a, -\mathbf{A}) \quad \text{and} \quad N(Q) = Q \cdot \bar{Q} = a^2 - \mathbf{A} \circ \mathbf{A}$$

$$a = \tfrac{1}{2}(q + \bar{q}) \quad \text{and} \quad b^2 = -N(Q - a). \tag{6.16}$$

A *polynomial* function of a quaternion variable is then expressed as[12]

$$f(Q) = \{\tfrac{1}{2}[f(a + b) + f(a - b)]; \quad (Q - a)[f(a + b) - f(a - b)]/2b\}. \tag{6.17}$$

In the event that $b = 0$ this becomes

$$f(Q) = \{f(a); (Q - a)f'(a)\}. \tag{6.18}$$

We take as our example the basic coordinate quaternion $Q_c = (t, x, y, z)$ then

$$b^2 = x^2 + y^2 + z^2, \qquad a = t.$$

In Eqn (6.17)

$$f(Q_c) = \left\{ \tfrac{1}{2}[f(t + r) + f(t - r)]; \frac{(x, y, z)}{2r} [f(t + r) - f(t - r)] \right\},$$

that is the transformation

$$t' = \tfrac{1}{2}\{f(t + r) + f(t - r)\},$$

$$x' = \frac{x}{2r} \{f(t + r) - f(t - r)\},$$

$$y' = \frac{y}{2r} \{f(t + r) - f(t - r)\}, \tag{6.19}$$

$$z' = \frac{z}{2r} \{f(t + r) - f(t - r)\}.$$

This is virtually identical to the second transformation of Bateman based upon the invariance of Maxwell's equations (Ref. 10, p. 481 Eqn B). Such difference that exists is that the reference gives the more general case with two independent functions f and ψ on the right-hand side of Eqn (6.19). This may well be the case had a more general function been operated on.

In pursuing this analysis this author has achieved a similar result by a direct approach the same as that used in ordinary complex variable. Defining

$$f(t, x, y, z) = (p, u, v, w)$$

in the manner $f(Q) = W$ of complex variable theory. Then

$$W' = \lim_{h \to 0} \frac{1}{h} [(p\{t + \delta t, x + \delta x, y + \delta y, z + \delta z\}; \quad u\{\ \}; \quad v\{\ \}; \quad w\{\ \})$$

$$- (p(t, x, y, z); u(\ ); v(\ ); w(\ )] \tag{6.20}$$

where for uniqueness W' has the same limit in whatever manner h tends to zero.

Therefore putting $h = \delta t$, $\alpha\delta x$, $\beta\delta y$, $\gamma\delta z$ in turn and making the other increments zero, we approach the limit along each of the *four* coordinate axes separately and we obtain

$$W' = \frac{\partial p}{\partial t} + \alpha \frac{\partial u}{\partial t} + \beta \frac{\partial v}{\partial t} + \gamma \frac{\partial w}{\partial t}$$

$$= \frac{1}{\alpha} \frac{\partial p}{\partial x} + \frac{\partial u}{\partial x} + \frac{\beta}{\alpha} \frac{\partial v}{\partial x} + \frac{\gamma}{\alpha} \frac{\partial w}{\partial x}$$

$$= \frac{1}{\beta} \frac{\partial p}{\partial y} + \frac{\alpha}{\beta} \frac{\partial u}{\partial y} + \frac{\partial v}{\partial y} + \frac{\gamma}{\beta} \frac{\partial w}{\partial y}$$

$$= \frac{1}{\gamma} \frac{\partial p}{\partial z} + \frac{\alpha}{\gamma} \frac{\partial u}{\partial z} + \frac{\beta}{\gamma} \frac{\partial v}{\partial z} + \frac{\partial w}{\partial z}. \tag{6.21}$$

By the uniqueness of the inverse we have $1/\alpha = -\alpha$, etc., and provided the right- or left-hand quotient is taken with consistency throughout $\beta/\alpha = -\alpha/\beta$ and, etc.† Thus equating coefficients of α, β, γ, α/β, etc., we derive an extended set of "Cauchy–Riemann" conditions

$$\frac{\partial p}{\partial t} = \frac{\partial u}{\partial x} = \frac{\partial v}{\partial y} = \frac{\partial w}{\partial z}$$

$$\frac{\partial p}{\partial x} = \frac{-\partial u}{\partial t}, \quad \frac{\partial p}{\partial y} = \frac{-\partial v}{\partial t}, \quad \frac{\partial p}{\partial z} = \frac{-\partial w}{\partial t}$$

$$\frac{\partial u}{\partial y} = \frac{-\partial v}{\partial x}, \quad \frac{\partial u}{\partial z} = \frac{-\partial w}{\partial x}$$

$$\frac{\partial v}{\partial z} = \frac{-\partial w}{\partial y} \tag{6.22}$$

in which the "original" Cauchy–Riemann conditions are underlined.

From these relations we find that u, v, and w each satisfy the wave equation in a separated form that is

$$\frac{\partial^2}{\partial x^2} + \frac{\partial^2}{\partial y^2} = 0 = \frac{\partial^2}{\partial t^2} - \frac{\partial^2}{\partial z^2}. \tag{6.23}$$

The left-hand side of this equation can be satisfied by the transformation of "ordinary" complex variable theory and the right-hand side is a one dimensional wave equation. The result is the transformation

$$u + iv = \phi(x + iy)$$

$$u - iv = \phi(x - iy)$$

$$w = F_1(z + t) + F_2(z - t)$$

$$p = F_1(z + T) - F_2(z - t) \tag{6.24}$$

for arbitrary functions ϕ, F_1 and F_2.

This result is found to be identical to the first of the Bateman transformations (loc. cit. p. 479 Eqn A).

Vector solutions to the wave equation can be obtained from the result

$$(0, u, v, w) = \{f(Q) - \overline{f(Q)}\}/2. \tag{6.25}$$

---

† Strictly one should define a W'_R and W'_L denoting whether the multiplication by the denominator is on the left- or right-hand side. Provided we avoid defining α/β or β/α we retain consistency by using the result $\alpha/\beta = -\beta/\alpha$ (Appendix II, p. 393) when the order is kept the same for both sides of the equality.

Finally we find that the Cauchy–Riemann conditions of (6.22) give rise to three pair-combinations of the kind

$$\frac{\partial u}{\partial x}\frac{\partial p}{\partial x} + \frac{\partial u}{\partial t}\frac{\partial p}{\partial t} = 0 = \frac{\partial v}{\partial y}\frac{\partial w}{\partial y} + \frac{\partial v}{\partial z}\frac{\partial w}{\partial z},$$

showing that the respective coordinate surfaces are orthogonal in the corresponding two spaces and demonstrating the conformality of the transformations. The particular transformation making $w = z$; $p = t$ in Eqn (6.24) is of course the "ordinary" conformal mapping in two dimensions.

It is of interest to note that the procedure used by Bateman in deriving the transformations of (6.11) and (6.12) involves the use of the bi-vector formulation of Maxwell's equation as did Kales (Ref. 5, Chapter 5) in his study of polarization. There is of course a well-known quaternion representation of Maxwell's equations by Silberstein.[13] The title of Ref. (12) from which the basic relation (6.17) was taken is Quaternions, bi-vectors and the Lorentz group. The connection with Lorentz transformations as a particular result of the transformation in (6.24) was also noted by Bateman and can be found in Ref. 12 (Eqn 79) in the same form. It is obtained by making the functions ϕ, F_1 and F_2 linear combinations of their arguments.

Once connected with the Lorentz group there appears to be no end to the applications provided by modern group theory which can form generalizations to the preceding and which should have a similar relevance to the transformation of optical systems. To take but one example (others are included in the bibliography to this chapter) and proceeding without unnecessary definitions, a group over a complex space has been investigated by Klotz.[14]

This is a group leaving the metric form

$$\mathrm{d}s^2 = \mathrm{d}z_0\,\overline{\mathrm{d}z_2} + \mathrm{d}z_2\,\overline{\mathrm{d}z_0} + \mathrm{d}z_1\,\overline{\mathrm{d}z_3} + \mathrm{d}z_3\,\overline{\mathrm{d}z_1} \qquad (6.26)$$

invariant, where the bar denotes complex conjugation (ordinary type). This is termed a connected symmetry group of twistor space and these have been shown to allow the formulation of conformally invariant relationships in a space of four dimensions. The relationships that are illustrated by this means are found to correspond to Lorentz transformations, translations, accelerations and dilatations each of which is a sub-group of the conformal group. As a result, every continuous transformation has as its generators, a Lorentz transformation, a translation, an acceleration, a dilatation and one other of four special mappings. These turn out to be the identity mapping and the two transformations of Bateman (1909) given in Eqns (6.11) and (6.12) (the latter with a $\pm$ possibility in the denominators). We further find that a particular form of the transformation in (6.19) is in fact a dilatation and was originally formulated by Lorentz himself in 1901. Cunningham[15] has shown that any electrodynamical field can be transformed into another by means of the transformation of (6.11).

6.5 GEOMETRICAL OPTICS IN FOUR DIMENSIONS

The reader might think that we have strayed a considerable way from our objective which is to design optical systems by a transformation process from known optical systems. We have illustrated that this transformation is most likely to be of a four dimensional character and may include some of the more exotic transformations that can be discovered in modern group theory.

That these transformations are mainly in use in the context of theoretical quantum physics is a further manifestation of the analogy that exists between these fields and ray optics. But since quantum theory itself is to particle trajectories in much the same way as diffraction theory is to geometrical optics, we may very well be in the process of including diffraction and aberration phenomena in with the more general of the transformations we have outlined.

We observe for instance that the transformations are used as much for the field variables as for the actual rays, wavefronts or optical surfaces. These latter always appear as real systems in a real space of three dimensions and thus it is apparent that a closer relationship will exist if these could be developed as the real projection of four-space phenomena.

An early application of this concept is found in the work of Pierpont[16] in 1928. In this study Fermat's principle is applied to the integral of the ray path in a hyperbolic space of four dimensions. A curved ray in real space is a straight line in such a Riemannian space and the standard results of reflection and refraction are deduced, including Bouguer's theorem, from the hyperbolic geometry (of the plane cross-section) that arises.

The definitive work on this subject is the volume by Synge[17] on Geometrical Mechanics and De Broglie Waves. This is based in the same way on a four dimensional version of the variational principle. The introduction which deserves to be copied here in full, outlines the entire situation regarding the particle-ray analogy and introduces for it the term "geometrical mechanics". The purpose of the book is to find the advantages of studying this problem in a four dimensional form so becoming a "relativistic geo-mectrical mechanics". This is in keeping with our expectations of the advantage to be derived from a "relativistic geometrical optics" since nearly all the transformations involved appear to have a connection with the Lorentz group.

The conclusions of the study are of interest but of course the details would require a repetition of the whole work. We can but survey the basic method and accept the results.

A function $f(x, \alpha)$ is defined on a space of four dimensions in which the position coordinates x_r ($r = 1, 2, 3, 4$) and direction cosines α_r ($\alpha_4/i > 0$) figure. These direction cosines, as we know, give the information required about the refractive index of the medium at any point. The variational principle is applied to obtain the space-time rays from the usual relation

$$\delta \int f(x, \alpha)\, \mathrm{d}s = 0 \tag{6.27}$$

along true ray paths. In this relation $\mathrm{d}s$ is an element of a line in a flat Minkowski space of four dimensions. The Euler–Lagrange equations and the definition of action follows exactly as in the three dimensional case (Ref. 3, p. 18). Gauge transformations can be made which transform the medium function f into a new function f^*, and the particular transformation studied is

$$f^*(x, \alpha) = f(x, \alpha) - \frac{\partial \phi}{\partial x_r}\alpha_r \tag{6.28}$$

where ϕ is a function of the *four* space coordinates x_r only. Hence where $x_4 = ict$ is involved, moving media or dispersion come into consideration also. Synge shows that this transformation leaves rays (in four space) invariant, (since if f satisfies Eqn (6.27) f^* can be shown to do so as well), but obviously changes the character of the medium throughout all space (–time). Some special types of medium functions are considered relating to homogeneous and non-homogeneous isotropic media, and the spherically symmetrical medium (yet again). The particle trajectories for this last case are shown to be the elliptical orbits of a particle in a central field and compares of course with the elliptical paths of the rays in a spherical medium (Chapter 2, p. 111). However, gauge transformations of and between these different functions are not investigated and this aspect is not proceeded with (loc. cit., p. 22).

The concluding section outlines methods of generalization. Its main finding is that it is as fruitful to generalize by increasing the number of dimensions in use as to go from the flat Minkowski spaces to curved spaces by the inclusion of a metric function. The latter subject is comparatively modern[18] and highly complex but the former can be recognised in the literature of differential geometry[19] and includes transformations of the solutions to the wave equation (the Laplace–Beltrami operator or D'Alembertian) by *rotations* of a higher dimensional sphere.

Herzberger[20] also discusses the transition from three to four dimensional optics and illustrates the universality of all physics through a fundamental relation from which, for optics, his fundamental optical invariant, Eqn (6.1), is derived. The generalization here is shown to have introduced diffraction into the subject. This therefore provides us with an avenue by which the transformation of *geometrical* optics can be extended to include the effects of physical optics. The use of the diffraction functions the circle polynomials and hyperspheroidal functions, in the solution of the wave equation in a higher dimensional space is also possible.

The most recent and basic method is that of Poeverlein.[21] This is based upon the transition to four dimensions of the Sommerfeld–Runge relation (Appendix I, Eqn (I.5)) also found derived in the work by Herzberger. As shown in Appendix I the ray path obeys the basic curl equation

$$\nabla \times \hat{t} = 0$$

in three dimensions which is extended by using the four dimensional operator $\square$, where

$$\square^2 = \frac{\partial^2}{\partial x^2} + \frac{\partial^2}{\partial y^2} + \frac{\partial^2}{\partial z^2} - \frac{\partial^2}{c^2 \partial t^2}$$

to

$$\square \times \hat{t} = 0. \tag{6.29}$$

Discontinuities in both space and time can be introduced as implying moving boundaries or dispersion relations and the introduction of physical optics effects such as diffraction, including the geometrical theory of diffraction is implied,[22] This last subject of course is the source of a considerable amount of current antenna research.

Finally a projective formulation of the problems of geometrical optics has been given by Cambi[23] which makes use of projective geometry, in essence a four dimensional geometry. This is to design intermediate refracting or reflecting surfaces specifically to apply to a particular degree of freedom, usually an aberration since this is basically an optical paper although mention is made of aperture illumination distributions for microwave purposes. The study is based on the fact that in a refraction process (Appendix I, p. 368) the incident ray, the refracted ray, the line bisecting them and the surface normal have a cross-ratio (anharmonic ratio) equal to the ratio $-\eta_2/\eta_1$ of the refractive indices. In a reflection, this cross ratio is -1 and the rays form an involution.

The cross ratio remains invariant under projective transformation (homographies) so if two rays are represented by two points in a plane the refraction connecting them is given by the cross ratio of magnitude η_2/η_1 on the line joining them. A system of rays becomes a curve of points but conformal mapping in this plane is definitely ruled out. The problem is of course simplest for the meridional plane of an axi-symmetric system. For skew rays and for the introduction of aberrations the projectivity is made complex and there results a complex homography on a straight line. The method is illustrated by reference to the Schmidt correcting plate for spherical aberration. We make mention of this method because complex homographies on a line are joined with quaternions and rotations, our other transformations in this chapter, in a work of that title by Duval.[24]

6.6 SUMMARY AND CONCLUSION

We can consider now the microwave antenna or optical system to be, in its most general sense, a particular case of the scattering problem of reflecting and refracting surfaces in which the source field is the system of rays from a point and the scattered field is specified. A special class of these scattering systems are those with axial symmetry including spherical symmetry and a sub-class are those for which the incident field is a spherical wave and the scattered field a plane wave.

The most general analysis will apply to the general cases which may then include the effects of diffraction and aberrations[25] and even time varying effects such as dispersion and scattering from moving surfaces. In the sub-classes of the systems, axi-symmetric or paraxial systems, the simpler analysis would be expected to apply. This should take the form of a transformation theory based on a one or two generator combination of the fundamental transformations. This would be the most fruitful field of investigation for generalized microwave antenna designs.

Summarizing the findings of this chapter as follows we see that:

(a) the work of Stavroudis establishes that a general lens group exists with a real matrix representation of the transformation between the incident field of rays and the transmitted (image or scattered) field of rays. This matrix, and its interpretation as an actual lens design, is subject to extremely involved algebraic analysis. Its application to the design of specific surfaces to correct particular optical effects can be expected to be as complex. The transformation belongs to the symplectic group and the generators are translations, rotations and semi-involutions.

(b) Applications of the transformations of Bateman have been shown to transform an entire physical problem into another physical problem. A particular demonstration transforms an optical lens into a different optical lens. The transformations were obtained by Bateman and Cunningham as a result of a very general analysis of the invariants under linear transformations of Maxwell's equations and the wave and eikonal equations. This process is more manageable than the previous one and operates in a four dimensional space.

(c) These same transformations have been shown in this chapter to be obtainable by even more direct means from the theory of the functions of a quaternion variable. This provides a conformal mapping in four dimensions in the same way that the complex variable provides conformal mappings in two. In a limited number of cases two dimensional mappings have been applied to geometrical optic systems and scattering problems. These transformations and those of Bateman add to the previous transformations the generators of dilatations, accelerations. Lorentz transformations and particular forms of inversions. These complete the total generators of the conformal group. The inversions in particular apply to the problem of lens transformations and arise also in the theory of Damien.

(d) Herzberger states that optics in a space of four dimensions is a model of all physics and Sachs[26] makes a similar claim for quaternion processes. (Hamilton himself was convinced of this fact and spent the last twenty years of his life in the attempt to prove it.) Synge shows that the transformations can with benefit be taken into higher dimensions where they are likely to be of a more simple form, rotations of a hypersphere for example. This continues the process already observed in going from the complicated three to the more amenable four dimensions. Some work has been done on such transformations of the wave equation on a space of higher dimension and on hyperspheres.

(e) Most of the groups and transformations involved and much of the associated algebra is paralleled in the transition from classical particle dynamics to quantum dynamics and even to relativistic quantum theory.

There is no doubt that a design process exists which can transform a given optical design with known properties into another with different properties. If these have the same number of surfaces an exchange may be effected between the action of the surfaces as for example "bending" a lens to convert it from an aplanatic lens to a lens with a specific illumination function. This still requires more knowledge of the precise relationship between the transformations of surfaces and their effects with regard to aberration, diffraction and other properties. Such transformations can be performed by the action of the elements of the conformal group in a space of four dimensions or by the quanternion algebra. Simpler transformations may result from a projection onto a higher dimensional space. This may asssist the more complex transformation processes involved with non-symmetrical scattering problems and time varying effects. Such projections are Cremona transformations.[27]

It is to be hoped that a study of the evidence as summarised here and in the bibliography will lead the reader to a similar conclusion and that the excursion into the higher realms of theoretical physics has been a worthwhile exercise in itself.

I have hesitated to caption the chapters with a quotation in the old style but in my reading for this book I have found that Leonardo da Vinci is said to have believed that "optics is the paradise of mathematicians". I would like to feel that I have shown a way for physicists and above all microwave antenna design engineers to have a share in that.

REFERENCES

1. E. Wolf (Ed.). "Progress in Optics", North Holland, Vol. 1 1960 and annually thereafter.
2. M. Herzberger. "Modern Geometrical Optics", Wiley Interscience, 1956; "Strahlenoptik", Springer-Verlag, 1931.
3. O. N. Stavroudis. "The Optics of Rays, Wavefronts and Caustics", Academic Press, London and New York, 1972, Chapter XV, p. 281.
4. S. Helgason. "Differential Geometry and Symmetric Spaces", Academic Press, London and New York, 1962, p. 340.
5. M. Mizushima. "Theoretical Physics", John Wiley, New York, 1972, Chapter II, p. 517.
6. R. K. Luneburg. "Mathematical Theory of Optics", University of California Press, 1964, p. 178.
7. V. P. Zaborov. The method of isometric transformation of radio lenses and Isometric transformation of constant thickness lenses, *Radiotek i Elekt.* 4 (4), (1959), 576 and 584.
8. H. Bateman. The transformations of coordinates which can be used to transform one physical problem into another, *Proc. Lond. Math. Soc.* (2) 8 (1910), 469; The conformal transformations of a space of four dimensions and their applications to geometrical optics, *Proc. Lond. Math. Soc.* (2) 7 (1908), 70.
9. J. Jeans. "Electricity and Magnetism", Cambridge University Press, 5th Edition, 1933, p. 279.
10. H. Bateman. The transformation of the electrodynamical equations, *Proc. London Math. Soc.* (2) 8 (1909), 223.
11. H. Bateman. "The Mathematical Analysis of Electrical and Optical Wave-motion", Dover Paperback No. S14, 1955, p. 31.
12. J. Ehlers, W. Rindler and I. Robinson. Quaternions, bi-vectors and the Lorentz Group. *In* "Perspectives in Geometry and Relativity" (B. Hoffmann, Ed.), Indiana University Press, 1966, p. 134.
13. L. Silberstein. "The Theory of Relativity", Macmillan, 1924, Chapter VIII, p. 205.
14. F. S. Klotz. Twistors and the conformal group, *J. Math. Phys.* 15 (12), (1974), 2242.

15. E. Cunningham. The principle of relativity in electrodynamics and an extension thereof, *Proc. Lond. Math. Soc.* (2) **8** (1909), 77.
16. J. Pierpont. Optics in hyperbolic space, *Trans. Amer. Math. Soc.* **30** (1928), 33.
17. J. L. Synge. "Geometrical Mechanics and de Broglie Waves", Cambridge University Press, (1954), p. 19.
18. F. G. Friedlander. "The Wave Equation on a Curved Space Time", Cambridge Monographs in Mathematical Physics, 1975.
19. R. Hermann. "Differential Geometry and the Calculus of Variations", Academic Press, London and New York, 1968, p. 390.
20. M. Herzberger. An optical model of physics, *Jour. Opt. Soc. Amer.* **40** (7), (1950), 424.
21. H. Poeverlein. The Summerfeld–Runge law and geometrical optics in four dimensions. *In* "Electromagnetic Theory and Antennas", (E. C. Jordan, Ed.), Pergamon Press, 1963, p. 261.
22. J. B. Keller. A geometrical theory of diffraction, *Jour. Opt. Soc. Amer.* **52** (1962), 116.
23. E. Cambi. Projective formulation of the problems of geometrical optics, *Jour. Opt. Soc. Amer.* **49** (1), (1959), 2 and 15.
24. P. Duval. "Quaternions, Homographies and Rotations", Oxford University Press, 1964.
25. H. A. Buchdahl. Symplectic formalism in the aberration theory of systems without symmetries, *Optik*, **37** (5), (1973), 571; also *Jour. Opt. Soc. Amer.* **62** (11), (1972), 1314.
26. M. Sachs. A new theory of elementary matter, *Int. Jour. Theor. Phy.* **4** (6), (1971), 433 and 453.
27. H. P. Hudson. "Cremona Transformations", Cambridge University Press, 1927.

BIBLIOGRAPHY

H. Bateman. A complete list of the published work is to be found in the commemorative article by E. T. Bell in *Quart. J. Appl. Maths* **4** (2) (1946), 105.

B. L. Beers and R. S. Millman. Analytic vector harmonic expansions on $SU(2)$ and S^2, *J. Math. Phys.* **16** (1) (1975), 11.

E. F. Bolinder. The classical electromagnetic equations expressed as four dimensional qualities, *J. Franklin Inst.* **264** (1975), 213.
A Survey of the use of non-Euclidean geometry in electrical engineering, *J. Franklin Inst.* **265** (1958), 169.
Geometric-analytic theory of transition in electrical engineering, *Proc. I.R.E.* **47** (1959), 1124.

L. G. Bossy. Le trace des rayons dans un milieu anisotrope, non permanent et legerment absorbant dans le cadre d'une theorie Hamiltonienne a quatre dimensions. Alta Frequenza, 1969, U.R.S. Symposium issue Stresa 1968, Vol. 38, p. 20.

H. R. Hassé. The equations of electrodynamics and the null influence of the earth's motion on optical and electrical phenomena, *Proc. Lond. Math. Soc.* (2), **8** (1909),

S. Hong and R. F. Goodrich. Application of conformal mapping to scattering and diffraction problems. *In* "Electromagnetic Wave Theory" (J. Brown, Ed.), Vol. 11. Pergamon Press, 1967, p. 907.

H. Jeger. "Transformation Geometry", Allen and Unwin, 1966.

D. S. Jones. High frequency refraction and diffraction in general media, *Phil. Trans. Roy. Soc.* **A255** (1963), 363.

R. Keskinen. Complex potentials in classical mechanics and geometrical optics, *Amer. J. Phys.* **40** (1972), 418.

P. W. Ketchum. Analytic functions of hypercomplex variables, *Trans. Amer. Math. Soc.* **30** (1928), 641.

J. Kronsbein. Kinematics, quaternions, spinors and the Pauli spin matrices. *Amer. Jour. Phys.* **35** (167) 335.

E. P. Lane. Hypergeodesic mapping of a surface on a plane, *Trans. Amer. Math. Soc.* **32** (1930), 558.

D. H. Mayer. Vector and tensor fields on conformal space, *J. Math. Phys.* **16** (4), (1975), 884.

Y. Miyazaki. Propagation properties of optical signal waves in perturbed dielectric waveguide by conformal mapping technique; Topical meeting on integrated optics paper M.2.2, Las Vegas, February, 1972.

T. C. Mo, C. H. Papas and C. E. Baum. General scaling method for electromagnetic fields with application to a matching problem, *J. Math. Phys.* **14** (4), (1973).

P. S. Modenov and A. S. Parkhomenko. "Geometric Transformations", Academic Press, London and New York, 1965.

M. Neviere and M. Cadilhac. Sur une nouvelle formulation de probleme de diffraction d'une onde plane par un reseau infiniment conducteur-cas general, *Optics Communications,* **3** (6), (1971), 379; Ibid, **4** (1) (1971), 13; Ibid, **2** (5) (1970), 235 with R. Petit, *Trans I.E.E.E.* **AP21** (1) (1973), 37.

J. H. Richter. Application of conformal mapping to earth-flattening procedures in radio propagation problems, *Radio Science* **1** (new series) (12) (1966), 1435.

Appendix 1

This appendix deals in the main with the fundamental laws of geometrical optics and the derivation of the particular forms used as the basis for reflector and lens design in the earlier chapters of this book. To do this one has of necessity to draw on the many and varied versions of these laws that have appeared in many and varied notations over the entire history of the subject. Many formulae will be given without derivation or proof where it can be more simply observed that the result is consistent with the generally accepted version. One or two applications will be introduced as illustrative examples as they are relevant to the general subject even if not exactly optical designs with direct application to microwave antennas. The resultant forms of the laws so obtained, besides being fundamental to optical design, do permit generalization and extension as will be indicated. For complete proof of the formula and in particular for the derivation of the most fundamental from Fermat's principle a standard text should be consulted.[1]

I.1 THE LAWS OF REFRACTION AND REFLECTION

We use the suffices i, r, and t throughout to denote the incident, reflected and transmitted (or refracted) sides of a surface of discontinuity between two media with refractive indices η_i and η_t accordingly. Unit vectors $\hat{s}$ will denote the directions of rays, which, for homogeneous isotropic media, will be straight lines. For non-homogeneous media, where the rays are not taken as piecewise straight increments, $\hat{s}$ will be tangential to a curved ray at any point on that ray. The increment of path length in such a case will be ds. Position vectors from an origin will be $\mathbf{r}$ or $\mathbf{\rho}$. Where the ray itself is a radius vector from the origin, as in a source distributions, $\hat{s}$ and ds may be replaced by $\hat{r}$ and dr. It is in this form that they are used in Chapter 1. Care must be exercised in differentiating between $\hat{r}$ and $\hat{s}$ whenever these do not derive from the same origin.

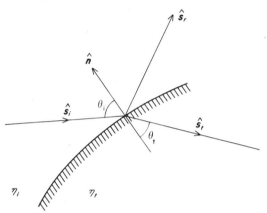

FIG. I.1. *The principle of reflection and refraction.*

The laws of reflection and refraction can be obtained rigorously from the solution of the infinite plane boundary value problem for the incident plane electromagnetic wave upon an interface between two refracting media.[2] The geometrical optics approximation enables one to dispense with the infinities and treat a local region as if it were part of a tangential infinitely plane discontinuity. This is a good approximation for regions with radius of curvature that is a large number of wavelengths and consequently where a uniquely defined normal $\mathbf{n}$ exists. The alternative derivation of the same law by Fermat's principle does not at first sight appear to suffer from this limitation, but, as observed by Pauli (Ref. 1, p. 7) the fact that the reflected ray does not appear in this formulation shows that Fermat's principle is essentially an approximation to reality.

The result with the notation of Fig. I. 1 is

$$\hat{\mathbf{n}} \times \hat{\mathbf{s}}_i = \hat{\mathbf{n}} \times \hat{\mathbf{s}}_r \qquad (\text{I.1a})$$

and

$$\hat{\mathbf{n}} \cdot \hat{\mathbf{s}}_i = -\hat{\mathbf{n}} \cdot \hat{\mathbf{s}}_r \qquad (\text{I.1b})$$

which are Snell's laws of reflection, and

$$\eta_i \hat{\mathbf{n}} \times \hat{\mathbf{s}}_i = \eta_t \hat{\mathbf{n}} \times \hat{\mathbf{s}}_t \qquad \text{(I.1c)}$$

the law of refraction.

The solution of the boundary value problem for a plane electromagnetic wave, shows, as does the above result, that the vectors are all coplanar.

Solving (I.1a) and (I.1b) gives for a reflected ray

$$\hat{\mathbf{s}}_r = \hat{\mathbf{s}}_i - 2\hat{\mathbf{n}}(\hat{\mathbf{s}}_i \cdot \hat{\mathbf{n}}), \qquad \text{(I.2)}$$

probably the most immediately useful to all the various forms of Snell's law of reflection.

The scalar form of Eqn (I.1c) gives the familiar law of refraction

$$\eta_i \sin \theta_i = \eta_t \sin \theta_t. \qquad \text{(I.3)}$$

To obtain a vector solution for the refracted ray we note that, since the incident and reflected rays are within the same medium, that is $\eta_i = \eta_r$ Eqns (I.1a) and (I.1c) combine to give

$$\hat{\mathbf{n}} \times \eta_i \hat{\mathbf{s}}_i = \hat{\mathbf{n}} \times \eta_r \hat{\mathbf{s}}_r = \hat{\mathbf{n}} \times \eta_t \hat{\mathbf{s}}_t. \qquad \text{(I.4)}$$

We define, in the light of this relation, the *directed* ray vector $\mathbf{t} = \eta \hat{\mathbf{s}}$ and thus relation 4 states that the vector $\hat{\mathbf{n}} \times \mathbf{t}$ is an invariant of the refraction at a plane surface. With this notation Eqn (I.1c) is

$$(\mathbf{t}_i - \mathbf{t}_t) \times \hat{\mathbf{n}} = 0 \qquad \text{(I.5)}$$

which is the Sommerfeld–Runge relation.[3] Hence the vector $\mathbf{t}_i - \mathbf{t}_t$ is parallel to $\hat{\mathbf{n}}$ and there exists a scalar γ such that

$$\mathbf{t}_t = \mathbf{t}_i + \gamma \hat{\mathbf{n}}. \qquad \text{(I.6)}$$

Taking the cross product of Eqn (I.4) with $\hat{\mathbf{n}}$ gives

$$\mathbf{t}_t = \mathbf{t}_i + \hat{\mathbf{n}}(\mathbf{t}_t \cdot \hat{\mathbf{n}} - \mathbf{t}_i \cdot \hat{\mathbf{n}}),$$

which on comparison with Eqn (I.6) shows that

$$\gamma = \mathbf{t}_t \cdot \hat{\mathbf{n}} - \mathbf{t}_i \cdot \hat{\mathbf{n}} = \eta_t \cos \theta_t - \eta_i \cos \theta_i. \qquad \text{(I.7)}$$

Using Eqn (I.3) to eliminate θ_t in Eqn (I.7) results in several equivalent forms for γ dependent only upon properties of the incident ray, which are

$$\gamma = \{\eta_t^2 - \eta_i^2 \sin^2 \theta_i\}^{\frac{1}{2}} - \eta_i \cos \theta_i \qquad \text{(I.8a)}$$

$$= \{\eta_t^2 - \eta_i^2(1 - \hat{\mathbf{n}} \cdot \hat{\mathbf{s}}_i)^2\}^{\frac{1}{2}} - \eta_i \hat{\mathbf{n}} \cdot \hat{\mathbf{s}}_i \qquad \text{(I.8b)}$$

$$= \{\eta_t^2 - \eta_i^2(\hat{\mathbf{n}} \times \hat{\mathbf{s}}_i)^2\}^{\frac{1}{2}} - \eta_i \hat{\mathbf{n}} \cdot \hat{\mathbf{s}}_i \qquad \text{(I.8c)}$$

$$= \{\eta_t^2 - \eta_i^2(\hat{\mathbf{n}} \times \hat{\mathbf{s}}_i)^2\}^{\frac{1}{2}} - \eta_i \hat{\mathbf{n}} \cdot \hat{\mathbf{s}}_i. \qquad \text{(I.8d)}$$

N

Taking now the differential form of (I.6) we obtain

$$\mathbf{dt}_t = \mathbf{dt}_i + \gamma \mathbf{d\hat{n}} + \hat{n}\, d\gamma. \qquad (I.9)$$

From Eqn (I.7) $d\gamma = -\eta_t \sin \theta_t\, d\theta_t + \eta_i \sin \theta_i\, d\theta_i$ and from Eqn (I.3) $\eta_i \cos \theta_i\, d\theta_i = \eta_t \cos \theta_t\, d\theta_t$ so that θ_t and $d\theta_t$ can be eliminated to give

$$d\gamma = f(\theta_i)\, d\theta_i.$$

Since $\hat{n}$ is a unit vector $\hat{n} \cdot d\hat{n} = 0$ hence taking the dot product of (I.9) with $\hat{n}$ there results

$$\hat{n} \cdot \mathbf{dt}_t - \hat{n} \cdot \mathbf{dt}_i = d\gamma. \qquad (I.10)$$

Comparing this with the differential of equation 7 leaves

$$\mathbf{t}_t \cdot d\hat{n} - \mathbf{t}_i \cdot d\hat{n} = 0$$

which since $d\hat{n}$ is tangential to the surface of refraction is a restatement of the original Snell's law.

In the reference given the following statement concludes this analysis:

"In essence formula (I.9) already contains the answer to all thinkable questions concerning the properties of the infinitesimal refracted pencil in terms of those of the incident one. It is enough to read it intelligently and to develop it appropriately in each particular case."

We can illustrate an application of Eqn (I.6) by the transmission of an oblique ray through a prism (Fig. I.2). Consider two consecutive plane refractions at a prism with surface normals $\hat{n}_1$ and $\hat{n}_2$ and let the intersecting edge have direction $\hat{e}$. Then since the refracted ray of the first surface becomes

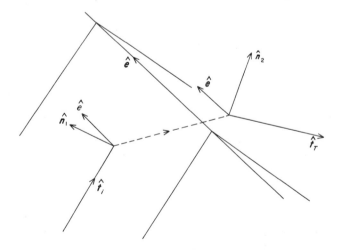

FIG. I.2. Double refraction at a plane prism.

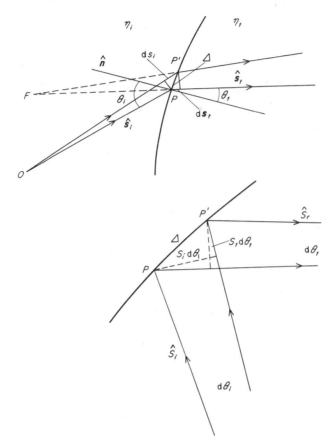

FIG. I.3. *Refraction and reflection of a ray pencil.*

the incident ray of the second we have from Eqn (I.6)

$$\mathbf{t}_t = \mathbf{t}_i + \gamma_1\hat{\mathbf{n}}_1; \qquad \mathbf{t}_T = \mathbf{t}_t + \gamma_2\hat{\mathbf{n}}_2,$$

so that the final ray is given by

$$\mathbf{t}_T = \mathbf{t}_i + \gamma_1\hat{\mathbf{n}}_1 + \gamma_2\hat{\mathbf{n}}_2.$$

Taking the scalar product with $\hat{\mathbf{e}}$

$$\mathbf{t}_T \cdot \hat{\mathbf{e}} = \mathbf{t}_i \cdot \hat{\mathbf{e}}$$

producing the well-known result that the incident and final rays make the same angle with the prism edge.

The same property can be shown to occur with a double mirror for a single reflection at each surface. In the case of a pure reflection we make $\eta_t = -\eta_i$ then from Eqn (I.7)

$$\gamma = -2\eta_i \cos \theta_i$$

and hence from Eqns (I.2) or (I.6) (with $\mathbf{t}_t$ replaced by $\mathbf{t}_r$)

$$\hat{\mathbf{s}}_r = \hat{\mathbf{s}}_i - 2\hat{\mathbf{n}} \cos \theta_i. \tag{I.11}$$

Consider now two adjacent rays of a flat pencil of rays as in Fig. I.3, issuing from the point O. After refraction the rays (or their extensions) intersect

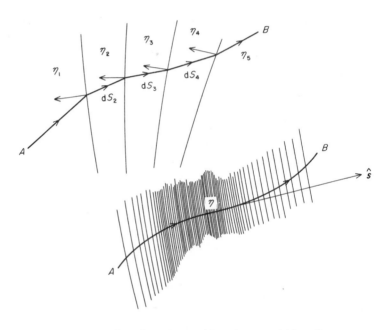

FIG. I.4. Ray paths in discontinuous and continuous variable media.

again at the point F, and, letting the adjacent points PP' on the surface be a distance Δ apart then

$$ds_i = \Delta \sin \theta_i \quad \text{and} \quad ds_t = \Delta \sin \theta_t \tag{I.12}$$

and hence by virtue of Eqn (I.4)

$$\eta_i \, ds_i = \eta_t \, ds_t. \tag{I.13}$$

An ambiguity in sign has to be included in these results to relate the changes between ds_i and ds_t which will be seen to determine whether the image formed is virtual (on the same side as O as shown) or real (on the opposite side).

Once again for a purely reflecting surface $\eta_i = -\eta_t$ and Eqn (I.13) becomes

$$\mathrm{d}s_i = \pm\ \mathrm{d}s_t \tag{I.14}$$

The two triangles with PP' as hypoteneuse become congruent and give the further relation for reflecting surfaces

$$s_i\,\mathrm{d}\theta_i = s_t\,\mathrm{d}\theta_t. \tag{I.15}$$

These are the relations upon which the design of fundamental focusing systems are based in Chapter 1.

The principle of Fermat, from which the above relations could also have been derived,[4] states that along a true ray path the function

$$\int \eta\,\mathrm{d}s \text{ is an extremum.}$$

Consider as in Fig. I.4 several refracting surfaces separating discrete media with refractive indices $\eta_1, \eta_2 \ldots$. Then from Fermat's principle

$$\delta\int_A^B \eta\,\mathrm{d}s = 0$$

for all ray paths between the end points A and B (one of which may be at infinity) we have

$$\int_A^B \eta\,\mathrm{d}s = \text{const,}$$

which for the piecewise straight ray in Fig. I.4 becomes

$$\sum_n \eta_n\,\mathrm{d}s_n = \text{const,} \tag{I.16}$$

with the correct sign attributed to each term.

In the case of a sequence of reflectors we can use Eqn (I.15) to give

$$\sum_n s_n\,\mathrm{d}\theta_n = \text{const,} \tag{I.17}$$

upon which the design of optical systems of more than one surface can be based.

I.2 CONTINUOUS NON-HOMOGENEOUS MEDIA

The basic form of Eqn (I.6) is

$$\eta_t \hat{s}_t - \eta_i \hat{s}_i = \gamma \hat{n} \tag{I.18}$$

at a single refracting surface. If this surface has the equation

$$\phi(x, y, z) = 0$$

then $\hat{n}$ is proportional to $\nabla \phi$ and hence the curl of Eqn (I.18) will be zero giving

$$\nabla \times (\eta_t \hat{s}_t) = \nabla \times (\eta_i \hat{s}_i). \tag{I.19}$$

Continuing this, first for a system of closely spaced refracting surfaces as in Fig. I.4 and then for a continuously varying refracting medium will give

$$\nabla \times (\eta \hat{s}) = \text{const along a ray path}, \tag{I.20}$$

or

$$\frac{d}{ds}(\nabla \times \eta \hat{s}) = 0 \text{ along a ray.} \tag{I.21}$$

Since for any ray with tangent vector $\hat{s}$ at a point P whose *radius vector* from an origin O is $\mathbf{r}$ then

$$\frac{d\mathbf{r}}{ds} = \hat{s}$$

and Eqn (I.21) becomes

$$\nabla \times \left[\frac{d}{ds} \left(\eta \frac{d\mathbf{r}}{ds} \right) \right] = 0, \tag{I.22}$$

which in turn gives

$$\frac{d}{ds} \left(\eta \frac{d\mathbf{r}}{ds} \right) = \text{the gradient of a space function.}$$

This gradient must lie parallel to the normal directions of the discrete surfaces in the stage shown in Fig. I.4 and thus is the gradient of the function describing the space variation of the refractive index, so giving finally the result which forms the basis of the ray method used for the study of the optics of non-homogeneous media in Chapter 2, namely

$$\frac{d}{ds} \left(\eta \frac{d\mathbf{r}}{ds} \right) = \text{grad } \eta. \tag{I.23}$$

A more rigorous derivation of this result can be found in many places in the literature of optics for example Ref. 2, p.121.

I.3 THE THEOREM OF MALUS AND DUPIN

This is given in Ref. 3, p. 130, as an application of Fermat's principle and in Ref. 4, p. 116, as a consequence of Eqn (I.19). That it derives from the latter is a point of great interest in the derivation of the four dimensional analysis. We will only give an adapted statement of the theorem here for reference purposes.

If a system of rays, issuing in the first instance from some source point and thus has a spherical wave front about that point, is refracted or reflected any number of times, and if equal optical paths are measured along each ray, from either the source itself or any of its original wave fronts, then the surface so formed by the end points of the equal optical path rays will be orthogonal to the system of rays and form a wave front.

This interprets the usual statement that a normal congruence $(q \cdot v)$ (or orthotomic system of rays) will remain a normal congruence after repeated reflection or refraction.

I.4 THE EIKONAL EQUATION

We take as starting point Eqn (I.20), and choosing the arbitrary constant to be zero we have

$$\nabla \times (\eta \hat{s}) = 0, \tag{I.24}$$

or

$$\eta \hat{s} = \nabla S,$$

where S is a function of the space coordinates, and also the consequence that the unit vectors $\hat{s}$ are perpendicular to the surfaces $S(x, y, z) = \text{const}$.

Since $\hat{s}$ is tangential to the ray at any point these surfaces are the normal surfaces referred to in the theorem of Malus and Dupin and are thus the wave-fronts of the propagating field.

Thus from Eqn (I.24) we have

$$\nabla S \cdot \nabla S = \eta^2 \qquad \text{or} \qquad |\nabla S| = \eta. \tag{I.25}$$

From the integral definition of curl, Eqn (I.24) implies

$$\int \eta \hat{\mathbf{s}} \cdot \mathbf{ds} = 0$$

for any closed path within the medium and hence

$$\int_A^B \eta \hat{\mathbf{s}} \cdot \mathbf{ds} = S_B - S_A \tag{I.26}$$

independently of the path taken. This can be seen to be a restatement of the fundamental principle of Fermat $\delta \int \eta \, ds = 0$ along a true ray path. S is the eikonal function (for example Ref. 4, p. 131) and its analogy with the potential function in the theory of conservative force fields demonstrated by Eqn (I.26) provides a further connection between the theory of ray paths and particle trajectories as discussed at more than one point in the text.

Equation (I.25) in a Cartesian coordinate system becomes

$$\left(\frac{\partial S}{\partial x}\right)^2 + \left(\frac{\partial S}{\partial y}\right)^2 + \left(\frac{\partial S}{\partial z}\right)^2 = \eta^2. \tag{I.27}$$

The same relation can be obtained by the substitution of an assumed solution into the scalar wave equation

$$\nabla^2 u + k^2 u = 0,$$

where $k = \omega\sqrt{\mu\varepsilon} = 2\pi/\lambda$ and μ, ε are the permeability and permittivity of the medium, λ the wave length of the radiated wave in the medium. Using the suffix 0 for the same parameters in a vacuum, we substitute

$$u = A(\exp -ik_0 S),$$

then since $\eta = k/k_0$ there results

$$\nabla^2 u + k^2 u = -k_0^2 u \left[\left(\frac{\partial S}{\partial x}\right)^2 + \left(\frac{\partial S}{\partial y}\right)^2 + \left(\frac{\partial S}{\partial z}\right)^2 - k^2/k_0^2\right]$$

$$+ 2ik_0 u \left[\tfrac{1}{2}\nabla^2 S + \frac{1}{A}\,\text{grad}\, A \cdot \text{grad}\, S\right] + \ldots \tag{I.28}$$

The remaining terms remain finite in the geometrical optics limit that is when $k_0 \to \infty$. We thus have an illustration of the approximate solution we have been using in that making the first term zero gives us Eqn (I.25) which derives from Fermat's principle. We further require the second term to be zero or

$$\frac{1}{A} \operatorname{grad} A . \operatorname{grad} S = -\tfrac{1}{2}\nabla^2 S. \tag{I.29}$$

Since $\nabla S = \eta\hat{\mathbf{s}}$ and grad $A \cdot \hat{\mathbf{s}} = \partial A/\partial s$,

$$\frac{dA}{ds} + \tfrac{1}{2}\frac{\nabla^2 S}{\eta}A = 0,$$

or

$$A(s) = A(0)\exp\left(-\tfrac{1}{2}\int_0^s \eta^{-1}\nabla^2 S \, ds\right). \tag{I.30}$$

The exponential factor can be expressed in terms of the Gaussian curvature of the wave front at the two chosen points with parameters[5] zero and s. If the principal radii of curvature are R and R and $K = 1/R_1R_2$ then

$$\exp\left(-\tfrac{1}{2}\int_0^s \eta^{-1}\nabla^2 S \, ds\right) = \{K(s)/K(0)\}^{\frac{1}{2}}. \tag{I.31}$$

These relations can be seen to be a statement of the conservation of energy. A full discussion is contained in Ref. 5 which concludes with the interesting result that the orthogonal triad of vectors associated with a curved ray in space, that is the tangent vector $\hat{\mathbf{s}}$, the normal $\hat{\mathbf{n}}$ and the bi-normal $\hat{\mathbf{b}}$, satisfy "pseudo-Maxwell" equations in which $\eta\hat{\mathbf{n}}$ and $\eta\mathbf{b}$ play the part of $\mathbf{E}$ and $\mathbf{H}$ and $\hat{\mathbf{s}}$ the Poynting vector $\mathbf{E} \times \mathbf{H}$.

I.5 REFRACTION AND REFLECTION OPERATORS

(a) *The reflection dyadic*
  The law of reflection given in Eqn (I.2) is

$$\hat{\mathbf{s}}_r = \hat{\mathbf{s}}_i - 2\hat{\mathbf{n}}(\hat{\mathbf{s}}_i \cdot \hat{\mathbf{n}}),$$

and can be written with a change in notation as

$$\hat{\mathbf{s}}_r = [I - 2(\hat{\mathbf{n}}\hat{\mathbf{n}})]\hat{\mathbf{s}}_i, \tag{I.32}$$

where I is an identity operator or idemfactor and $(\hat{\mathbf{n}}\hat{\mathbf{n}})$ is a dyadic. The properties of the reflection dyadic

$$\Omega = I - 2(\hat{\mathbf{n}}\hat{\mathbf{n}})$$

can be obtained by the comparison between these two equations. It can be shown that this operator has the necessary algebraic properties of distributivity and associativity required for the following but not (Ref. 3, p. 25) commutativity. Then to determine the direction of a final ray after a number of intermediate reflections we have

$$\hat{s}_{r_n} = \Omega_n \Omega_{n-1} \dots \Omega_1 \hat{s}_i. \tag{I.33}$$

By the associative property these may be bracketed in any way (without disturbing the order of course) that affords simplification of the result.

The angle between the incident and final rays will then be ψ where

$$\cos \psi = \hat{s}_i \Omega \hat{s}_i, \tag{I.34}$$

where Ω stands for the product of all the operators in Eqn (I.33). For example the double mirror system with normals $\hat{n}_1$ and $\hat{n}_2$ has

$$\Omega = \Omega_2 \Omega_1 = \left[I - 2(\hat{n}_2 \hat{n}_2) \right]\left[I - 2(\hat{n}_1 \hat{n}_1) \right]$$
$$= I - 2(\hat{n}_2 \hat{n}_2) - 2(\hat{n}_1 \hat{n}_1) + 4\hat{n}_1 \cdot \hat{n}_2(\hat{n}_1 \hat{n}_2),$$

so that

$$\hat{s}_{r_2} = \hat{s}_i - 2\hat{s}_i \cdot \hat{n}_2 \hat{n}_2 - 2\hat{s}_i \cdot \hat{n}_1 \hat{n}_1 + 4\hat{n}_1 \cdot \hat{n}_2 \hat{s}_i \cdot \hat{n}_1 \hat{n}_2. \tag{I.35}$$

If the line of the intersection of the mirrors has direction $\hat{e}$ then $\hat{e} \cdot \hat{n}_1 = \hat{e} \cdot \hat{n}_2 = 0$ and so $\hat{e} \cdot \hat{s}_{r_2} = \hat{e} \cdot \hat{s}_i$ and the ray undergoes a rigid rotation about $\hat{e}$ as was the case for the prism.

Since $\Omega_1 \Omega_2 \neq \Omega_2 \Omega_1$ in general, a beam of rays wide enough to illuminate both sides will separate into two reflected beams. If the mirrors are at right angles the operator Ω becomes self-conjugate and the order of reflections is immaterial. However, no such double mirror can reverse the direction of *every* incident ray as is the case with (some) triple mirrors.

(b) *The refraction dyadic*[6]

With regard to refraction the situation is more complex from Eqns (I.6) and (I.8) we obtain

$$\hat{s}_t = \frac{\eta_i}{\eta_t} \hat{s}_i - \frac{\eta_i}{\eta_t} \hat{n}(\hat{n} \cdot \hat{s}_i) + \hat{n}\left(1 - \frac{\eta_i^2}{\eta_t^2} + \frac{\eta_i^2}{\eta_t^2} (\hat{n} \cdot \hat{s}_i)^2 \right)^{\frac{1}{2}}, \tag{I.36}$$

which can be seen to be Snell's law by taking the scalar product with $\hat{n}$ to give

$$\cos \theta_t = \left(1 - \frac{\eta_i^2}{\eta_t^2} \sin^2 \theta_i \right)^{\frac{1}{2}}$$

Putting

$$\eta_i/\eta_t = \beta; \hat{\mathbf{n}} \cdot \hat{\mathbf{s}}_i = -p \quad \text{and} \quad \frac{\eta_i}{\eta_t}(\hat{\mathbf{n}} \cdot \hat{\mathbf{s}}_i - \cos\theta_t) = \alpha$$

Eqn (I.36) becomes

$$\alpha = \beta p \left\{ 1 - \left(1 + \frac{1 - \beta^2}{\beta^2 p^2} \right)^{\frac{1}{2}} \right\},$$

or

$$\alpha = \beta p q. \tag{I.37}$$

Then Eqn (I.6) is simply

$$\hat{\mathbf{s}}_t = \beta[I - q\hat{\mathbf{n}}\hat{\mathbf{n}}]\hat{\mathbf{s}}_i, \tag{I.38}$$

whence $R = \beta[I - q\hat{\mathbf{n}}\hat{\mathbf{n}}]$ is the refraction dyadic. As noted by Silberstien (Ref. 3), q contains p explicitly and this fact prevents R from being distributive. It does however have a defined inverse which is found to be (Ref. 6)

$$R^{-1} = \beta^{-1}\left[I - \frac{q}{q-1}\hat{\mathbf{n}}\hat{\mathbf{n}} \right]. \tag{I.39}$$

This is the reversal of the ray and hence is obtained from R by the transformation $\eta_i \leftrightarrow \eta_t$ and $\theta_i \leftrightarrow \theta_t$.

It is thus limited by the condition for total internal reflection since

$$\frac{q}{q-1} = \left\{ 1 - \left(1 + \frac{\beta^2 - 1}{1 - \beta^2 + \beta^2 p^2} \right) \right\}^{\frac{1}{2}},$$

which becomes imaginary when $\eta_i/\eta_t < \sin\theta_t$.

(c) The reflection quaternion

Reverting to the basic relations of Snell's law for reflection given in Eqn (I.1) which are (noting change of cross product)

$$\hat{\mathbf{n}} \cdot \hat{\mathbf{s}}_i = -\hat{\mathbf{n}} \cdot \hat{\mathbf{s}}_r; \quad \hat{\mathbf{n}} \times \hat{\mathbf{s}}_r = -\hat{\mathbf{s}}_i \times \hat{\mathbf{n}}$$

and adding we obtain

$$-\hat{\mathbf{n}} \cdot \hat{\mathbf{s}}_r + \hat{\mathbf{n}} \times \hat{\mathbf{s}}_r = \hat{\mathbf{n}} \cdot \hat{\mathbf{s}}_i - \hat{\mathbf{s}}_i \times \hat{\mathbf{n}}$$

$$= -[-\hat{\mathbf{s}}_i \cdot \hat{\mathbf{n}} + \hat{\mathbf{s}}_i \times \hat{\mathbf{n}}]. \tag{I.40}$$

This combination of dot and cross product typifies the product of two quaternions which have zero scalar elements (Appendix II).

Calling then $N \equiv (0, \hat{\mathbf{n}})$, $R \equiv (0, \hat{\mathbf{s}}_r)$ and $I = (0, \hat{\mathbf{s}}_i)$ Eqn (I.40) is $NR = -IN$ where quaternion algebra has to be used.

Consequently $R = N^{-1}IN$ for a single reflection. Again in a system of multiple reflections we have after m reflections

$$R_m = (-1)^m N_m^{-1} N_{m1}^{-1} \ldots N_1^{-1} I N_1 \ldots N_m, \tag{I.41}$$

or

$$R_m = (-1)^m Q^{-1} I Q,$$

where Q is the single quaternion, the product of $N_1 \ldots N_m$.

As always happens with such condensed versions of operator algebras and tensors, there is as much complexity in calculating the final result as in other longhand versions. The procedure is however straightforward though complicated vector algebra.

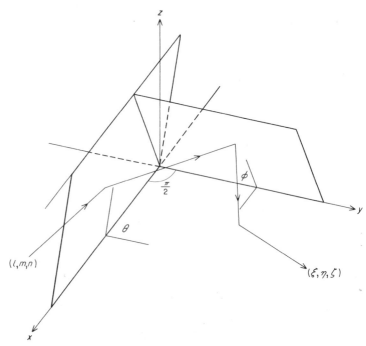

FIG. I.5. *Reflection in a triple mirror.*

We give as illustration the result for the three mirror system illustrated in Fig. I.5. The mirror in the x, y plane has a 90° corner and two mirrors intersect it along the x and y axes with inclinations θ and ϕ respectively. Due to the non-commutative property of quaternion products the order of the mirror reflections affects the final result (as is physically obvious).

Taking a reflection first from the "θ" mirror, then the "ϕ" mirror and finally the horizontal mirror, we find, for a ray with original direction cosines (l, m, n) that the reflected ray has direction cosines given by

$$\xi = l\cos 2\phi + m\sin 2\theta \sin 2\phi - n\cos 2\theta \sin 2\phi,$$

$$\eta = m\cos 2\theta + n\sin 2\theta, \qquad (1.42)$$

$$\zeta = -l\sin 2\phi + m\sin 2\theta \cos 2\phi - n\cos 2\theta \cos 2\phi.$$

For example with $\theta = \phi = \pi/2$ we have the usual 90° corner reflector and $(\xi, \eta, \zeta) = (-l, -m, -n)$ that is the ray is returned along its original direction. Other interesting results can be obtained by putting $\theta = \frac{1}{2}\pi$, $\phi = \frac{1}{4}\pi$ and $\theta = \frac{1}{4}\pi$, $\phi = \frac{1}{2}\pi$ in turn in Eqn (1.42), giving reflections from the semioctant corner reflector. For a full appraisal however, the result is required for all combinations of the order of reflections. If, in the example given, the incidence was first upon the "ϕ" mirror and second upon the "θ" mirror the result is a transformation of Eqns (1.42) by $l \leftrightarrow m$; $\xi \leftrightarrow \eta$; $\theta \leftrightarrow \phi$.

A study of the more complex refraction quaternion has been made by Wagner.[7]

(d) The reflection matrix

A quaternion is an algebraic form of a matrix or tensor operator in much the same way as a complex number is an algebraic form of vector. Hence the results of the previous paragraph will have a matrix version and this is to be found in the work of Beggs.[8]

For a plane mirror whose surface is given by the equation

$$Ax + By + Cz + D = 0$$

the direction cosines of the reflected ray is related to those of the incident ray by the matrix M given by

$$\begin{pmatrix} \xi \\ \eta \\ \zeta \end{pmatrix} = \begin{pmatrix} 1 - 2A^2/F^2 & -2AB/F^2 & -2AC/F^2 \\ -2AB/F^2 & 1 - 2B^2/F^2 & -2BC/F^2 \\ -2AC/F^2 & -2BC/F^2 & 1 - 2C^2/F^2 \end{pmatrix} \begin{pmatrix} l \\ m \\ n \end{pmatrix}, \qquad (1.43)$$

where $F^2 = A^2 + B^2 + C^2$.

Successive reflections are then calculated via the successive multiplication of reflection matrices.

The main role played by this combination of plane mirrors in the field of microwaves is in the design of corner reflectors or clusters of corner reflectors.[9] Such arrangements are necessary to smooth out some of the blind spots which may occur with the single 90° trihedral corner. The connection between such systems and the crystal symmetries is fairly obvious and thus it is of no surprise that some of the methods used here make their appearance in those subjects also and have been studied by authors more renowned in those fields.

By regarding the reflections of a ray from a system of mirrors as a single rigid rotation of the direction of the ray by some angle about an axis, both of which can be determined by the methods given, shows again the connection that exists between the quaternions for example, the rotations in three (and four) space and the symmetry groups.[10]

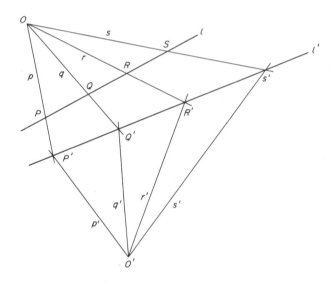

FIG. I.6. *Geometry of the projectivity.*

(e) *The refractive projectivity*

Four coplanar straight lines through a common point O, as shown in Fig. I.6, having a transversal l which intersects them in points PQR and S have the elementary geometrical property

$$PR/OP = \sin{(pr)}/\sin{(ORP)}$$

indicating by (pr) the angle between the orientated lines p and r (taken counter clockwise for positive angles). So for a single ratio

$$\frac{PR}{QR} = \frac{OP}{OQ}\frac{\sin{(pr)}}{\sin{(qr)}}.$$

Indicating this ratio by the symbol (PQR) then for a ratio of ratios

$$(PQRS) = \frac{PQR}{PQS} = \frac{\sin(pr)}{\sin(qr)}\bigg/\frac{\sin(ps)}{\sin(qs)} = \frac{pqr}{pqs} = (pqrs). \qquad (I.44)$$

This expression defines the anharmonic or cross-ratio between the points $PQRS$ or the lines $pqrs$. As can be seen it is dependent only upon the angles between the lines and not upon the position of the transversal l. It forms a condition between the four entities, so that for a given cross-ratio *any* three points or lines, will determine the unique fourth member.

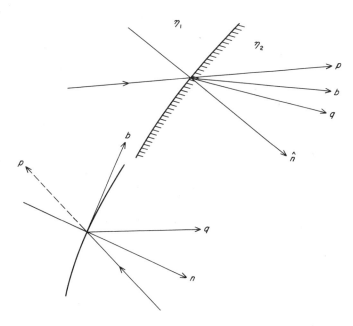

FIG. I.7. *Refraction and reflection projectivities.*

The property remains constant for any other transversal l' and by the uniqueness property for any other point of common origin O' whether in the same plane or not. The proof of this fundamental property comes from an application of Desargue's theorem.[11] The relation in Eqn (I.44) is said to be invariant under the projection from a point, or the points (or lines) are a projectivity.

In the case of refraction of a ray between two media with refractive indices η_1 and η_2 let the incident ray be p and the refracted ray q, the line bisecting the angle between them b and the normal to the surface n (Fig. I.7(b)) then as Cambi[12] shows

$$\frac{\sin{(pn)}}{\sin{(qn)}} = \frac{\eta_2}{\eta_1} \quad \text{and} \quad \frac{\sin{(pb)}}{\sin{(qb)}} = -1$$

and hence $(pqnb) = -\eta_2/\eta_1$.

Thus Snell's law defines a projectivity on the refracting surface which thus remains invariant under projective (and involutary) transformations of the plane (or line) containing the rays (or points) and its cross section of the refracting surface.

In the case of a reflector

$$\sin{(pn)} = \sin{(qn)} \quad \text{and} \quad (pb) = -(qb)$$

thus

$$(pqnb) = -1.$$

The ascent in dimension from points and lines to lines and planes follows from the principle of duality and as Ref. 12 shows to higher dimensionality. Therefore the refractive process defined in this way illustrates the relevance of some of the transformations, for example the involutions, conjectured as a design process in Chapter 6.

I.6 THE FOCAL LINE OF A REFLECTOR

If a system of rays from a point source is reflected from a specified reflecting surface given by the equation

$$f(x, y, z) = \text{const}$$

then the rays will meet any specified surface at points which can be obtained by a direct application of Snell's law. For those cases where these points all lie upon a single curve this curve will be a focal line for the reflector and surface given. Such is the case in particular for cylindrical reflectors and the perpendicular plane through the source. The focal line in such cases is also a cross-section of the zero distance wave-front which is required for the application of Damien's theorem (q.v.).

Normals to the surface at a point $P(a, b, c)$ have the direction

$$\hat{\mathbf{n}} = \frac{\operatorname{grad} f_P}{\left|\operatorname{grad} f\right|_P}.$$

The incident ray $\hat{\mathbf{s}}_i$ from a point $(d, 0, 0)$ will have direction

$$\hat{\mathbf{s}}_i = \frac{(a - d)\hat{\mathbf{i}} + b\hat{\mathbf{j}} + c\mathbf{k}}{\{(a - d)^2 + b^2 + c^2\}^{\frac{1}{2}}}. \tag{I.45}$$

Hence the reflected ray will be in the direction

$$\hat{\mathbf{s}}_r = \frac{1}{\{(a - d)^2 + b^2 + c^2\}^{\frac{1}{2}}} \left\{ (a - d)\hat{\mathbf{i}} + b\hat{\mathbf{j}} + c\mathbf{k} \right.$$

$$\left. - \frac{2\left(\dfrac{\partial f}{\partial x}\hat{\mathbf{i}} + \dfrac{\partial f}{\partial y}\hat{\mathbf{j}} + \dfrac{\partial f}{\partial z}\mathbf{k}\right)\left((a - d)\dfrac{\partial f}{\partial x} + b\dfrac{\partial f}{\partial y} + c\dfrac{\partial f}{\partial z}\right)}{\left(\dfrac{\partial f}{\partial x}\right)^2 + \left(\dfrac{\partial f}{\partial y}\right)^2 + \left(\dfrac{\partial f}{\partial z}\right)^2} \right\} \tag{I.46}$$

where all the differentials are evaluated at the point P.

Letting

$$\left\{ \left(\frac{\partial f}{\partial x}\right)^2 + \left(\frac{\partial f}{\partial y}\right)^2 + \left(\frac{\partial f}{\partial z}\right)^2 \right\} = D,$$

and

$$\left\{ (a - d)\frac{\partial f}{\partial x} + b\frac{\partial f}{\partial y} + c\frac{\partial f}{\partial z} \right\} = B,$$

the direction cosines of $\hat{\mathbf{s}}_r$ are thus proportional to

$$l \equiv (a - d)D^2 - 2\frac{\partial f}{\partial x}B,$$

$$m \equiv bD^2 - 2\frac{\partial f}{\partial y}B, \tag{I.47}$$

$$n \equiv cD^2 - 2\frac{\partial f}{\partial z}B.$$

The equation of the ray is therefore

$$\frac{x - a}{l} = \frac{y - b}{m} = \frac{z - c}{n}. \tag{I.48}$$

This line meets a specified surface in a point which varies as the point (a, b, c) is moved on the reflector. One variable may be eliminated in these relations therefore by the condition $f(a, b, c) = \text{const}$. For certain surfaces and reflectors two of the variables can be eliminated resulting in a single parameter system of points which is the focal line.

I.7 REFLECTION AND REFRACTION AT A LOSSY MEDIUM

The boundary value solution giving rise to the form of Snell's law for refraction so far considered was expressly for the incidence of a wave from free space or medium with refractive index η_i upon a semi-infinite lossless medium with refractive index η_t. The following result due to Bell *et al.*[13] generalizes this result for the situation where the second medium has a loss mechanism and is thus characterised by a complex refractive index

$$\eta_t = \eta_2 + iK. \tag{I.49}$$

In the notation of this appendix the generalized Snell's law is given as

$$\sin \theta_t = [\eta_i/\eta_2 \sin \theta_i] F^{\frac{1}{2}}, \tag{I.50}$$

where F is given as

$$F = \frac{1}{2} \left\{ \frac{-(\eta_2^2 - K^2 + \eta_i^2 \sin^2 \theta_i) + [(\eta_2^2 - K^2 + \eta_i^2 \sin^2 \theta_i) + 4(\eta_2^2 K^2 - \eta_i^2 \sin^2 \theta_i(\eta_2^2 - K^2)]}{K^2 - \eta_i^2 \sin^2 \theta_i(1 - K^2/\eta_2^2)} \right\}.$$

when $K = 0$ this reduced to the law for lossless media.

The Fresnel coefficients for this case for two conditions of reflection from a lossy interface are

(a) for polarization perpendicular to the plane of incidence

$$R_\perp = \frac{\sin^2 (\theta_i - \theta_t) + \left(\dfrac{\eta_2 K}{\eta_i}\right)^2 \left(\dfrac{\sin^2 \theta_t}{\sin \theta_i \cos \theta_t}\right)^2}{\sin^2 (\theta_i + \theta_t) + \left(\dfrac{\eta_2 K}{\eta_i}\right)^2 \left(\dfrac{\sin^2 \theta_t}{\sin \theta_i \cos \theta_t}\right)^2}. \tag{I.52}$$

(b) for polarization parallel to the plane of incidence

$$R_\parallel = \frac{N \text{ with upper signs}}{N \text{ with lower signs}}$$

where

$$N = \tan^2 (\theta_i \mp \theta_t) + \left\{ \frac{\sin^4 \theta_t \sin^2 2\theta_i}{\eta_i^4 \sin^4 \theta_i (\cos 2\theta_i + \cos 2\theta_t)^2} \right\} G,$$

and

$$G = \eta_2^4 - \eta_i^4 \frac{\sin^4 \theta_i}{\sin^4 \theta_t} \mp 2\eta_i^2 \frac{\cos \theta_t \sin \theta_i}{\cos \theta_i \sin \theta_t} \left(\eta_2^2 - \eta_i^2 \frac{\sin^2 \theta_i}{\sin^2 \theta_t} \right)$$

$$\pm K^4 \pm 2\eta_i^2 K^2 \frac{\sin \theta_i \cos \theta_t}{\cos \theta_i \sin \theta_t} + 2\eta_2^2 K^2 \left(\frac{1 + 2 \sin^2 \theta_t}{\sin^2 2\theta_i \cos^2 \theta_t} \mp \frac{4 \sin \theta_t}{\sin 2\theta_i \cos \theta_t} \right), \tag{I.53}$$

when $K = 0$ these reduce to the Fresnel equations for lossless media.

I.8 THE DIFFERENTIAL EQUATION OF A REFLECTOR

If the reflector has a curved profile shown as the curve AB in Fig. I.8(a) and is illuminated by rays from a source at the origin then from the geometry of the figure we have

$$\psi = \psi' - \tfrac{1}{2}\pi,$$
$$\psi' = \alpha + \theta,$$

and

$$2\alpha = \phi - \theta.$$

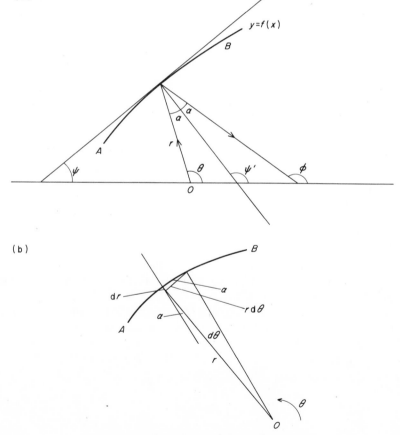

FIG. I.8. *Differential geometry of a reflector.*

In terms of a Cartesian geometry the reflector will have a profile given by

$$y = f(x) \qquad \text{then} \qquad \tan \psi = f'(x) = -\cot\left(\frac{\theta}{2} + \frac{\phi}{2}\right). \qquad (I.54)$$

Alternatively, if the profile is given in polar coordinates (r, θ) then from the law of reflection (Fig. I.8(b)) we have

$$\tan \alpha = dr/r \, d\theta$$

and hence

$$\frac{dr}{r} = \tan\left(\frac{\phi}{2} - \frac{\theta}{2}\right) d\theta. \qquad (I.55)$$

If, as was shown in Section 13.2 of Chapter 1, ϕ can be obtained explicitly as a function $g(\theta)$ of θ Eqn (I.55) can be integrated numerically to give

$$\log (r/r_0) = \int_{\theta_0}^{\theta} \tan \tfrac{1}{2}\{g(\theta) - \theta\} \, d\theta, \qquad (I.56)$$

where (r_0, θ_0) refers to the starting point A of the process.

I.9 ABBE'S SINE CONDITION AND THE HERSCHEL CONDITION

In all optical systems with more than a single surface, the additional surfaces allow the application of a further condition for each surface, upon the properties of the focusing ray pencils. For a fundamental two surface axi-symmetric system one such condition is that the system of rays focuses a second focal source point into a second image point near to the first. Two basic situations exist, that where the second foci are displaced transversely to the optical axis of the system and where the displacement is axially. The situation is as shown in Fig. I.9(a).

If P and P' are the axial foci and Q and Q' the displaced foci and using primes to indicate properties of the image space, we consider rays $\hat{s}$ and t which are subtended by the semi-aperture first at P and then at Q. These appear in the image space as $\hat{s}'$ and $\hat{t}'$ subtended at P' and Q' respectively.

From Eqn (I.26) we have for any closed true ray path

$$I \equiv \oint \eta \hat{s} \cdot ds = 0. \qquad (I.56)$$

Taking this closed path to be the route $PP'Q'QP$ first along the s rays then along the t rays, and writing

$$\int_A^B \ldots \equiv \{AB\}$$

for simplicity, then

$$\delta I \equiv \{PP' + \mathrm{d}l' + Q'Q + \mathrm{d}l\} \text{ along } s \text{ rays}$$
$$= \{PP' + \mathrm{d}l' + Q'Q + \mathrm{d}l\} \text{ along } t \text{ rays}.$$

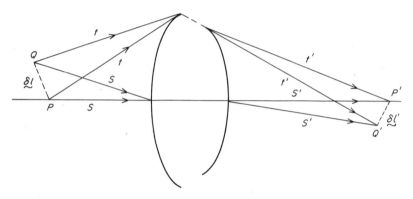

FIG. I.9(a). *Ray paths for small displacement of source.*

Since $\{PP'\}$ along s rays $= \{PP'\}$ along t rays and we require $\{QQ'\}$ along s rays to equal $\{QQ'\}$ along t rays, then this relation will be satisfied if for $\mathrm{d}l$ and $\mathrm{d}l'$

$$\int_P^Q \eta \hat{\mathbf{s}} \cdot \mathrm{d}\mathbf{s} = \int_{P'}^{Q'} \eta \hat{\mathbf{s}}' \cdot \mathrm{d}\mathbf{s}',$$

and

$$\int_P^Q \eta \hat{\mathbf{t}} \cdot \mathrm{d}\mathbf{s} = \int_{P'}^{Q'} \eta \hat{\mathbf{t}}' \cdot \mathrm{d}\mathbf{s}. \tag{I.57}$$

For *small* displacements $\mathrm{d}l$ equation I.57 can be approximated by

$$\eta(\hat{\mathbf{s}} - \hat{\mathbf{t}}) \cdot \mathrm{d}\mathbf{l} = \eta'(\hat{\mathbf{s}}' - \hat{\mathbf{t}}') \cdot \mathrm{d}\mathbf{l}' \tag{I.58}$$

which is Brun's law (Ref. 1, p. 11).

For a displacement $\mathrm{d}\mathbf{l}$ perpendicular to the axis of this system the second of the relations in equation I.57 gives (approximately)

$$\eta \, dl \sin \theta = \eta' \, dl' \sin \theta' \qquad (I.59)$$

which is the Abbé sine condition (Ref. 2, p. 165).

In the case where the image focus is at infinity the integral I along the s ray from P must equal the integral along any t ray from P up to the point where it intersects its corresponding t' ray. Thus the sine rule is said to be obeyed if the intersection of the incident rays and the finally transmitted rays

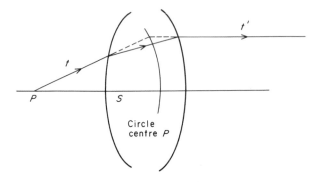

FIG. I.9(b). *The sine condition with focus at infinity.*

all lie on a circle centred on the source of rays (Fig. I.9(b)). This can be derived as a limiting condition on equation I.59 (Ref. 2, p. 167). If we now take dr along the axis of symmetry Eqn (I.58) gives the relation

$$\eta \, dr(1 - \cos \theta) = \eta' \, dr'(1 - \cos \theta'). \qquad (I.60)$$

This is Herschel's condition for an axial displacement of focus.

It is obviously incompatible with the sine condition to which it approximates only for small angles θ, that is for paraxial rays only. Thus no optical system can simultaneously produce perfect foci at local transversely and axially displaced points from a point source similarly displaced from its true position.

I.10 THE FLUX DENSITY FOR RAY PROPAGATION IN GEOMETRICAL OPTICS

We give here a summary of the method of Burkhard and Shealy[14] relating the scalar energy density in a reflected or refracted (unfocused) beam with the incident energy and the shape of the deflecting surface. This summary is taken in its entirety from the reference given with a slight change in notation.

In a Cartesian coordinate system a point on the irradiated surface is defined in terms of curvilinear coordinates (u, v) by

$$\mathbf{r} = x(u, v)\hat{\mathbf{i}} + y(u, v)\hat{\mathbf{j}} + z(u, v)\hat{\mathbf{k}},$$

and a point on the receiving surface by coordinates (U, V) where

$$\mathbf{R} = X(U, V)\hat{\mathbf{i}} + Y(U, V)\hat{\mathbf{j}} + Z(U, V)\hat{\mathbf{k}}.$$

A ray with incident direction $\hat{\mathbf{a}}$ is reflected by the surface into the direction

$$\hat{\mathbf{A}} = \hat{\mathbf{a}} - 2\hat{\mathbf{n}}(\hat{\mathbf{a}} \cdot \hat{\mathbf{n}}),$$

where $\hat{\mathbf{n}}$ is the reflector normal.

In the case of refraction

$$\hat{\mathbf{A}} = \frac{\eta_i}{\eta_t} \hat{\mathbf{a}} + \left[\cos \theta_t - \frac{\eta_i}{\eta_t} \cos \theta_i \right] \hat{\mathbf{n}}$$

as in Eqns (I.6) and (I.7).

The direction cosines of the line joining points (x, y, z) on the first surface to (X, Y, Z) on the second can be obtained from these relations and are designated (l, m, n) each being a function of the parameters u and v.

We define, as required by the reference,

$$\rho = (Z - z)/n,$$

$$I_0 = \hat{\mathbf{A}} \cdot \left(\frac{\partial \mathbf{r}}{\partial u} \times \frac{\partial \mathbf{r}}{\partial v} \right),$$

$$I_1 = \hat{\mathbf{A}} \cdot \left[\left(\frac{\partial \mathbf{r}}{\partial u} \times \frac{\partial \hat{\mathbf{A}}}{\partial v} \right) + \left(\frac{\partial \hat{\mathbf{A}}}{\partial u} \times \frac{\partial \mathbf{r}}{\partial v} \right) \right],$$

$$I_2 = \hat{\mathbf{A}} \cdot \frac{\partial \hat{\mathbf{A}}}{\partial u} \times \frac{\partial \hat{\mathbf{A}}}{\partial v},$$

and

$$\cos \psi = \hat{\mathbf{A}} \cdot \frac{\dfrac{\partial \mathbf{R}}{\partial U} \times \dfrac{\partial \mathbf{R}}{\partial V}}{\left| \dfrac{\partial \mathbf{R}}{\partial U} \times \dfrac{\partial \mathbf{R}}{\partial V} \right|}.$$

The flux incident upon an element dS_1 of the reflector or refractor surface is given by

$$F_{inc} = \sigma \cos \theta_i \, dS_1$$

where σ is the flux density of the incident beam, which could be a function of θ_i. The flux per unit area on the receiving surface is then

$$F_{rec} = \rho\sigma \cos \theta_i \, dS_1/dS_2 \qquad (I.61)$$

where dS_2 is the area mapped on the receiving surface by rays through dS_1 on the deflecting surface and ρ is a transmissity or reflectivity factor as required.

Then since

$$dS_1 = \left| \frac{\partial \mathbf{r}}{\partial u} \times \frac{\partial \mathbf{r}}{\partial v} \right| du \, dv,$$

and

$$dS_2 = \left| \frac{\partial \mathbf{R}}{\partial U} \times \frac{\partial \mathbf{R}}{\partial V} \right| dU \, dV, \qquad (I.62)$$

$$F_{rec} = \frac{\rho\sigma \cos \theta_i \cos \psi \left| \dfrac{\partial \mathbf{r}}{\partial u} \times \dfrac{\partial \mathbf{r}}{\partial v} \right|}{\left| I_0 + \rho I_1 + \rho^2 I_2 \right|}, \qquad (I.63)$$

in which the required ratio $du \, dv/dU \, dV$ has been obtained by an exercise in differential geometry.

The equation for the caustic surface is the loci of the singularities of Eqn (I.63) and is therefore given by

$$I_0 + \rho I_1 + \rho^2 I_2 = 0. \qquad (I.64)$$

Solving this as a quadratic in ρ since $I_0 I_1$ and I_2 are functions of u and v only, the equation of the caustic surface is found to be

$$\mathbf{R}_{caust} = \mathbf{r}(u, v) + \rho(u, v)\hat{\mathbf{A}}(u, v). \qquad (I.65)$$

REFERENCES

1. C. P. Enz. Pauli Lectures on Physics Vol .2, "Optics and the Theory of Electrons", M.I.T. Press, 1972, p. 4.
2. M. Born and E. Wolf. "Principles of Optics", Pergamon Press, 1959, p. 36.
3. L. Silberstein. "Simplified Method of Tracing Rays Through Any Optical System of Lenses Prisms or Mirrors", Longmans Green & Co., London, 1918.
4. O. N. Stavroudis. "The Optics of Rays, Wavefronts and Caustics", Academic Press, New York and London, 1972, Chapter 2.

5. R. E. Collin and F. J. Zucker. "Antenna Theory", Part II, McGraw Hill Inter University Electronics Series, Vol. 7, 1969, p. 5 and p. 31.

6. F. D. Bennett. Refraction operators and ray tracing through cones of constant refractive index, *Jour. Opt. Soc. Amer.* **47** (1) (1957), p. 85.

7. H. Wagner. Zur mathematischen behandlung von Spiegelungen *Optik,* **8** (10) (1951), p. 456.

8. J. S. Beggs. Mirror image kinematics. *Jour. Opt. Soc. Amer.* **50** (4) (1960), p. 388.

9. G. C. Southworth. Principles and Applications of Waveguide Transmission, Van Nostrand, 1966, p. 475.

10. J. B. Keller. Parallel reflection of light by plane mirrors, *Quart. J. App. Maths.* **11** (1953), p. 216.

    L. B. Tuckerman. Multiple reflections by plane mirrors, *Quart. J. App. Maths.* **5** (2) (1947), p. 133.

    J. L. Synge. Reflection in a corner formed by three plane mirrors, *Quart. J. App. Maths.* **4** (2) (1946), p. 116.

    H. S. M. Coxeter. The product of three reflections, *Quart. J. App. Maths.* **5** (2) (1947), p. 217.

    A. J. Montgomery. Analysis of two-tilt compensating interferometers, *Jour. Opt. Soc. Amer.* **57** (9) (1967), p. 1121.

    J. C. Polasek. Matrix analysis of gimballed mirror and prism systems, *Jour. Opt. Soc. Amer.* **57** (10) (1967), p. 1193.

11. H. F. Baker. "Principles of Geometry", Vol. 1, Cambridge University Press, 1929.

12. E. Cambi. Ref. 23, Chapter 6.

13. R. J. Bell, K. R. Armstrong, C. S. Nichols and R. W. Bradley. Generalized laws of refraction and reflection. *Jour. Opt. Soc. Amer.* **59** (2) (1969), p. 187.

14. See Ref. 43, Chapter 1.

Appendix II

II.1 THE CIRCLE POLYNOMIALS

The circle polynomials of Zernike[1] as applied to the theory of optical aberrations are the polynomials

$$R_n^m(r) = \sum_{k=0}^{(n-m)/2} (-1)^k \frac{(n-k)!}{[(n+m-2k)/2]!\,[(n-m-2k)/2]!} \frac{r^{n-2k}}{k!} \quad \text{(II.1)}$$

for $(n-m)$ an even positive integer ($R_n^m(r) = 0$ otherwise). Other descriptions include

$$R_n^m(r) = r^{-m} \frac{1}{\Gamma[(n-m+2)/2]} \left(\frac{d}{d(r^2)}\right)^{(n-m)/2} \left(r^{(m+n)/2}(r^2-1)^{(n-m)/2}\right) \quad \text{(II.2)}$$

and more particularly

$$R_n^m(r) = (-1)^{(n-m)/2} \binom{(m+n)/2}{m} r^m \,_2F_1\{(m+n+2)/2, -(n-m)/2;$$

$$m+1; r^2\}. \quad \text{(II.3)}$$

A full derivation of these polynomials including the differential equation satisfied by them can be found in Chako.[2] A table of $R_n^m(r)$ ($n < 10$ $m < 10$) is given in Table II.1.

TABLE II.1
Zernike Polynomials $R_n^m(r)$
$[R_n^m(r) = 0$ if $(n - m)$ is not an even integer$_7]$

$R_0^0 = 1$	$R_7^1 = 35r^7 - 60r^5 + 30r^3 - 4r$
	$R_7^3 = 21r^7 - 30r^5 + 10r^3$
$R_1^1 = r$	$R_7^5 = 7r^7 - 6r^5$
	$R_7^7 = r^7$
$R_2^0 = 2r^2 - 1$	
$R_2^2 = r^2$	$R_8^0 = 70r^8 - 140r^6 + 90r^4 - 20r^2 + 1$
	$R_8^2 = 56r^8 - 105r^6 + 60r^4 - 10r^2$
$R_3^1 = 3r^3 - 2r$	$R_8^4 = 28r^8 - 42r^6 + 15r^4$
$R_3^3 = r^3$	$R_8^6 = 8r^8 - 7r^6$
	$R_8^8 = r^8$
$R_4^0 = 6r^4 - 6r^2 + 1$	
$R_4^2 = 4r^4 - 3r^2$	$R_9^1 = 126r^9 - 280r^7 + 210r^5 - 60r^3 + 5r$
$R_4^4 = r^4$	$R_9^3 = 84r^9 - 168r^7 + 105r^5 - 20r^3$
	$R_9^5 = 36r^9 - 56r^7 + 21r^5$
$R_5^1 = 10r^5 - 12r^3 + 3r$	$R_9^7 = 9r^9 - 8r^7$
$R_5^3 = 5r^5 - 4r^3$	$R_9^9 = r^9$
$R_5^5 = r^5$	
	$R_{10}^0 = 252r^{10} - 630r^8 + 560r^6 - 210r^4 + 30r^2 - 1$
$R_6^0 = 20r^6 - 30r^4 + 12r^2 - 1$	$R_{10}^2 = 210r^{10} - 504r^8 + 420r^6 - 140r^4 + 15r^2$
$R_6^2 = 15r^6 - 20r^4 + 6r^2$	$R_{10}^4 = 120r^{10} - 252r^8 + 168r^6 - 35r^4$
$R_6^4 = 6r^6 - 5r^4$	$R_{10}^6 = 45r^{10} - 72r^8 + 28r^6$
$R_6^6 = r^6$	$R_{10}^8 = 10r^{10} - 9r^8$
	$R_{10}^{10} = r^{10}$

Comparison of Eqn (II.3) with one of the possible definitions of the Jacobi polynomials[3]

$$P_s^{(\alpha, \beta)}(x) = \binom{s + \alpha}{s} {}_2F_1 \left\{s + \alpha + \beta + 1, -s; \alpha + 1; \frac{1 - x}{2}\right\} \quad \text{(II.4)}$$

shows that

$$R_n^m(r) = (-1)^{\frac{1}{2}(n - m)} r^m P_{(n - m)/2}^{(m, 0)}(1 - 2r^2). \quad \text{(II.5)}$$

With $m = 0$ and n therefore an even positive integer the Jacobi polynomials reduce to the Legendre polynomials and Eqn (II.5) becomes

$$R_{2n}^0(r) = P_n(2r^2 - 1). \quad \text{(II.6)}$$

From the orthogonality properties of the Jacobi and Legendre polynomials we obtain the orthogonality relation of the circle polynomials

$$\int_0^1 R_n^m(r)R_p^m(r)r\,dr = \delta_{n,\,p}/(2n + 2), \tag{II.7}$$

where $\delta_{n,\,p}$ is the Kronecker symbol.

From the Hankel transform of the Jacobi polynomial as given in Erdelyi[4] we have

$$\int_0^a \frac{r^\nu}{a^{\nu+2}} P_n^{(\nu,o)}(1 - 2r^2/a^2) J_\nu(ur)r\,dr = \frac{J_{\nu+2n+1}(au)}{au}. \tag{II.8}$$

In terms of the circle polynomials (with $a = 1$) we obtain Eqn (3.37)

$$\int_0^1 R_n^m(r) J_m(ur)r\,dr = (-1)^{(n-m)/2} J_{n+1}(u)/u. \tag{II.9}$$

The Fourier–Bessel inverse of this is

$$(-1)^{(n-m)/2} R_n^m(r) = \int_0^\infty J_{n+1}(ru) J_m(u)\,du \qquad r < 1$$
$$= 0 \qquad r > 1. \tag{II.10}$$

Recurrence relations can be obtained from the recurrence relations for Legendre and Jacobi polynomials (Ref. 3, p. 782). These include, from Chako[2]

$$2(n + 1)rR_n^m(r) = (n + m + 2)R_{n+1}^{m+1}(r) + (n - m)R_{n-1}^{n+1}(r), \tag{II.11}$$

and from Myrick[5] (with correction in the case of Eqn (II.12))

$$(2n + 4)(n + m)(n - m)R_{n-2}^m(r) + \{2n(n - m + 2)^2 + (2n + 4)(n + m)^2$$
$$- 2nr^2(2n + 4)(2n + 2)\}R_n^m(r) + 2n(n - m + 2)(n + m + 2)R_{n+2}^m(r) = 0, \tag{II.12}$$

and

$$(n - m + 2)(n + m)R_n^{m-2}(r) - \{2n^2 + 2m^2 - 2m(2m - 2)/r^2\}R_n^m(r)$$
$$+ (n + m + 2)(n - m)R_n^{m+2}(r) = 0 \tag{II.13}$$

Chako also gives the relation (corrected)

$$\frac{d}{dr} R_n^0(r) = 2nR_{n-1}^1(r) + 2(n - 2)R_{n-3}^1(r) + \ldots\ldots + 4R_1^1(r). \tag{II.14}$$

The coefficients a_n for *uniform* illumination are particularly elementary for all orders of Hankel transform. They are given by Eqn (3.42) as

$$a_{nm} = 2(n + 1)\int_0^1 f(r)R_n^m(r)r\,dr,$$

which for uniform illumination, $f(r) = 1$, become simply

$$a_{nm} = (-1)^{(n-m)/2} \frac{2m(n+1)}{n(n+2)}. \tag{II.15}$$

For zero order Hankel tranforms of polynomial functions $f(r)$ the following integral is useful in obtaining the coefficients α_{2s+1} in Eqn (3.41)

$$\int_0^1 r^{2(n-1)} R_{2s}^0(r) r \, dr = \frac{\{\Gamma(n)\}^2}{2\Gamma(n+s)\Gamma(n-s+1)} \tag{II.16}$$

$$n > 0$$

(Ref. 4, 278 corrected).

Finally, in the light of the discussion in Chapter 6, it is interesting to observe that both Chako and Myrick (quoting Zernike) consider as natural the extension of the circle polynomial theory to solutions of Laplace's equation in higher dimensional polar coordinates.

II.2 BESSEL FUNCTIONS, ASSOCIATED BESSEL FUNCTIONS AND SERIES

We list below basic series of Bessel and associated functions and fundamental relations which have been found to be of value in deriving Hankel and Lommel transforms in the form of infinite series by the method of circle polynomials. Page references are those of Watson[6]

(1) $J_n(z) = \sum_{m=0}^{\infty} \frac{(-1)^m (\frac{1}{2}z)^{n+2m}}{m!(n+m)!}$

(2) $J_{n-1}(z) + J_{n+1}(z) = 2n/z \, J_n(z)$

(3) $J_{n-1}(z) - J_{n+1}(z) = 2 J_n'(z)$

(4) $J_n(z) = \frac{1}{2\pi} \int_0^{2\pi} \cos(n\theta - z \sin \theta) \, d\theta$

(5) $J_n(y + z) = \sum_{m=-\infty}^{\infty} J_m(y) J_{n-m}(z)$

$$\left\{\begin{array}{l} \cos(z\sin\theta) = J_0(z) + 2\sum_{n=1}^{\infty} J_{2n}(z)\cos 2n\theta \\[2mm] \sin(z\sin\theta) = \quad\quad 2\sum_{n=1}^{\infty} J_{2n+1}(z)\sin(2n+1)\theta. \\[2mm] \cos(z\cos\theta) = J_0(z) + 2\sum_{n=1}^{\infty}(-1)^n J_{2n}(z)\cos 2n\theta \\[2mm] \cos(z\sin\theta) = \quad\quad 2\sum_{n=1}^{\infty}(-1)^n J_{2n+1}(z)\cos(2n+1)\theta \end{array}\right.$$

(6)

$$\left\{\begin{array}{l} \dfrac{\sin z}{z} = \dfrac{2}{z^2}\{2^2 J_2(z) - 4^2 J_4(z) + 6^2 J_6(z)\dots\} \\[3mm] \cos z = \dfrac{2}{z}\{1^2 J_1(z) - 3^2 J_3(z) + 5^2 J_5(z)\dots\} \end{array}\right.$$

(7)

(8)
$$\left\{\begin{array}{ll} 1 = J_0(z) + 2\sum_{m=1}^{\infty} J_{2n}(z) & \text{Watson p. 34} \\[3mm] z^m = 2^m \sum_{n=0}^{\infty} \dfrac{(m+2n)(m+n-1)!}{n!} J_{m+2n}(z) & \end{array}\right.$$

(9) $\quad J_\nu(\lambda z) = \lambda^\nu \sum_{m=0}^{\infty} \dfrac{(-1)^m(\lambda^2 - 1)^m(\frac{1}{2}z)^m}{m!} J_{\nu+m}(z)$ $\qquad$ Watson p. 142

(10) $\quad E_\nu(z) = -\Omega_\nu(z) = \dfrac{1}{\pi}\int_0^\pi \sin(\nu\theta - z\sin\theta)\,d\theta$ $\qquad$ the Anger–Weber function
Watson, p. 308

(11) $\quad \Omega_n(z) = \dfrac{2}{\pi}\sum_{s=0}^{\infty} \dfrac{2(2s+1)}{(2s+1)^2 - n^2} J_{2s+1}(z)$ $\qquad$ *n* even

$\qquad\qquad\quad = \dfrac{2}{\pi}\left[-\dfrac{1}{n}J_0(z) + \sum_{s=1}^{\infty} \dfrac{2n}{4s^2 - n^2} J_{2s}(z)\right]$ $\qquad$ *n* odd

(12) $\quad \Omega_0(z) + 2\Omega_2(z) + 2\Omega_4(z) + \dots = 0$

$\qquad\quad \Omega_0(z) - 2\Omega_2(z) + 2\Omega_4(z) - \dots = \sin z$

$\qquad\quad 2[\Omega_1(z) - \Omega_3(z) + \Omega_5(z) - \dots] = -\cos z$

(13) $\quad J_n^2(z) = \sum_{m=0}^{\infty} \dfrac{(-1)^m(2n+2m)!\,(\frac{1}{2}z)^{2n+2m}}{m!(2n+m)!\,[(n+m)!]^2}$ $\qquad$ Watson, p. 32

(14) $\quad \frac{1}{2} + \sum_{m=1}^{\infty}(-1)^m J_0(mx) = 0 \qquad x > 0$ $\qquad$ Watson, p. 634

(15) $(\tfrac{1}{2}az)^{\mu - \nu} J_\nu(az) = \dfrac{1}{\Gamma(\nu + 1)} \displaystyle\sum_{t=0}^{\infty} \dfrac{(\mu + 2t)\Gamma(\mu + t)}{t!}$

$$\times \ _2F_1\{-t; \mu + t; \nu + 1; a^2\} J_{\mu + 2t}(z)$$

(Erdelyi: "Higher Transcendental Functions", Vol. 2, 19, p. 64). This relation can also be derived from a result by MacRobert[7] and agrees with that obtained in Chapter 4 Eqn (4.9). Hence for $\nu = n$ and $\mu - \nu = 1$

$$\tfrac{1}{2} J_n(z) = \frac{1}{z} \sum_{t=0}^{\infty} (-1)^t (n + 2t + 1) J_{n+2t+1}(z)$$

and thus

$$J_0(z) = \frac{2}{z} \sum_{t=0}^{\infty} (-1)^t (2t + 1) J_{2t+1}(z),$$

$$J_1(z) = \frac{2}{z} \sum_{t=0}^{\infty} (-1)^t (2t + 2) J_{2t+2}(z), \text{ etc.}$$

(16) The modified Anger–Weber function [8]

From (10) we define the modified Anger–Weber function

$$\Phi_\nu(z) = \frac{1 - \cos \nu\pi}{\nu\pi} + \Omega_\nu(z)$$

then for integer values

$$\Phi_{2n}(z) = \Omega_{2n}(z) \quad \text{but} \quad \Phi_{2n+1}(z) = \Omega_{2n+1}(z) + \frac{2}{(2n + 1)\pi}.$$

Most relevant properties can then be obtained from those of $E_\nu(z)$ and in particular the two expansions

$$\Phi_{2n}(z) = \frac{2}{\pi} \sum_{t=0}^{\infty} \frac{2(2t + 1)}{(2t + 1)^2 - (2n)^2} J_{2t+1}(z)$$

$$\Phi_{2n+1}(z) = \frac{2}{(2n + 1)\pi}[1 - J_0(z)] + \frac{1}{\pi} \sum_{t=0}^{\infty} \frac{2(2n + 1)}{(2t)^2 - (2n + 1)^2} J_{2t}(z)$$

(17) $\Omega_n(z) = -\dfrac{1}{\pi} \displaystyle\int_{-\infty}^{\infty} \dfrac{J_n(u)\, du}{u - z}$ 　　　　Moss

(Chapter 3, Ref. 14)

$$= -\frac{1}{\pi} \int_{-\infty}^{\infty} \frac{J_n(z + u)\, du}{u}$$

$$= \sum_{s=-\infty}^{\infty} \Omega_{n-s}(0) J_s(z)$$

(18) $\Omega_0(z) = \dfrac{4}{\pi}\{J_1(z) + \tfrac{1}{3}J_3(z) + \tfrac{1}{5}J_5(z) + \ldots\}$

(19) $\Phi_{2n+1}(z) = (-1)^n \displaystyle\sum_{s=1}^{\infty} \dfrac{(-1)^s(z/2)^{2s}}{\Gamma(s-n-\tfrac{1}{2})\Gamma(s+n+3/2)}$

(20)   From $J_n(Nxy) = \displaystyle\sum_{s=0}^{\infty} c_s R_{n+2s}^n(y)$,

$$\int_0^1 J_n(Nxy)R_{n+2t}^n(y)y \, dy = \dfrac{c_t}{2(n+2t+1)},$$

hence

$$J_n(Nxy) = \sum_{s=0}^{\infty} \dfrac{2(n+2s+1)}{Nx}(-1)^s J_{n+2s+1}(Nx)R_{n+2s}^n(y)$$

(cf. relation (9))

(21)   Similarly

$$(-1)^t \int_0^1 R_{n+2s}^n(y)J_{n+2t+1}(Ny) \, dy$$

$$= (-1)^s \int_0^1 R_{n+2t}^n(y)J_{n+2s+1}(Ny) \, dy$$

(22) $\displaystyle\int_0^{2\pi} \exp\left(i\{-n\theta + \alpha r \cos\theta + \beta r \sin\theta\}\right) d\theta$

$$= 2\pi \exp\left[in\tan^{-1}(\alpha/\beta)\right] J_n(\sqrt{\alpha^2 + \beta^2}r)$$

(23) $\displaystyle\int J_1(ur)r \, dr = \sum_{s \text{ odd}} \dfrac{2(s+1)}{s(s+2)}\dfrac{J_{s+1}(u)}{u}$

$\displaystyle\int J_2(ur)r \, dr = \sum_{s \text{ even}} \dfrac{4(s+1)}{s(s+2)}\dfrac{J_{s+1}(u)}{u}$

$\displaystyle\int J_3(ur)r \, dr = \dfrac{8}{5}\dfrac{J_4(u)}{u} + \dfrac{36}{35}\dfrac{J_6(u)}{u} + \dfrac{16}{21}\dfrac{J_8(u)}{u} + \dfrac{60}{99}\dfrac{J_{10}(u)}{u}\cdots$

$\displaystyle\int J_4(ur)r \, dr = \dfrac{10}{6}\dfrac{J_5(u)}{u} + \dfrac{14}{12}\dfrac{J_7(u)}{u} + \dfrac{18}{20}\dfrac{J_9(u)}{u} + \dfrac{22}{30}\dfrac{J_{11}(u)}{u}$

(24) $\dfrac{e^{ikR}}{R} = \displaystyle\int_0^{\infty} \dfrac{u \, du}{(u^2-k^2)^{\frac{1}{2}}} J_0(ur) \exp\left[\mp(u^2-k^2)^{\frac{1}{2}}z\right]$ Watson, p. 416

$R^2 = z^2 + r^2$ upper sign for $+$ ve z
                      lower sign for $-$ ve z.

II.3 GREEN'S FUNCTIONS—PARTICULAR INTEGRALS OF THE WAVE EQUATION

In keeping with the concepts of the latter part of this book, we present a four dimensional scalar analysis for *deriving* Green's functions[9]. These are usually defined in three-dimensional analysis and Green's theorem in a vector or dyadic form then utilises their properties (Refs (5) and C. T. Tai, Chap. 3).

The four dimensional divergence theorem applied to a four component entity W is

$$\int_\tau \left(\frac{\partial W_x}{\partial x} + \frac{\partial W_y}{\partial y} + \frac{\partial W_z}{\partial z} + \frac{\partial W_\xi}{\partial \xi} \right) d\tau = \int_H (l_x W_x + l_y W_y + l_z W_z + l_\xi W_\xi) \, dH$$

(II.17)

where τ is a hypervolume of three real space and one complex ξ dimension and H a hypersurface of two space and one ξ dimension. l_x, l_y, l_z and l_ξ are direction cosines of the outward "normal" to H and we use the abbreviation $\xi = \text{ict}$. Then in a precise analogue of the three dimensional theory we use the four dimensional operator $\square$ of Chapter 6 (p. 347) and the wave equation with a source field $f(x', y', z', t')$ becomes

$$\square^2 \psi = f(x', y', z', \xi').$$

(II.18)

Substituting $W = U \square V - V \square U$ into equation 1 we obtain

$$\int_\tau (U \square^2 V - V \square^2 U) \, d\tau = \int_H \left(U \frac{\partial V}{\partial N} - V \frac{\partial U}{\partial N} \right) dH,$$

(II.19)

which is Green's theorem in four dimensions.

Putting R as the four dimensional position vector

$$R^2 = \Sigma(x - x')^2,$$

then a solution to $\square^2 V = 0$ is

$$V = \frac{1}{R^2}, \qquad R \neq 0$$

Isolating the field point $P(x, y, z, \xi)$ by a small hypersphere of radius $R = a$, and putting $U = \psi$; $V = 1/R^2$ in Eqn (II.19), then since

$$\square^2 U = f(x' \, y' \, z' \, \xi'),$$

$$\int_\tau \frac{f(x'y'z'\xi')}{R^2} \, d\tau = \int_H \left(\frac{1}{R^2} \frac{\partial \psi}{\partial N} - \psi \frac{\partial}{\partial N} \left(\frac{1}{R^2} \right) \right) dH,$$

(II.20)

o

where H is now the combined surface of

(a) the small hypersphere about $P(x, y, z, \xi)$
(b) the hypersphere at infinity
(c) any other surface + time in between

(a) For the small hypersphere we have

$$\frac{\partial \psi}{\partial N} = -\frac{\partial \psi}{\partial R} \quad \text{and} \quad \left| \frac{\partial}{\partial N} \left(\frac{1}{R^2} \right) \right| = \frac{-2}{a^3}$$
$$R = a$$

Using hyperspherical coordinates

$$x = R \cos \theta_1 \qquad\qquad \theta_1 \text{ from 0 to } \pi,$$
$$y = R \sin \theta_1 \cos \theta_2 \qquad \theta_2 \text{ from 0 to } \pi,$$
$$z = R \sin \theta_1 \sin \theta_2 \cos \phi \qquad \phi \text{ from 0 to } 2\pi,$$
$$\xi = R \sin \theta_1 \sin \theta_2 \sin \phi,$$

the "area" of the small hypersurface over which ψ is constant $\psi = \psi_p$ is

$$\int_0^\pi \int_0^\pi \int_0^{2\pi} a^3 \sin^2 \theta_1 \sin \theta_2 \, d\theta_1 \, d\theta_2 \, d\phi = 2\pi^2 a^3.$$

Therefore as the small hypersphere shrinks to zero at P the integral over its surface

$$\int_a \left(\frac{1}{R^2} \frac{\partial \psi}{\partial N} - \psi \frac{\partial}{\partial N} \left(\frac{1}{R^2} \right) \right) dH \rightarrow 2\pi^2 \left(\frac{-a^3}{a} \frac{\partial \psi}{\partial R} - 2\psi_p \right)_{\substack{a \to 0 \\ \text{at } r = a}}$$

$$= -4\pi^2 \psi_p.$$

This leaves Eqn (II.20) as

$$\int_\tau \frac{f}{R^2} \, d\tau = -4\pi^2 \psi_p + \int_H \left(\frac{1}{R^2} \frac{\partial \psi}{\partial N} - \psi \frac{\partial}{\partial N} \left(\frac{1}{R^2} \right) \right) dH. \qquad \text{(II.21)}$$

The integral over the infinite hypersphere can be shown to tend to zero if in the limit the fields obey the radiation condition at infinity. Hence H in Eqn (II.21) need only be considered as the hypersurface of all local surfaces and time.

We now integrate the left hand side of Eqn (II.21) with respect to the time coordinate. We have

$$I(x, y, z, \xi) = \int_\tau \left(\frac{f(x'y'z'\xi')}{R^2} \, dx' \, dy' \, dz' \right) d\xi',$$

where

$$R^2 = (x - x')^2 + (y - y')^2 + z - z')^2 + (ic)^2(t - t')^2.$$

We take in the first instance the time t at the field point P to be zero, that is $t = \xi = 0 \; P = P(x, y, z, 0)$. Then

$$R = (x - x')^2 + (y - y')^2 + (z - z')^2 + \xi'^2$$
$$= |\mathbf{r} - \mathbf{r}'|^2 + \xi',$$

where $\mathbf{r}$ is the real three space position vector from the source point $\mathbf{r}$ to the field point $P \equiv \mathbf{r}'$. Then

$$I(t = 0) = \int_{-\infty}^{\infty} \frac{f(x'y'z'\xi')}{|\mathbf{r}-\mathbf{r}'|^2 + \xi'^2} \, d\xi' = \int_C \frac{f}{\rho^2 + \mu^2} \, d\mu, \qquad \text{(II.22)}$$

where (ρ, μ) is the complex plane and (Fig. II.1) the contour C includes the real axis. The poles of the integrand are at $\mu = \pm i\rho$. This contrasts with the usual three dimensional analysis where the poles are on the contour of integration and various methods have to be resorted to to obtain the principal value. If the contour is closed by an infinite semi-circle in the lower half plane we only enclose the pole at $\mu = -i\rho$. This as we shall see gives retarded potentials. The analysis is however identical if the contour were closed in the upper half plane containing the pole at $\mu = i\rho$ and giving advanced potentials. Then from equation 6

$$I(t = 0) = 2\pi i \, \{\text{residue of } f/(\rho^2 + \mu^2) \text{ at } \mu = -i\rho\}$$

$$= 2\pi i \lim_{\mu \to -i\rho} \left[\frac{(\mu + i\rho)}{\mu^2 + \rho^2} f \right]$$

$$= -\frac{2\pi i}{2i\rho} f \text{ (evaluated at } \mu = -i\rho).$$

Noting that when $\mu = i\rho$, $-i|\mathbf{r} - \mathbf{r}'| = \xi' = ict'$

$$t' = \frac{-|\mathbf{r} - \mathbf{r}'|}{c}.$$

Therefore

$$I(t = 0) = \frac{-\pi}{|\mathbf{r} - \mathbf{r}'|} f\left(x', y', z', t' = \frac{-|\mathbf{r} - \mathbf{r}'|}{c} \right).$$

For a time t at P the same procedure gives

$$I(x, y, z, t) = \frac{-\pi}{|\mathbf{r} - \mathbf{r}'|} f\left(x', y', z', t - \frac{|\mathbf{r} - \mathbf{r}'|}{c}\right), \qquad \text{(II.23)}$$

that is the retarded value of f. We note that the remaining factor π "converts" the four dimensional factor $4\pi^2$ in equation II.21 to the three dimensional 4π.

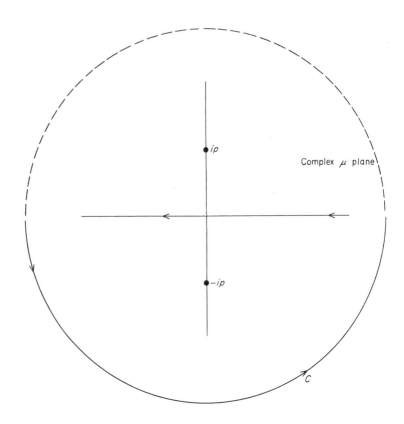

FIG. II.1. *Contour for integration of Eqn. (II.22).*

Some of the complications of an extended distribution of sources where the field point P could be within the range of $\mathbf{r}'$ and hence $|\mathbf{r} - \mathbf{r}'|$ become zero, can be seen in this treatment. In such a case both poles coincide at the origin which is on the path of integration and the principal value (in essence the average of the advanced and retarded solution) has to be taken.

$I(x, y, z, t)$ now *defines* the Green's function for specified sources $f(x', y', z', t')$.

For example

f = impulse function at the "point" $(\mathbf{r}', t')$

$$I = \frac{-\pi}{|\mathbf{r} - \mathbf{r}'|} \delta\left\{ t - t' - \frac{|\mathbf{r} - \mathbf{r}'|}{c} \right\} \overset{\text{def}}{=} -4\pi^2 G_{\text{impulse}} \qquad (\text{II.24})$$

f = steady state harmonic = $A \exp(\pm i\omega t)$

$$I = \frac{-\pi}{|\mathbf{r} - \mathbf{r}'|} \; A \exp(\pm i\omega t') \mp |\mathbf{r} - \mathbf{r}'|/c)]$$

$$= \frac{-\pi A \exp(\mp ik|\mathbf{r} - \mathbf{r}'|)}{|\mathbf{r} - \mathbf{r}'|} \; \exp(\pm i\omega t' \overset{\text{def}}{=} -4\pi^2 G_{\text{harmonic}}.$$

$$(\text{II.25})$$

Substituting this result back into Eqn (II.21) gives

$$4\pi^2 \psi_P = -\pi \int_v \frac{[f]_{\text{ret}}}{\mathbf{r} - \mathbf{r}'} \, dx' \, dy' \, dz' + \int_H \left(\frac{1}{R^2} \frac{\partial \psi}{\partial N} - \psi \frac{\partial}{\partial N}\left(\frac{1}{R^2} \right) \right) dH, \quad (\text{II.26})$$

where v is the real three space containing the sources f, and the sign has been changed to account now for the *inward* pointing normal in the surface integral.

Ideally one would continue to show how the surface integral in equation (II.26) gives the standard form shown in the table in Chapter 3. This does not appear to be simple. The result below can be "seen" in Eqn (II.26) since we have shown that time integration of terms containing the factor $1/R^2$ converts them to Green's functions, but in fact we can only state that since the first integral in (II.26) alone contains the source function f, it is the particular integral of Eqn (II.18) and hence the second integral must be the solution of

$$\square^2 \psi = 0$$

and proceed in the standard manner. Then for the steady state solution we can ignore both the time integrations that remain and $\partial/\partial t'$ terms to give the standard result

$$\psi_p = \frac{-1}{4\pi} \int_v \frac{[f]_{\text{ret}} \, dx' \, dy' \, dz'}{|\mathbf{r} - \mathbf{r}'|} + \int_s \left(G \frac{\partial \psi}{\partial n} - \psi \frac{\partial G}{\partial n} \right) ds. \qquad (\text{II.27})$$

II.4 QUATERNIONS

Quaternions have a long and interesting history, the most intriguing aspect of which being the manner that they have begun to make an immense impact in the field of modern theoretical physics[10] after being forgotten for nearly a hundred years since their discovery as a pure algebraic-geometrical concept. This history can be traced in many ways, by concentration on the algebra leading back to Hamilton and Grassmann and even to Euler, the geometry to Clifford, Servois and Cayley and the physics in more recent terms to Pauli and Dirac. Such a history and the recent advent of a spate of applications may well be the concern of a separate volume on the subject. The basic algebra that is all that is required at this time for the material in this book can be discussed with comparative brevity. This unfortunately hides the enormous beauty and potency of the subject illustrated as it has been here by comparatively simple examples.

To pose just one question, that in fact formed this author's own introduction to the subject and apparently was first asked by Servois in 1813, that is shortly after Argand and De Moivre; "if $re^{i\theta}$ is descriptive of a circle in a complex plane, what form of number would describe a sphere". The answer found by Hamilton and forming the basis of the entire subject was that no such number could describe a sphere in a real three space but such could readily be found in a four space (and later in an eight space).

To do this we require to label each of the three space coordinate axes with a complex number in the manner that $i = \sqrt{-1}$ labels the y axis in the Argand diagram. In common with the concept that this complex i is also a rotation of $\pi/2$ about the z axis, the other complex labels will then give rotations about the other axes as well. We label a rotation of $\pi/2$ about the x axis as the complex number α with $\alpha^2 = -1$, about the y axis as β, and about the z axis as γ. Thus the first elements of the algebra are $\alpha^2 = \beta^2 = \gamma^2 = -1$.

The next step is algebraic and is to find the equivalent in four dimensions of the product of scalar invariants of the two dimensional field. Given two complex numbers $z_1 = (a + ib)$; $z_2 = (c + id)$ one can find by the product of the norms

$$|z_1|^2 |z_2|^2 = |z_1 z_2|^2,$$

that

$$(a^2 + b^2)(c^2 + d^2) = A^2 + B^2,$$

where

$$A = ac - bd,$$
$$B = ad + bc.$$

Performing this in four dimensions (and proving that there is no solution in three dimensions) was Hamilton's great discovery. It takes the form[11]

$$(a^2 + b^2 + c^2 + d^2)(t^2 + x^2 + y^2 + z^2) = A^2 + B^2 + C^2 + D^2,$$

with

$$A = at - bx - cy - dz,$$
$$B = ax + bt + cz - dy,$$
$$C = ay - bz + ct + dz,$$
$$D = az + by - cx + dt.$$

This is now called a skew (or symplectic) field or a division ring.

Performing the same product $|Q_1|^2 |Q_2|^2 = |Q_1 Q_2|^2$ with the hyper-complex numbers

$$Q_1 = (a + \alpha b + \beta c + \gamma d),$$
$$Q_2 = (t + \alpha x + \beta y + \gamma z),$$

requires the additional non-commutative property

$$\alpha\beta = \gamma, \qquad \beta\gamma = \alpha, \qquad \gamma\alpha = \beta$$

and

$$\alpha\beta = -\beta\alpha, \text{ etc.}$$

The numbers Q are the quaternions. Hamilton originally labelled the axes i, j and k and this led on to vector analysis and the non-commutative cross product. The absorption of vector theory into the science of the time led to the neglect of quaternion theory. The essential three dimensionality of the former had greater appeal it seems than the four dimensionality of the latter. This is in the process of being corrected.

A unique inverse is defined by

$$Q^{-1}Q = QQ^{-1},$$

whence we have $\alpha\alpha^{-1} = \alpha^{-1}\alpha = 1$ or $\alpha = -\alpha^{-1}$. In the quotient α/β we have the alternatives of forming the left hand quotient $\beta^{-1}\alpha$ or the right hand quotient $\alpha\beta^{-1}$ and these are different (by -1). To avoid the need for labelling we can observe the following

$$\frac{\alpha}{\beta} = \alpha\beta^{-1} = -\alpha\beta \qquad \text{right hand quotient,}$$

$$\frac{\beta}{\alpha} = \beta\alpha^{-1} = -\beta\alpha \qquad \text{right hand quotient.}$$

Therefore, if *both* are taken right quotients $\alpha/\beta = -\beta/\alpha$. The same occurs if *both* are taken as left hand quotients, thus $\alpha/\beta = -\beta/\alpha$ is independent of the order of division.

A quaternion $Q = (a + \alpha b + \beta c + \gamma d)$ has a conjugate, the quaternion

$$\bar{Q} = (a - \alpha b - \beta c - \gamma d)$$

We have the square

$$Q^2 = (a + \alpha b + \beta c + \gamma d)^2$$
$$= \{a^2 - b^2 - c^2 - d^2, 2\alpha ab, 2\beta ac, 2\gamma ad\},$$

and the norm

$$Q\bar{Q} = (a + \alpha b + \beta c + \gamma d)(a - \alpha b - \beta c - \gamma d)$$
$$= \{a^2 + b^2 + c^2 + d^2, 0, 0, 0\}.$$

From the rule of multiplication we find

$$Q_1 Q_2 = (a + \alpha b + \beta c + \gamma d)(t + \alpha x + \beta y + \gamma z)$$
$$= A + \alpha B + \beta C + \gamma D$$

with A, B, C and D defined above.

Then *if* b, c, d and x, y, z were *considered* as the components of three-space vectors **V** and **W**

$$Q_1 = (a, \mathbf{V}), \qquad Q_2 = (t, \mathbf{W})$$

and

$$Q_1 Q_2 = (at - \mathbf{V} \bigcirc \mathbf{W}, a\mathbf{W} + t\mathbf{V} + \mathbf{V} \otimes \mathbf{W})$$

where $\bigcirc$ and $\otimes$ refer to the inner product and outer product of **V** and **W** in the "ordinary" vector sense. This last result can be seen directly by expanding the terms and comparing with A, B, C and D. It follows that if

$$Q = (a, \mathbf{V}) \qquad \bar{Q} = (a, -\mathbf{V})$$

The norm is $Q\bar{Q} = a^2 - \mathbf{V} \bigcirc \mathbf{V}$, the norm of $(Q - a)$ is $(Q - a)\overline{(Q - a)} = -\mathbf{V} \bigcirc \mathbf{V}$ and the square of $(Q - a)$ is $(Q - a)(Q - a) = +\mathbf{V} \bigcirc \mathbf{V}$. These are the relations required in chapter 6, section 4.

Using the ordered notation we have

$$(a, b, c, d)^2 = (a^2 - b^2 - c^2 - d^2, 2ab, 2ac, 2ad)$$

$$(0, x, y, z)^2 = -(x^2 + y^2 + z^2)$$
$$(0, x, y, z)^3 = -(0, x, y, z)(x^2 + y^2 + z^2)$$
$$(0, x, y, z)^4 = (x^2 + y^2 + z^2)^2, \text{ etc.}$$

Hence

$$\exp\{0, x, y, z\} = 1 + (0, x, y, z) - \frac{x^2 + y^2 + z^2}{2!} - \frac{(0, x, y, z)(x^2 + y^2 + z^2)}{3!}$$

$$= \cos\{x^2 + y^2 + z^2\}^{\frac{1}{2}} + \frac{(0, x, y, z)}{\{x^2 + y^2 + z^2\}^{\frac{1}{2}}} \sin\{x^2 + y^2 + z^2\}^{\frac{1}{2}}$$

Therefore

$$\left|\exp\{0, x, y, z\}\right| = 1$$

Putting $\{x^2 + y^2 + z^2\} = \rho^2$

$$x = \rho \cos\phi,$$

$$y = \rho \sin\phi \cos\psi,$$

$$z = \rho \sin\phi \sin\psi$$

(polar coords w.r.t. the x axis).

then

$$\exp\{\rho(\alpha \cos\phi + \beta \sin\phi \cos\psi + \gamma \sin\phi \sin\psi\}$$

$$= \cos\rho + \sin\rho\{\alpha \cos\phi + \beta \sin\phi \cos\psi + \gamma \sin\phi \sin\psi\}.$$

Hence

$$\alpha \cos\phi + \beta \sin\phi \cos\psi + \gamma \sin\phi \sin\psi$$

is the three dimension analogue of $\sqrt{-1}$ as can be verified by the direct product.
In hyperspherical coordinates we can put

$$a = \mu \cos\rho.$$

$$b = \mu \sin\rho \cos\phi,$$

$$c = \mu \sin\rho \sin\phi \cos\psi,$$

$$d = \mu \sin\rho \sin\phi \sin\psi,$$

then

$$\mu = \{a^2 + b^2 + c^2 + d^2\}^{\frac{1}{2}}.$$

Any quaternion is then

$$Q = \mu[\cos\rho + \sqrt{-1} \sin\rho]$$

where $\sqrt{-1}$ is given above and a generalized De Moivre's theorem states

$$\{\mu \cos\rho + \mu \sin\rho(\alpha \cos\phi + \beta \sin\phi \cos\psi + \gamma \sin\phi \sin\psi)\}^{\nu}$$

$$= \mu^{\nu} \cos(\nu\rho + 2\nu n\pi) + \mu^{\nu}\sqrt{-1} \sin(\nu\rho + 2\nu n\pi).$$

The last sample we take of the variety of forms that this entity can assume is the symplectic decomposition of the quaternion, that is its matrix representation. From the matrix representation of the complex number

$$z = x + iy \quad \text{namely} \quad z = \begin{pmatrix} x & y \\ -y & x \end{pmatrix}$$

which can be confirmed by taking the product

$$z_1 z_2 = \begin{pmatrix} a & b \\ -b & a \end{pmatrix} \begin{pmatrix} c & d \\ -d & c \end{pmatrix},$$

we form

$$Q = \begin{pmatrix} t + ix & y + iz \\ -y + iz & t - ix \end{pmatrix},$$

that is the complex matrix

$$Q = \begin{pmatrix} w & u \\ -\bar{u} & \bar{w} \end{pmatrix},$$

with

$$w = t + ix \qquad u = y + iz$$

This matrix, it will be found, conforms in all respects with the product rules and the definitions of the product factors A, B, C and D required by a quaternion.

Thus we find

$$Q = t \begin{pmatrix} 1 & 0 \\ 0 & 1 \end{pmatrix} + x \begin{pmatrix} i & 0 \\ 0 & -i \end{pmatrix} + y \begin{pmatrix} 0 & 1 \\ -1 & 0 \end{pmatrix} + z \begin{pmatrix} 0 & i \\ i & 0 \end{pmatrix}.$$

Then if Q also is $(t + \alpha x + \beta y + \gamma z)$

$$\alpha = \begin{pmatrix} i & 0 \\ 0 & -i \end{pmatrix} = i\sigma_z,$$

$$\beta = \begin{pmatrix} 0 & 1 \\ -1 & 0 \end{pmatrix} = i\sigma_y,$$

$$\gamma = \begin{pmatrix} 0 & i \\ i & 0 \end{pmatrix} = i\sigma_x,$$

where σ_x, σ_y and σ_z are the Pauli spin matrices. The reversal of the order is to be noted.

Putting each complex number into its 2×2 matrix form results in the 4×4 representation of the quaternion basis

$$
E = \begin{pmatrix} 1 & 0 & 0 & 0 \\ 0 & 1 & 0 & 0 \\ 0 & 0 & 1 & 0 \\ 0 & 0 & 0 & 1 \end{pmatrix}, \quad
I = \begin{pmatrix} 0 & 1 & 0 & 0 \\ -1 & 0 & 0 & 0 \\ 0 & 0 & 0 & -1 \\ 0 & 0 & 1 & 0 \end{pmatrix}, \quad
J = \begin{pmatrix} 0 & 0 & 1 & 0 \\ 0 & 0 & 0 & 1 \\ -1 & 0 & 0 & 0 \\ 0 & -1 & 0 & 0 \end{pmatrix},
$$

$$
K = \begin{pmatrix} 0 & 0 & 0 & 1 \\ 0 & 0 & -1 & 0 \\ 0 & 1 & 0 & 0 \\ -1 & 0 & 0 & 0 \end{pmatrix}
$$
in which E, I, J, K has the identical algebra to $1, \alpha, \beta, \gamma$

The matrix J is the kernel matrix of the lens symmetry transformation given in Chapter 6 (Section 6, p. 334). Complex forms of E, I, J, K are bi-quaternion and relate to the Clifford numbers (q.v.).

The physical applications are as many and as varied. In some cases they require the even more general form the bi-quaternion $Q_1 + iQ_2$ ("ordinary" i) as for example the transmission matrices of Chapter 2.

A real direction is given by the quaternion $P = (0, x, y, z)$ and a rotation by $P' = QPQ^{-1}$ where Q has to be unitary. If Q is of the form $(\cos \rho, \hat{u} \sin \rho)$ this is a rotation of an angle 2ρ about the direction $\hat{u}$. This can be separated to give the Euler rotations and the Cayley–Klein parameters.

If Q has the form

$$
\left(\sqrt{\frac{1 + \lambda}{2}}, \quad \hat{u}\sqrt{\frac{1 - \lambda}{2}} \right)
$$

where $\lambda = (1 - v^2/c^2)^{-\frac{1}{2}}$ then QPQ^{-1} operating on a physical quaternion P (a four vector) is the Lorentz transformation. Other forms of Q applied in this way are rotations in four dimensions and will derive the transformations indicated in Chapter 6.

This is anticipated to be the centre of the lens transformation theory conjectured in chapter 6. More complex transformations may involve biquaternions or even octonions (the last of the line). These should by the account given introduce aberration or diffraction effects.

It is unnecessary to enter here into the dramatic involvement that quaternion theory now has with fundamental particle theory, spinor theory except in so far as they may relate eventually to the ray analogue of the particle trajectory. The theory seems to be more related to the spin properties associated with polarization than with dynamical problems. In this respect the optical transformations may be of value in quantum mechanics.

II.5 RAY INTEGRAL IN A SPHERICAL MEDIUM

We present here the standard treatment[12] deriving the refractive index law $\eta(r)$ from the integral equation defining it (Eqn (2.24), Chapter 2). For any plane curve in a polar coordinate systems (r, θ) the tangent to the curve makes an angle with the radius vector where

$$r\frac{d\theta}{dr} = \pm \tan \phi \tag{II.28}$$

the ambiguity in sign depending upon the increase or decrease of r with respect to θ. We have from Bouguer's theorem in a spherically non-homogeneous medium

$$r\eta(r) \sin \phi = \kappa,$$

from which

$$\tan \phi = \frac{-\kappa}{[r^2\eta^2(r) - \kappa^2]^{\frac{1}{2}}},$$

and hence

$$\int d\theta = \int \frac{-\kappa \, dr}{r[r^2\eta^2(r) - \kappa^2]^{\frac{1}{2}}}, \tag{II.29}$$

the limits of the integral being the end points of the path being considered. In a spherically symmetrical medium the path between the two points at the same radial distance is symmetrical with respect to a diameter bisecting the chord joining them. So for a ray beginning and ending on the surface of a spherical lens the angle turned through by the radius vector θ is double the angle turned through by the radius vector between the beginning point P on the sphere and the central point N at which it bisects the chord PP' as shown in the accompanying Fig. II.2(a). This takes into account the change in sign of $\tan \phi$ in Eqn I as the ray goes through its point of closest approach to the origin $r_{\min}$. At this central point the ray is perpendicular to the radius vector and hence

$$r_{\min} \eta(r_{\min}) = \kappa$$

and κ is also known by the condition at the point of entry of the ray into the

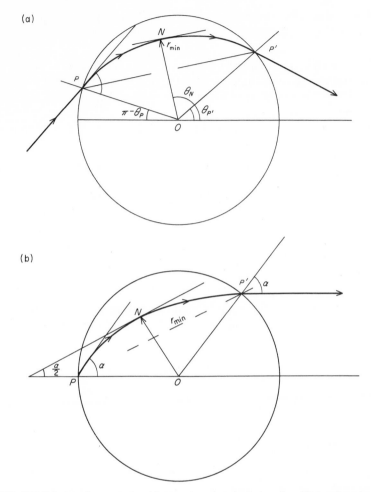

FIG. II.2. *Solution of ray equation. (a) general ray in spherical medium. (b) Luneburg lens.*

lens

$$r_{\text{surface}}\,\eta(r_{\text{surface}})\sin\phi = \kappa \qquad (\text{II.30})$$

ϕ being the angle the ray makes with r at P.

Thus for a lens of unit radius the angle turned through by the radius vector

$$\theta_P - \theta_N = \int_1^{r_{\min}} \frac{-\kappa\,dr}{r(\eta^2 r^2 - \kappa^2)^{\frac{1}{2}}} \qquad (\text{II.31})$$

and it is this integral equation for a given value of θ (as required by the specified focusing property) that has to be solved for the unknown function $\eta(r)$.

If as shown in figure II.2(b), the ray makes an angle α with the radius vector at the surface then at the mid-point N the ray tangent is parallel to the chord PP' and by the argument of Chapter 2 p. 149 makes an angle $\alpha/2$ with the *original* radius vector OP.

This determines the left hand side of equation 4 which then becomes

$$\frac{\pi}{2} - \frac{\alpha}{2} = \int_1^{r_{min}} \frac{-\kappa \, dr}{r(\eta^2 r^2 - \kappa^2)^{\frac{1}{2}}} \tag{II.32}$$

At the surface point P we require $\eta(1) = 1$ and hence from Eqn (II.30) $\kappa = \sin \alpha$. As we shall see later it is the absence of this definitive condition that prevents the complete solution for rays of a hyperbolic form. Substituting for α and putting $r\eta(r) = u$ in Eqn (II.32) results in

$$\pi - \sin^{-1} \kappa = 2 \int_1^{\kappa} \frac{-\kappa (1/r) (dr/du) \, du}{(u^2 - \kappa^2)^{\frac{1}{2}}}. \tag{II.33}$$

This is Abel's integral[13] and its method of solution can be found in the literature of mathematical analysis. Multiplying both sides by

$$\frac{d\kappa}{(\kappa^2 - z^2)^{\frac{1}{2}}}$$

where z is a completely general (well-behaved) function as yet, and integrating with respect to κ from z to 1

$$\int_z^1 \frac{\pi - \sin^{-1} \kappa}{(\kappa^2 - z^2)^{\frac{1}{2}}} \, d\kappa = 2 \int_z^1 \frac{d\kappa}{(\kappa^2 - z^2)^{\frac{1}{2}}} \int_{\kappa}^1 \frac{\kappa(1/r)(dr/du)\,du}{(u^2 - \kappa^2)^{\frac{1}{2}}}. \tag{II.34}$$

Interchanging the order of integration by Dirichlet's formula (Ref. 13, p. 77) gives on the right hand side

$$2 \int_z^1 \frac{1}{r} \frac{dr}{du} \, du \int_z^u \frac{\kappa \, d\kappa}{(\kappa^2 - z^2)^{\frac{1}{2}}(u^2 - \kappa^2)^{\frac{1}{2}}}. \tag{II.35}$$

Abel's identity states

$$\int_z^u \frac{\kappa \, d\kappa}{(\kappa^2 - z^2)^{\frac{1}{2}} (u^2 - \kappa^2)^{\frac{1}{2}}} = \frac{\pi}{2},$$

and then equation (II.34) is

$$\int_z^1 \frac{d\kappa}{(\kappa^2 - z^2)^{\frac{1}{2}}} - \frac{1}{\pi} \int_z^1 \frac{\sin^{-1} \kappa \, d\kappa}{(\kappa^2 - z^2)^{\frac{1}{2}}} = \int_z^1 \frac{1}{r} \frac{dr}{du} du, \tag{II.36}$$

and

$$\frac{1}{\pi} \int_z^1 \frac{\sin^{-1} \kappa}{(\kappa^2 - z^2)^{\frac{1}{2}}} d\kappa = \tfrac{1}{2} \log \left[1 + (1 - z^2)^{\frac{1}{2}} \right],$$

Thus putting $z = u = r\eta(r)$

$$\log \left[\frac{1 + (1 - z^2)^{\frac{1}{2}}}{z} \right] - \tfrac{1}{2} \log \left[1 + (1 - z^2)^{\frac{1}{2}} \right] = -2 \log \frac{z}{\eta},$$

or

$$\frac{1 + (1 - r^2\eta^2)^{\frac{1}{2}}}{r^2\eta^2} = \frac{1}{r^2},$$

whence

$$\eta^2 = 2 - r^2.$$

We can now see in the case of the hyperbolic rays that the condition for *refraction* at the surface gives

$$\sin \theta_{p'} = \eta_1 \sin \alpha = \kappa, \tag{II.37}$$

where η_1 is the unknown refractive index at the surface of the lens $r = 1$ and has to comply with the eventual result obtained for $\eta(r)$ when $r = 1$ is inserted.

Then Eqn (II.33) has on the left hand side

$$\pi - \sin^{-1}\left(\frac{\kappa}{\eta_1} \right)$$

The required scaling of the integral limits on the right hand side has so far not produced a form capable of solution by the more simple mathematical functions. This then remains a further challenge for the antenna theorist.[14]

The same does not occur if $\theta_{p'}$ is made a constant multiple of α instead of the sine law in Eqn (II.37). The result in this case is the lens of Toraldo di Francia as given in chapter 2.

REFERENCES

1. F. Zernike. *Physica* **1** (1934), 687.
2. N. Chako. Characteristic curves in image space, The McGill Symposium on Microwave Optics, ASTIA No. AD 211499, 1959, p. 67.
3. M. Abramowitz and I. A. Stegun (Eds). "Handbook of Mathematical Tables", Dover, 1965, p. 779, equation (22.5.42).

4. A. Erdelyi (Ed). "Tables of Integral Transforms", Vol. II, McGraw Hill, 1954, p. 47.
5. D. R. Myrick. A generalization of the radial polynomials of F. Zernike (and references therein), *S.I.A.M. Jour. App. Maths.* **14** (2) (1966), p. 476.
6. G. N. Watson. "The Theory of Bessel Functions", Cambridge University Press, 1941.
7. T. M. MacRobert. Expression of an E-function as a finite series of E-functions, *Math. Rev.* **22** (7A) (1961), 967, and *Math. Ann.* **140** (1966), 414.
8. G. D. Bernard and A. Ishimaru. "Tables of Anger and Lommel–Weber Functions", AFCRL Report No. 53 Univ. of Washington Press 1962. P. Brauer and E. Brauer. Über unvollständiger Anger-Webersche Funktionen, *Z. Agnew. Math. Mech.* **21** (3) 1941.
9. J. A. Stratton. "Electromagnetic Theory", McGraw Hill, 1941.
10. A. Kyrala. Applications of vectors, matrices, tensors and quaternions, W. B. Saunders Co., 1967, Chapters 8 and 9.
11. H. Halberstam and R. E. Ingram. "The Mathematical Papers of Sir William Rowan Hamilton", vol III "Algebra" Cambridge University Press, 1967.
12. Ref. (7) of Chapter 2.
13. E. T. Whittaker and G. N. Watson. "Modern Analysis" Cambridge University Press, 1958, p. 229.
14. D. K. Cheng. Modified Luneburg lens for defocussed source. *Trans. I.R.E.* **AP8** (1) (1960), 110.
    W. R. Wing and R. V. Neidigh. A rapid Abel inversion. *Amer. Jour. of Phys.* **39** (1971), 760.

Author Index

Page numbers in italic refer to bibliographical references.

Subject Index

PURE AND APPLIED PHYSICS

A Series of Monographs and Textbooks

Consulting Editors

H. S. W. Massey
University College, London, England

Keith A. Brueckner
University of California, San Diego
La Jolla, California

1. F. H. Field and J. L. Franklin, Electron Impact Phenomena and the Properties of Gaseous Ions. (Revised edition, 1970.)

2. H. Kopfermann, Nuclear Moments. English Version Prepared from the Second German Edition by E. E. Schneider.

3. Walter E. Thirring, Principles of Quantum Electrodynamics. Translated from the German by J. Bernstein. With Corrections and Additions by Walter E. Thirring.

4. U. Fano and G. Racah, Irreducible Tensorial Sets.

5. E. P. Wigner, Group Theory and Its Application to the Quantum Mechanics of Atomic Spectra. Expanded and Improved Edition. Translated from the German by J. J. Griffin.

6. J. Irving and N. Mullineux, Mathematics in Physics and Engineering.

7. Karl F. Herzfeld and Theodore A. Litovitz, Absorption and Dispersion of Ultrasonic Waves.

8. Leon Brillouin, Wave Propagation and Group Velocity.

9. Fay Ajzenberg-Selove (ed.), Nuclear Spectroscopy. Parts A and B.

10. D. R. Bates (ed.), Quantum Theory. In three volumes.

11. D. J. Thouless, The Quantum Mechanics of Many-Body Systems. (Second edition, 1972.)

12. W. S. C. Williams, An Introduction to Elementary Particles. (Second edition, 1971.)

13. D. R. Bates (ed.), Atomic and Molecular Processes.

14. Amos de-Shalit and Igal Talmi, Nuclear Shell Theory.

15. Walter H. Barkas. Nuclear Research Emulsions. Volume I.
    Nuclear Research Emulsions. Volume II.

16. Joseph Callaway, Energy Band Theory.

17. John M. Blatt, Theory of Superconductivity.

18. F. A. Kaempffer, Concepts in Quantum Mechanics.

19. R. E. Burgess (ed.), Fluctuation Phenomena in Solids.

20. J. M. Daniels, Oriented Nuclei: Polarized Targets and Beams.

21. R. H. Huddlestone and S. L. Leonard (eds.), Plasma Diagnostic Techniques.

22. Amnon Katz, Classical Mechanics, Quantum Mechanics, Field Theory.

23. Warren P. Mason, Crystal Physics in Interaction Processes.

24. F. A. Berezin, The Method of Second Quantization.

25. E. H. S. Burhop (ed.), High Energy Physics. In five volumes.

26. L. S. Rodberg and R. M. Thaler, Introduction to the Quantum Theory of Scattering.

27. R. P. Shutt (ed.), Bubble and Spark Chambers. In two volumes.

28. Geoffrey V. Marr, Photoionization Processes in Gases.

29. J. P. Davidson, Collective Models of the Nucleus.

30. Sydney Geltman, Topics in Atomic Collision Theory.

31. Eugene Feenberg, Theory of Quantum Fluids.

32. Robert T. Beyer and Stephen V. Letcher, Physical Ultrasonics.

33. S. Sugano, Y. Tanabe, and H. Kamimura, Multiplets of Transition-Metal Ions in Crystals.

34. Walter T. Grandy, Jr., Introduction to Electrodynamics and Radiation.

35. J. Killingbeck and G. H. A. Cole, Mathematical Techniques and Physical Applications.

36. Herbert Überall, Electron Scattering from Complex Nuclei. Parts A and B.

37. Ronald C. Davidson, Methods in Nonlinear Plasma Theory.

38. O. N. Stavroudis, The Optics of Rays, Wavefronts, and Caustics.

39. Hans R. Griem, Spectral Line Broadening by Plasmas.

40. Joseph Cerny (ed.), Nuclear Spectroscopy and Reactions. Parts A, B, C, and D.

41. Sidney Cornbleet, Microwave Optics: The Optics of Microwave Antenna Design.